T0396710

GPU Computing and Applications

Yiyu Cai • Simon See
Editors

GPU Computing and Applications

Editors
Yiyu Cai
Nanyang Technological University
Singapore

Simon See
Nvidia
Singapore

ISBN 978-981-287-133-6 ISBN 978-981-287-134-3 (eBook)
DOI 10.1007/978-981-287-134-3
Springer Singapore Heidelberg New York Dordrecht London

Library of Congress Control Number: 2014955661

Printed on acid-free paper

Springer is part of Springer Science+Business Media (www.springer.com)

Preface

Graphics Processing Unit (GPU) technology is playing an important role in computing today. This book presents a collection of state-of-the-art research on GPU computing and their applications. The major part of this book is selected from the work submitted to the 2013 Symposium on GPU Computing and Applications jointly organized by the Institute for Media Innovation of Nanyang Technological University, and NVIDIA Corporation (South East Asia).

The book addresses the fundamental issues in GPU computing with a focus on big data processing. Three major domains of GPU applications are covered in the book including (1) Engineering design and simulation; (2) Biomedical Sciences; and (3) Interactive & Digital Media.

This book has 17 chapters. Each chapter is carefully selected to present ideas and techniques with insight in a specific area. In Chap. 1, Panpan Cai et al. will report a GPU-enabled parallel genetic algorithm for path planning. In Chap. 2, Alexandre Kaspar and Bailin Deng will introduce real-time deformation of constrained meshes using GPU. In Chap. 3, Yanlin Luo et al. will investigate GPU-based real-time volume interaction for scientific visualization education. In Chap. 4, Petros Papanikoloaou and George Papagiannakis will illustrate real-time separable subsurface scattering for animated virtual characters. In Chap. 5, Yusha Li et al. will describe adaptive NURBS tessellation on GPU. In Chap. 6, Huagen Wan et al. will discuss a graphics native approach to identifying surface atoms of macromolecules. In Chap. 7, Farhoosh Alghabi et al. will explain their scalable software framework for stateful stream data processing on multiple GPUs. In Chap. 8, Tananan Pattanangkur et al. will share their solution for high performance mobile medical imaging. In Chap. 9, David Mainzer and Gabriel Zachmann will showcase their collision detection based on fuzzy scene subdivision. In Chap. 10, Philip Boyer et al. will present the smoothed particle hydrodynamics applied to cartilage deformation. In Chap. 11, Kyrylo Shegeda and Pierre Boulanger will describe a GPU-based real-time algorithm for virtual viewpoint rendering from multi-video. In Chap. 12, Ettikan K. Karuppiah et al. will illustrate a middleware framework for programmable multi-GPU based big data applications. In Chap. 13, Byungjoon Chang et al. will talk on the efficient implementation of a real-time Kd-tree construction algorithm. In Chap. 14, Niko Lukac and Borut Zalik will discuss fast approximate k-nearest neighbors search using GPGPU. In Chap. 15, Shafaatunnur Hasan et al. will share their soft computing methods for big data problems. In Chap. 16, Martin Němec and Lumír Janošek will show a numerical solution of BVP on GPU with application to path-planning. And in Chap. 17, Amirul Abdullah et al. will investigate fast multi-keyword range search in GPGPU.

Readers will benefit from this book which is contributed by experienced GPU researchers and educators. The book may also motivate researchers and developers to develop new possible applications of GPU technology in various areas.

Singapore Yiyu Cai
Singapore Simon See

Contents

List of Contributors

Amirul Abdullah MIMOS Berhad, Kuala Lumpur, Malaysia

Tiranee Achalakul Computer Engineering, King Mongkut's University of Technology Thonburi, Bangkok, Thailand

Farhoosh Alghabi Institute for Vision and Graphics, University of Siegen, Siegen, Germany

Katchaguy Areekijseree Computer Engineering, King Mongkut's University of Technology Thonburi, Bangkok, Thailand

Pierre Boulanger Computing Science Department, University of Alberta, Edmonton, AB, Canada

Philip Boyer Department of Systems and Computer Engineering, Carleton University, Ottawa, ON, Canada

Panpan Cai School of Mechanical & Aerospace Engineering, Nanyang Technological University, Singapore, Singapore

Yiyu Cai Nanyang Technological University, Singapore, Singapore

Institute for Media Innovation, Nanyang Technological University, Singapore, Singapore

Indhumathi Chandrasekaran School of Mechanical & Aerospace Engineering, Nanyang Technological University, Singapore, Singapore

Byungjoon Chang Digital Media & Communications R&D Center, Samsung Electronics, Suwon-si, Gyeonggi-do, South Korea

Bailin Deng Computer Graphics and Geometry Laboratory, École Polytechnique Fédérale de Lausanne, Lausanne, Switzerland

Radoslav Fasuga Department of Computer Science, VŠB-Technical University, Ostrava, Czech Republic

Yunqing Guan Institute for Media Innovation, Nanyang Technological University, Singapore, Singapore

Shafaatunnur Hasan Soft Computing Research Group, Faculty of Computing, Universiti Teknologi Malaysia, Skudai, Johor, Malaysia

Wan Huagen State Key Lab of CAD&CG, Zhejiang University, Hangzhou, China

Insung Ihm Department of Computer Science and Engineering, Sogang University, Seoul, South Korea

Lumír Janošek Department of Computer Science, VŠB-Technical University, Ostrava, Czech Republic

Chris Joslin Department of Systems and Computer Engineering, Carleton University, Ottawa, ON, Canada

Ettikan K. Karuppiah MIMOS Berhad, Kuala Lumpur, Malaysia

Alexandre Kaspar Computer Graphics and Geometry Laboratory, École Polytechnique Fédérale de Lausanne, Lausanne, Switzerland

Yong Keh Kok MIMOS Berhad, Kuala Lumpur, Malaysia

Andreas Kolb Institute for Vision and Graphics, University of Siegen, Siegen, Germany

Sean LeBlanc Department of Systems and Computer Engineering, Carleton University, Ottawa, ON, Canada

Yusha Li School of Computer Engineering, Nanyang Technological University, Singapore, Singapore

Noel Lopes UDI, Institute of Guarda, Guarda, Portugal

CISUC, University of Coimbra, Coimbra, Portugal

Xingjiang Lu Department of Mathematics, Zhejiang University, Hangzhou, China

Niko Lukač Faculty of Electrical Engineering and Computer Sience, University of Maribor, Maribor, Slovenia

Yanlin Luo The College of Information Science and Technology, Beijing Normal University, Beijing, China

Zuying Luo The College of Information Science and Technology, Beijing Normal University, Beijing, China

Yanhong Luo The College of Electrical Engineering, Northwest University for Nationalities, Lanzhou, China

The School of Nuclear Science and Technology, Lanzhou University, Lanzhou, China

David Mainzer Clausthal University, Clausthal-Zellerfeld, Germany

Soo Saw Meng MIMOS Berhad, Kuala Lumpur, Malaysia

Amril Nazir MIMOS Berhad, Kuala Lumpur, Malaysia

Martin Němec Department of Computer Science, VŠB-Technical University, Ostrava, Czech Republic

G. Papagiannakis Computer Science Department, University of Crete, Heraklion, Greece

Foundation for Research and Technology Hellas, Heraklion, Greece

P. Papanikolaou Computer Science Department, University of Crete, Heraklion, Greece

Foundation for Research and Technology Hellas, Heraklion, Greece

Tananan Pattanangkur Computer Engineering, King Mongkut's University of Technology Thonburi, Bangkok, Thailand

Sarunya Pumma Computer Engineering, King Mongkut's University of Technology Thonburi, Bangkok, Thailand

Ulrich Schipper Institute for Vision and Graphics, University of Siegen, Siegen, Germany

Mohanavelu Senapan MIMOS Berhad, Kuala Lumpur, Malaysia

Woong Seo Department of Computer Science and Engineering, Sogang University, Seoul, South Korea

Siti Mariyam Shamsuddin UDI, Institute of Guarda, Guarda, Portugal

Kyrylo Shegeda Computing Science Department, University of Alberta, Edmonton, AB, Canada

Keeratpal Singh MIMOS Berhad, Kuala Lumpur, Malaysia

Sikana Tanupabrungson Computer Engineering, King Mongkut's University of Technology Thonburi, Bangkok, Thailand

Guozhao Wang Department of Mathematics, Zhejiang University, Hangzhou, China

Zhongke Wu The College of Information Science and Technology, Beijing Normal University, Beijing, China

Gabriel Zachmann University of Bremen, Bremen, Germany

Borut Žalik Faculty of Electrical Engineering and Computer Science, University of Maribor, Maribor, Slovenia

Wenjing Zhang School of Computer Engineering, Nanyang Technological University, Singapore, Singapore

Jianmin Zheng School of Computer Engineering, Nanyang Technological University, Singapore, Singapore

Chapter 1
A GPU-Enabled Parallel Genetic Algorithm for Path Planning of Robotic Operators

Panpan Cai, Yiyu Cai, Indhumathi Chandrasekaran, and Jianmin Zheng

Abstract Genetic algorithm (GA) is a class of global optimization algorithm inspired by the Darwinian biological evolution. It is widely applied in the field of robotic path planning. Parallel GA (PGA) is a subclass of GA which is able to achieve good solutions in a short time. This chapter discusses the utilization of a PGA in determining collision-free path for robotic operators. GPU-style genetic operators are designed to speed up the GA process while improving the quality of solutions. GPU parallelization for a master–slave parallel GA (MSPGA) is implemented by parallelizing the selection, crossover and mutation operators.

Keywords Genetic Algorithm • Parallel GA • GPU • Master-slave Parallel GA

1.1 Introduction

1.1.1 Motivation

Genetic algorithms (GA) [1–3] are promising in achieving globally optimized solutions for path planning. The process of GA requires a large number of iterations with intensive computations. Thus, it is difficult to have fast GA optimization in serial platform. GPUs provide a highly parallel computing structure which enables various types of data processing. By embedding GA into the GPU platform, it is possible to achieve significant performance improvements. The structure of GPU is

P. Cai • I. Chandrasekaran
School of Mechanical & Aerospace Engineering, Nanyang Technological University, Singapore, Singapore
e-mail: pcai2@e.ntu.edu.sg

Y. Cai (✉)
Nanyang Technological University, Singapore, Singapore

Institute for Media Innovation, Nanyang Technological University, Singapore, Singapore
e-mail: myycai@ntu.edu.sg

J. Zheng
School of Computer Engineering, Nanyang Technological University, Singapore, Singapore

Y. Cai, S. See (eds.), *GPU Computing and Applications*,
DOI 10.1007/978-981-287-134-3_1

quite different with single-core and multi-core CPUs [4]. It performs in a massively parallel pattern with minimum communication between stream processors. As such the design of parallel GA also needs to be adapted to the GPU architecture.

Automated path planning for robotic operators is highly desired in many applications. Path planning is an optimization problem with non-explicitly represented objective function and multiple hard constraints. Because of the high degrees of freedom (DOF) of robotic operators, the search spaces of the optimization problems are extremely huge. Therefore designing a fast and effective GA-based path planning algorithm becomes challenging.

1.1.2 Objectives

The objectives of this work include:

- Designing an efficient GPU-based genetic algorithm
- Implementing a GPU-enabled parallel genetic algorithm
- Developing an automated path planning system using GPU-based parallel GA (GPUPGA) with fast convergence and good solution quality

1.1.3 Organization of the Chapter

The rest of the chapter is organized as follows: Sect. 1.2 presents the prior arts. Firstly a comprehensive review of the concept and theories of GA is given. Then commonly used parallel genetic algorithms will be introduced. Section 1.3 discusses about the GPU-based genetic algorithm. Detailed algorithm design and GPU implementation will be discussed in this section. In Sect. 1.4, the application of the GPUPGA algorithm in the path planning problem of robotic operators is presented with graphic and statistical results. Conclusions and discussions are presented at the end of the chapter in the final section.

1.2 Prior Arts

1.2.1 GA Study

GA is a class of combinatorial optimization algorithm first put forward by John Henry Holland in the 1960s. It is described as a computational abstraction of Darwinian biological evolution [3, 5]. Selection, crossover and mutation are the three most important components in GA which are noted as genetic operators. The evolutionary procedure formed by specific designs of genetic operators is called the "adaptive plan".

Many types of genetic operator designs exist, and some are tested in different applications. For crossover operators, the most commonly used are one point, two point, multi-point and parameter based. Popular mutation strategies include single-point mutation, multi-point mutation and parameter-based mutation. Major selection criteria include roulette wheel selection, rank selection [6, 7], tournament selection [2] and so on. Different selection scheme will cause different selection pressure in the population. Some of them assert higher pressure at start stage, and others will increase pressures when the evolutionary process goes on.

One important component in the selection is calculating fitness values for chromosomes (strings). This process is called fitness evaluation. The fitness function is actually the objective function of the optimization problem. It will directly affect the chance of reproduction of chromosomes. Thus, it is the major guiding force of the evolutionary direction of the population. For constrained problems, constraints usually perform as part of the fitness function.

A genetic algorithm equipped with standard operators and selection schemes is called a *simple genetic algorithm* (SGA). It usually has the following features [1, 2]:

- Binary bits
- Fixed length linear chromosomes, and
- With simple GA procedure

Parameter setting has substantial influence on the performance of SGA. These parameters include the *size of population*, *severity of selection*, *crossover rate* and *mutation rate*. The trade-off between the selection pressure and crossover rate [2] is most important. If no beneficial crossover happens before the population being conquered by a single chromosome, the process would stop before global convergence. On the other hand, if crossovers happen too frequently, building blocks (BB) [8, 9] will be easily destroyed and thus hinder the convergence.

1.2.2 Parallel GA (PGA) Study

Parallel genetic algorithm (PGA) is one kind of GA making use of the power of parallel computing to achieve better performance. Major types of parallel GA include master–slave parallel GA, coarse-grained parallel GA and fine-grained parallel GA. Coarse-grained parallel GA divides the set of chromosomes into multiple populations with minor communications. Parallelism is implemented among subpopulations. In order to achieve shorter execution time, the sizes of subpopulations need to be decreased, which have negative effects on the supply of building blocks in the initial populations. Fine-grained parallel GA assigns topological structure to the single population and restricts global communications to better fit parallel computational structures. Lower communication ability in populations hinders the spread of building blocks. Thus, the convergence speed of fine-grained GAs is usually slower than MSPGA.

Master–slave parallel GA is a straightforward parallel version of serial GAs. As the MSPGAs preserve the original procedure of serial GA, their behaviours are more predictable. The master processor takes care of the overflow of the GA process while handling simple components [10]. Functional parts where intensive computation is required are pulled into the slave processor. The most typical component handled by the slave processor is fitness evaluation where computations like hard constraint evaluation, distance calculation and fitness calculation are done [11]. Other operators like crossover and mutation may also be parallelized [12].

Researchers have started investigating the adaptation of parallel GAs in GPU in recent years. Pospíchal et al. [13] investigated the general GPU implementation of a coarse-grained parallel GA. Feier et al. [14] applied the GPU coarse-grained PGA in optimizing NP-complete problems. Jaros [15] implemented the algorithm in solving the knapsack problem. In the field of hybrid PGAs, Munawar et al. [16] implemented an adaptive resolution PGA in dealing with Minimization Linear Programming (MINLP) problems and reported good results.

MSPGA is the most popular type of PGA investigated these years. Arora et al. [17] discussed the GPU implementation of both binary-coded and real-coded MSPGA and reported a significant speed-up. Oiso et al. [18] tested a steady-state MSPGA for function optimization in GPU. Wang and Shen [19] applied a GPU-based MSPGA in generating daily activity plans. Some researchers like Fujimoto and Tsutsui [12] specified in the parallel design of single genetic operators.

1.3 GPU-Enabled PGA

1.3.1 GPU Fundamentals

GPUs are equipped with tremendous computational horsepower and high memory bandwidth which can bring substantial speed-ups in a variety of applications [20]. General purpose GPUs (GPGPUs) are a new generation of GPUs which is aiming at handling more general, complex and intensive processing. GPGPU provides a complete functional set of operations which work on arbitrary length data. A GPGPU contains several streaming multiprocessors (SMs) which can run hundreds of threads concurrently. The SMs are equipped with caches and control units which are shared by internal threads.

CUDA C [20] is a typical GPU accessing APIs designed by nVIDIA as an extension of the standard C language. It allows programmers to allocate GPU memories and run kernels on parallel threads in a C/C++ like style [4, 20]. A variety of types of access to GPU memory is provided in CUDA C/C++. Global memory, constant memory and texture memory lie in the global physical memory, while shared memory resides inside SMs. Local memory and register memory are only usable for the threads who allocated them.

CUDA has a hierarchical thread structure reflecting the hierarchical hardware architecture. Each launched kernel is handled by one thread grid. The thread grid consists of an array or matrix of thread blocks, and the blocks contain a similar matrix of threads.

1.3.2 GPU PGA Design and Implementation

We investigate the GPU parallelization of a master–slave parallel GA. The master processor is the CPU and the slave processor is GPU. We aim to do the GA search using a single CPU and a single GPU. The overall flow of MSPGA process is shown in Fig. 1.1. Highly parallelized 3D collision detection is done in iterations. Thus, no preprocessing (such as configuration space generation) is required for the algorithm.

In the starting phase of an MSPGA, an initial population is firstly generated by random approaches. Hard constraints are applied to the random generation process to make all initial chromosomes within the feasible space. Then the MSPGA proceeds into an evolutionary iteration loop where four functional components are performed in sequence (Fig. 1.2).

The first procedure in the loop is to evaluate the fitness value of chromosomes which are proportionally scaled into unified selection rates later. Next, a roulette wheel selection operator is applied in the population. In the selection process, chromosomes with higher fitness value have better chances to be pulled into the mating pool for reproduction.

The mating pool is represented as indexes of chromosomes instead of actual chromosomes. In order to achieve global communications in parallel processors, we use a specially designed generating process (as denoted in Fig. 1.3). Firstly a CUDA kernel is launched with kernel parameters specified as $<<<1,N,1>>>$ where N is the population size. A uniform random float number $r \in [0, 1)$ is then generated in each thread and is compared between the array of selection rates of chromosomes. If the selection rate array is denoted by $(c_0, c_1, \ldots, c_{N-1})$, then the task of threads is to find the index i which satisfies $c_{i-1} \leq r < c_i$. Here c_{-1} is counted as zero. The index generated by this process is then stored into the corresponding position in the mating pool.

After mating pool data are settled, the crossover operator is applied immediately (Fig. 1.4). The CUDA kernel parameters are specified as $< <<N, L, 1>> >$ where N is the population size and L is the chromosome length. In the GPU kernel, the j th thread block B_j ($j \neq 0$) handles a pair of chromosomes $\left(s_{i_{j-1}}, s_{i_j} \right)$ where i_j stands for the index in the j th position of the mating pool B_0 is reserved for elitism. Data of the two chromosomes are pulled into the shared memory of the block for future use.

Next, gene pairs of parents are assigned accordingly into the thread arrays. The k th thread in the j th thread blocks retrieves g_k^{j-1} and g_k^j into its local memory. Then the following work is the generation of the offspring gene from parent genes.

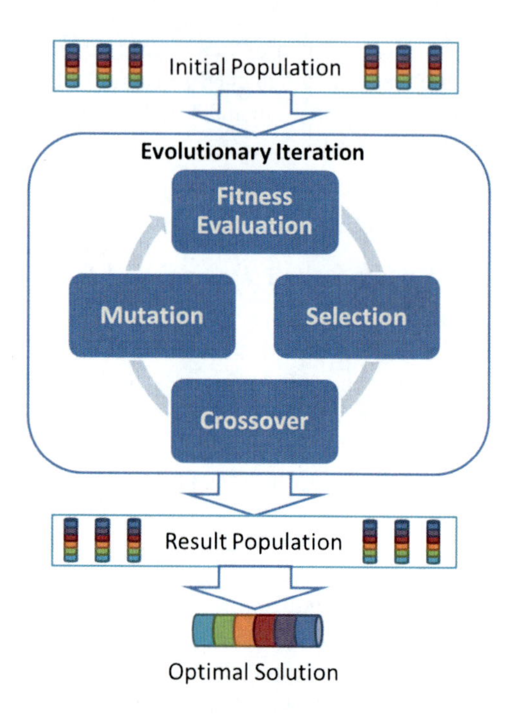

Fig. 1.1 Overall flow of the master–slave genetic algorithm process

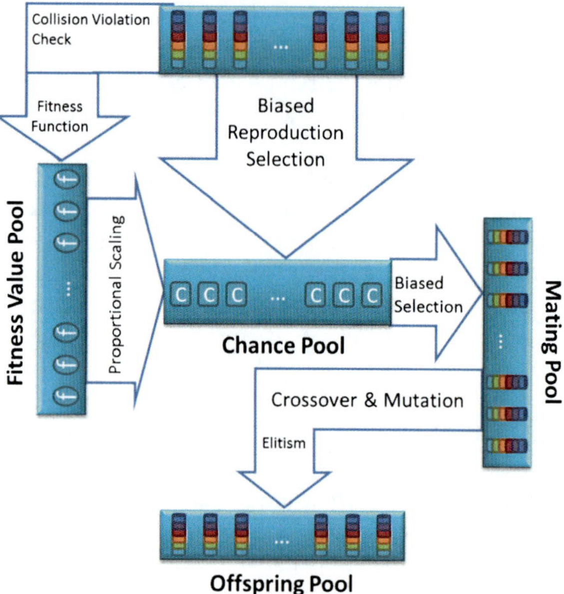

Fig. 1.2 Indication of the procedure inside the evolutionary iteration

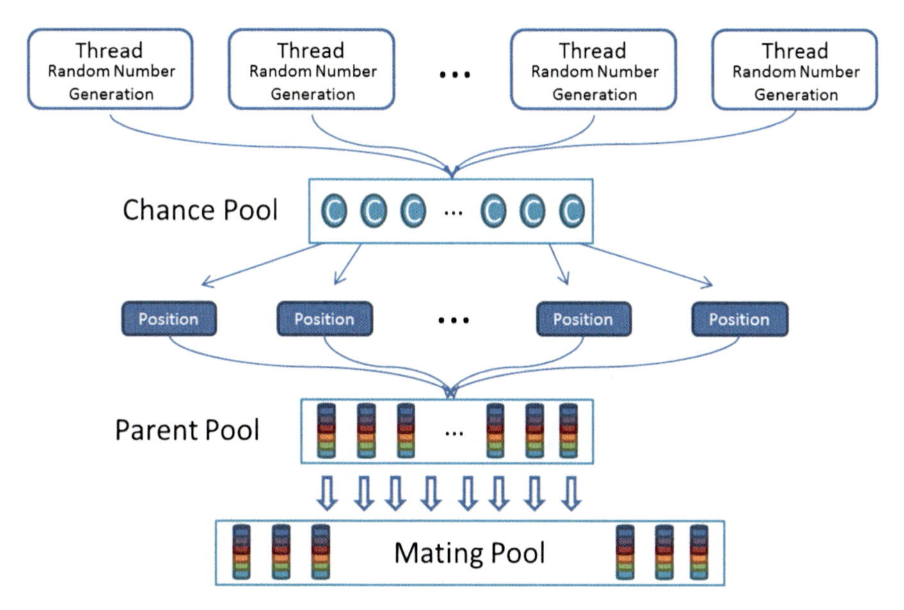

Fig. 1.3 GPU implementation details of mating pool generation

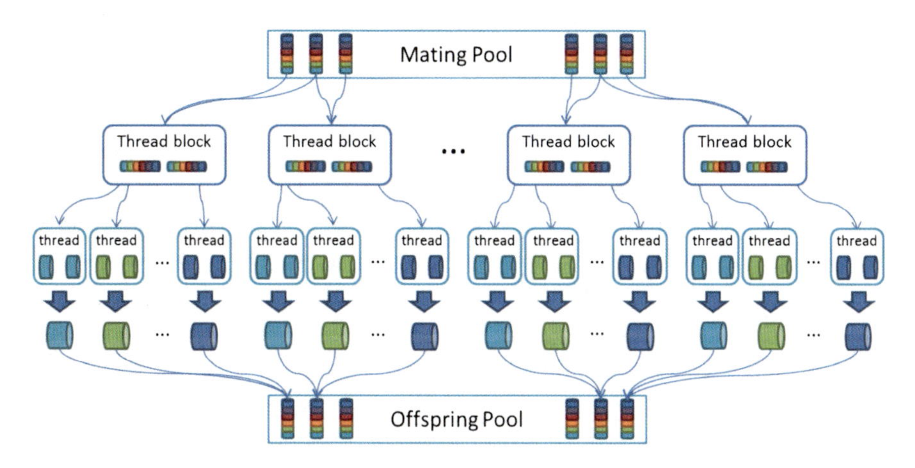

Fig. 1.4 GPU implementation details of the parameter-based crossover operator

Several generation strategies can be applied. In this work, we select a better one from the parents to be the offspring. After all gene positions in all threads are determined, the kernel will terminate and the offspring remain in the GPU global memory.

1.4 GPU-Enabled PGA Application for Collision-Free Path Planning

1.4.1 Collision Avoidance Path Planning

The chromosome structure of the GPUPGA for the path planning problem is designed as a linear structure with each gene representing a configuration node. Some operators have restrictions on neighbouring configurations. Some operations are not allowed to happen simultaneously. Hence the final structure of the chromosome is an interrelated linear string.

For path planning optimization problems, collision avoidance performs as a hard constraint. In our algorithm, this constraint performs as a part of the fitness function. To make the collision avoidance constraint "hard", the gradient of fitness function along the collision violation should be large enough. Another component in the fitness function would be the distance cost of the path which is the weighted sum of the motional cost of all operations. Combining all the considerations, the fitness function is written as:

$$f(s_i) = \begin{cases} \dfrac{c}{N_i}, N_i > 0 \\ c\left(1 + \dfrac{c}{d_i}\right), N_i = 0 \end{cases}$$

Where c is a constant scaling factor and d_i is the distance cost of the i th string.

The kernel parameters of the fitness evaluation kernel are written as < <<*N*, *L*, 1>>>. GPU implementation is shown in Fig. 1.5. Each chromosome in the population is assigned into a thread block, while genes in the string are handled by separate threads. Within each thread, continuous collision detection is done for genes. A synchronized shared memory unit is used to count the number of collision violations in the chromosome (wrote as N_i in the formula). This counted number is further used in the fitness function. The kernel returns when the fitness values are settled in the GPU global memory, and the values will be used to generate the mating pool later on.

Another try to fulfil collision avoidance is through adding constraints to the crossover operator. Collision violation can be taken into consideration when two parent genes are compared. If one of the parent genes is colliding with the environment, the non-colliding one will be chosen. By this approach, the survival chances of invalid genes are largely reduced.

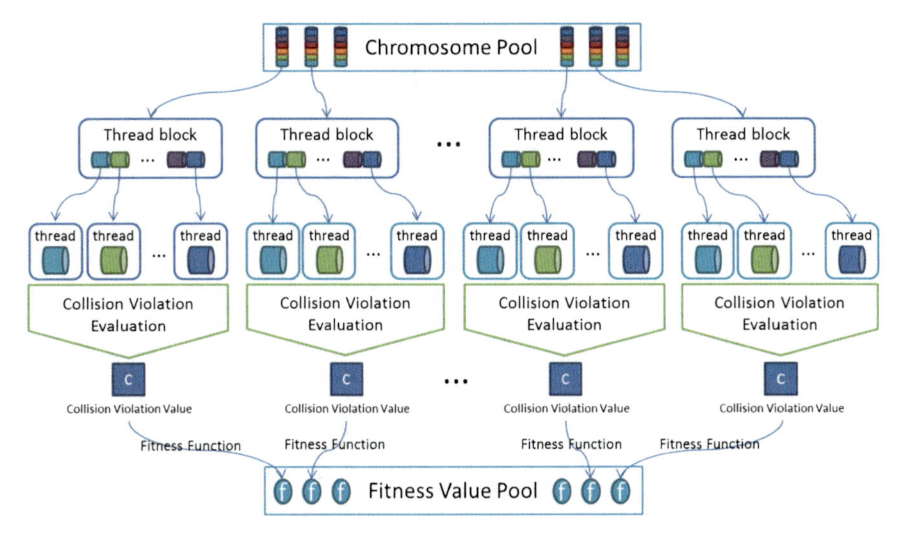

Fig. 1.5 GPU implementation details of the fitness evaluation process

1.4.2 Terminal Conditions

As GA is an optimization process with many random factors, it is hard to determine the precise point of termination. The algorithm uses multiple clues to decide when to terminate the search. These criteria [21, 22] include:

- When the iterations do not bring improvements for several consecutive steps
- When the best fitness value in the population reaches some user defined value
- When the latest optimal solution shows a satisfactory performance

1.4.3 Global Optimization

Unlike path planning algorithms such as the A* or potential field method, our algorithm aims to achieve globally optimized solutions. To show that the designed GPUPGA could achieve globally optimized solutions for path identification, we build a site environment using CAD tools which contains eight boxes, three cylinders, one cone and a Z-shaped road. A robot is designed to simulate a mobile crane with 3 degrees of freedom for lifting jobs in industrial sites. All objects are within the working area of the crane. The experiment is done on a PC with a GeForce GTX 660 graphic card and an Intel i7-3770 CPU. It takes 31.5 s to achieve the result shown in Fig. 1.6.

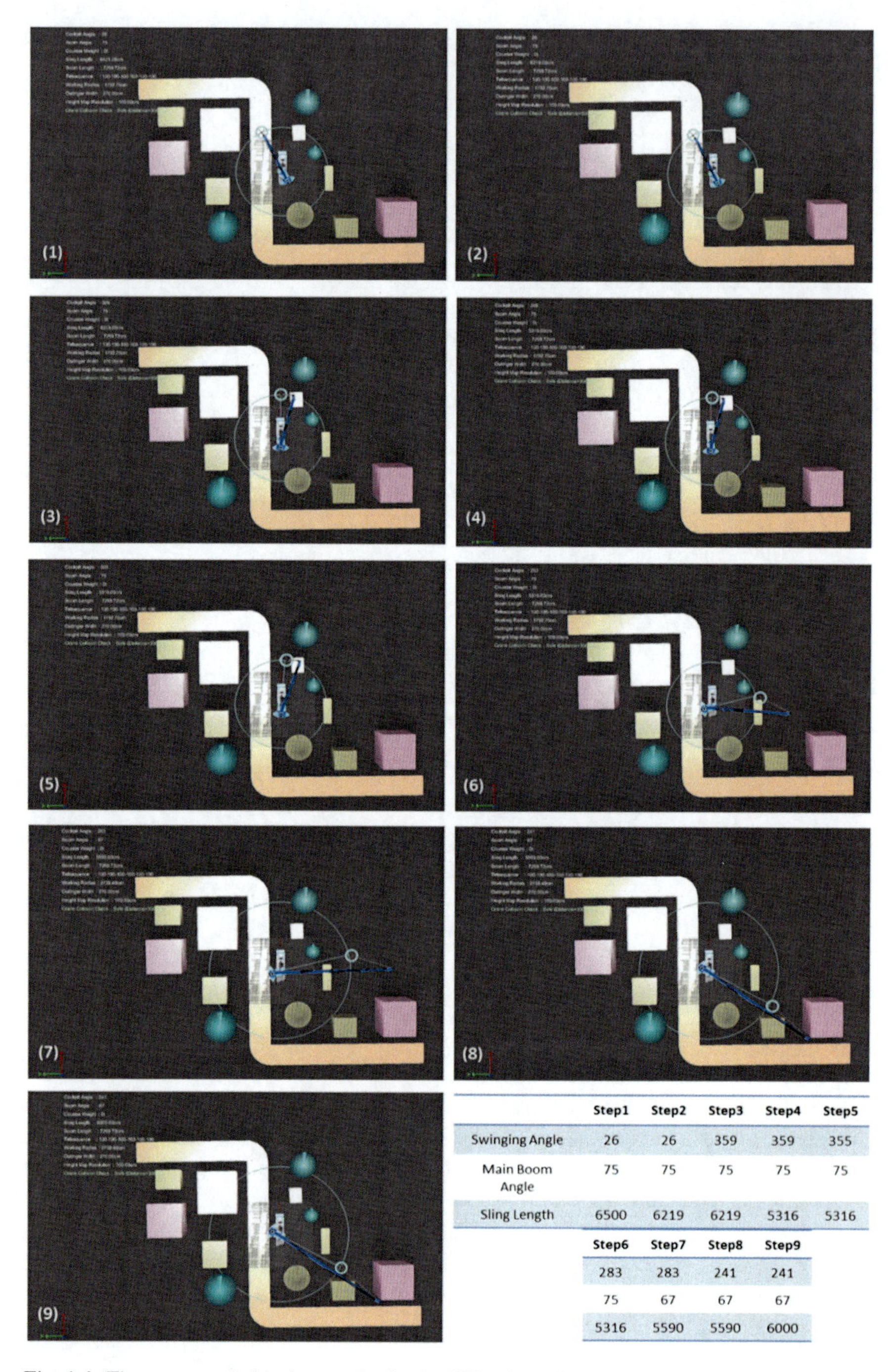

	Step1	Step2	Step3	Step4	Step5	Step6	Step7	Step8	Step9
Swinging Angle	26	26	359	359	355	283	283	241	241
Main Boom Angle	75	75	75	75	75	75	67	67	67
Sling Length	6500	6219	6219	5316	5316	5316	5590	5590	6000

Fig. 1.6 The automated planning results for the lifting job of a mobile crane. The nine snapshots stand for the nine configuration steps in the final path. The result is done within an execution time of 31.5 s. Parameters of the search are set as: population size-50, chromosome length-5, mutation rate-25 % and crossover rate-75 %

Table 1.1 Execution time in CPU and GPU for plants with various complexities

	Plant 1	Plant 2	Plant 3	Plant 4
Estimated CPU execution time	30 s with 960 cores	31.5 s with 960 cores	32 s with 960 cores	34 s with 960 cores
Triangle number	48	1,342	34,791	104,202
Estimated CPU time per triangle	600 s	22.53 s	0.88 s	0.31 s
Execution time (wall time) per triangle	625 ms	23.5 ms	0.9 ms	0.32 ms

1.4.4 Performance

To test the performance of the algorithm, we used four plant models with various complexities. The least complex one contains 48 triangles, while the most complex one contains 104,202 triangles. Here we estimate the CPU time of the algorithm by multiplying the execution time with the number of parallel cores. It would be an upper bound of the actual CPU time. The results are shown in Table 1.1 and Fig. 1.7.

From Fig. 1.7, we can see that, for environments with different numbers of triangles, the execution time is similar. When considering the average execution time and CPU time, they are actually fast decreasing when triangle number increases. The result shows that the algorithm has good scalability and is highly suitable for complex environments.

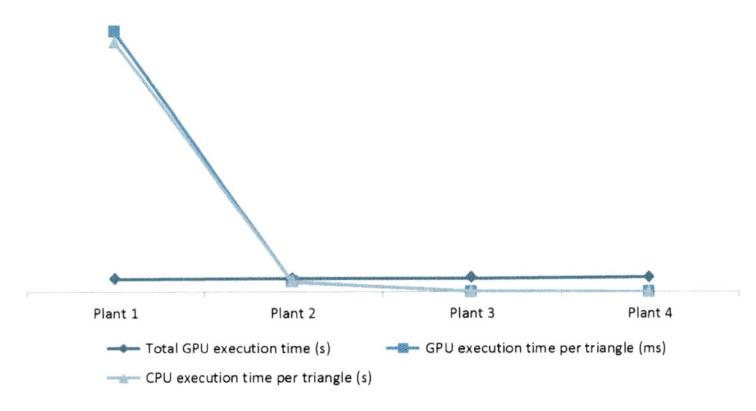

Fig. 1.7 The data chart for execution times for plant models

Conclusions

Contributions

A GPU implementation scheme of MSPGA is designed in this chapter which considers the hardware architecture of GPUs. The algorithm parallelizes four functional components in GA including fitness evaluation, mating pool generation, crossover and mutation. The GPU MASPGA algorithm is further tuned to solve the path planning problem by applying proper chromosome structure, collision detection and termination criteria.

The algorithm manages to identify global optima for path planning problems in random 3D environments. It is able to provide a clear and zigzag-free path with high operability for human operators. The algorithm is fast and has promising potential in further speed-up.

Limitations

The fast performance of GPU hardware is still not fully exploited. Balance among different memory types needs to be further analysed to reach a maximum use of the memory bandwidth. Communication between the CPU and GPU is still necessary in the algorithm which may limit the speed-up of the parallelization.

Future Improvements

To further improve the performance and solution quality of the GPU MAPGA algorithm, we will optimize the program design in a more GPU-friendly style taking full consideration of the memory usage and multiprocessor occupation. Selection and crossover operators will be improved to enable more effective search and higher convergence speed.

Acknowledgements The authors would like to express their sincere thanks to those who help this work in one way or another.

References

1. Goldberg, D. E.: Simple genetic algorithms and the minimal, deceptive problem. In: Davis, L. (ed.) Genetic algorithms and simulated annealing, pp. 74–88. Pitman, London (1987)
2. Goldberg, D.E.: Genetic Algorithms in Search, Optimization, and Machine Learning. Addison-wesley Reading, Menlo Park, CA (1989)
3. Holland, J.H.: Adaptation in Natural and Artificial Systems: An Introductory Analysis with Applications to Biology, Control, and Artificial Intelligence. MIT Press, Cambridge, MA (1992). ISBN 0262581116
4. Sanders, J., Kandrot E.: CUDA by Example: An Introduction to General-Purpose GPU Programming. Addison-Wesley Professional (2010). ISBN: 0132180138
5. Holland, J.H.: Genetic algorithms. Sci. Am. **267**(1), 66–72 (1992)

6. Goldberg, D. E., Deb, K.: A comparative analysis of selection schemes used in genetic algorithms. In: Foundations of Genetic Algorithms, vol. 51, 61801–62996 (1991)
7. Miller, B.L., Goldberg, D.E.: Genetic algorithms, selection schemes, and the varying effects of noise. Evol. Comput. **4**(2), 113–131 (1996)
8. Goldberg, D.E., et al.: Messy genetic algorithms: Motivation, analysis, and first results. Complex Syst. **3**(5), 493–530 (1989)
9. Goldberg, D. E., et al.: On the supply of building blocks. In: Proceedings of the Genetic and Evolutionary Computation Conference, Citeseer, pp.336–342 (2001)
10. Nowostawski, M. and Poli R.: Parallel genetic algorithm taxonomy. In: IEEE Third International Conference on Knowledge-Based Intelligent Information Engineering Systems, pp. 88–92 (1999)
11. Ismail, M. A.: Parallel genetic algorithms (PGAs): master slave paradigm approach using MPI. In: IEEE E-Tech 2004, pp. 83–87 (2004)
12. Fujimoto, N., Tsutsui. S.: Parallelizing a Genetic Operator for GPUs. In: 2013 I.E. Congress on Evolutionary Computation (CEC), pp. 1271–1277 (2013)
13. Pospíchal, P., et al. Parallel genetic algorithm on the cuda architecture. In: Applications of Evolutionary Computation, pp. 442–451. Springer, Heidelberg (2010)
14. Feier, M. C., et al.: Solving NP-Complete Problems on the CUDA Architecture Using Genetic Algorithms. In: IEEE 2011, 10th International Symposium on Parallel and Distributed Computing (ISPDC), pp. 278–281 (2011)
15. Jaros, J.: Multi-GPU island-based genetic algorithm for solving the knapsack problem. In: 2012 I.E. Congress on Evolutionary Computation (CEC), pp. 1–8 (2012)
16. Munawar, A., et al.: Advanced genetic algorithm to solve minlp problems over GPU. In: 2011 I.E. Congress on Evolutionary Computation (CEC), pp. 318–325 (2011)
17. Arora, R., et al.: Parallelization of binary and real-coded genetic algorithms on GPU using CUDA. In: 2010 I.E. Congress on Evolutionary Computation (CEC), pp. 1–8 (2010)
18. Oiso, M., et al.: Accelerating steady-state genetic algorithms based on CUDA architecture. In: 2011 I.E. Congress on Evolutionary Computation (CEC), pp. 687–692 (2011)
19. Wang, K., Shen, Z.: A GPU-based parallel genetic algorithm for generating daily activity plans. IEEE Trans. Intell. Transp. Syst. **13**(3), 1474–1480 (2012)
20. NVidia, C.: C programming guide version 3.2. NVIDIA Corporation, Santa Clara, CA (2010)
21. Renders, J.M., Flasse, S.P.: Hybrid methods using genetic algorithms for global optimization. IEEE Trans. Syst. Man. Cybern. B Cybern. **26**(2), 243–258 (1996)
22. Safe, M., et al.: On stopping criteria for genetic algorithms. In: Advances in Artificial Intelligence–SBIA, 405–413 (2004)

Chapter 2
Real-Time Deformation of Constrained Meshes Using GPU

Alexandre Kaspar and Bailin Deng

Abstract Constrained meshes play an important role in free-form architectural design, as they can represent panel layouts on free-form surfaces. It is challenging to perform real-time manipulation on such meshes, because all constraints need to be respected during the deformation while the shape quality needs to be maintained. This usually leads to nonlinear constrained optimization problems, which are challenging to solve in real time. In this chapter, we present a GPU-based shape manipulation tool for constrained meshes, using the parallelizable algorithm proposed in Deng et al. (Computer-Aided Design, 2014). We discuss the main challenges and solutions for the GPU implementation and provide timing comparison against CPU implementations of the algorithm. Our GPU implementation significantly outperforms the CPU version, allowing real-time handle-based deformation for large constrained meshes.

2.1 Introduction

With the advances in computer-aided design tools, complex free-form shapes are becoming more and more popular in architectural design nowadays. While digital models can be easily created using a computer, the construction of such shapes remains a challenge, due to the limitation of fabrication technologies. To realize free-form architectural designs at a reasonable cost, the design surfaces usually need to be decomposed into panels of simple shapes that facilitate manufacturing. This process is called *rationalization*, which amounts to approximating the NURBS-based design surface using a set of panels subject to requirements such as approximation tolerance, panel types, aesthetics of panel layouts, etc. Rationalization usually involves nonlinear optimization with a large number of variables and is therefore computationally expensive [1].

A. Kaspar (✉) • B. Deng
Computer Graphics and Geometry Laboratory, École Polytechnique Fédérale de Lausanne, Lausanne, Switzerland
e-mail: alexandre.kaspar@a3.epfl.ch; bailin.deng@epfl.ch

Y. Cai, S. See (eds.), *GPU Computing and Applications*,
DOI 10.1007/978-981-287-134-3_2

From a designer's point of view, it is important to explore different design shapes and their corresponding panel layouts. One possible way is to modify the NURBS design and perform rationalization for each new shape. Due to the heavy computational cost of rationalization, it is time-consuming to explore designs via this approach. An alternative approach is to directly manipulate the panel shapes and layouts while respecting the shape requirements for panel types and maintaining the aesthetics of the overall shape. In this way, the user only explores panel layouts that satisfy all the requirements, with intuitive feedback about what modifications are possible under the given requirements. Such *fabrication-aware* shape exploration methods for free-form architecture have been a popular research topic recently [2–7].

Usually, a panel layout can be represented by a polygonal mesh, with mesh faces representing the panels and mesh edges representing the panel boundaries. The shape requirements for panel layout induce geometric constraints for mesh elements. For example, a layout of planar panels corresponds to a polygonal mesh where the vertices of each face are required to be coplanar (see Fig. 2.1). Therefore, manipulating the panel layout reduces to deforming the mesh while satisfying certain geometric constraints and maintaining the shape quality. This usually leads to a nonlinear constrained optimization problem for mesh vertex positions. Due to the difficulty of the optimization, it is a challenging task to perform real-time manipulation, especially for large meshes.

Bouaziz et al. [3] proposed a general framework for handle-based deformation of meshes subject to soft constraints, formulated as a nonlinear least-squares problem. Utilizing projections of individual mesh elements onto their feasible configurations, they propose an iterative solver that alternates between global linear system solving and local mesh element projections. The projections are independent and can be executed in parallel, thus achieving significant speedup on multi-core processors. When run on a multi-core CPU, the method achieves interactive results for meshes with about 1K vertices, but is still unable to handle large meshes. Recently, this method was extended in [8] to allow both hard and soft constraints. The proposed numerical solver consists of a series of simple subproblems similar to those in [3], enabling speedup from parallelism. In this chapter, we present an implementation of the method in [8] on GPU using CUDA, which provides many more computational cores than CPU. By carefully optimizing for performance, our implementation allows real-time deformation of constrained meshes with up to 20 K vertices and 20 K constraints.

2.1.1 Related Work

Besides [3] and [8], other handle-based deformation methods for constrained meshes have been developed in recent years. Zhao et al. [5] extended the shape space exploration approach in [2], using curve handles to control target shapes. Vaxman [4] proposed a method to deform polyhedral meshes while keeping their

Fig. 2.1 Panel layouts can be represented by polygonal meshes subject to geometric constraints. *Left*: Yas Viceroy Hotel in Abu Dhabi, designed by Asymptote Architecture (image courtesy of Asymptote Architecture). *Right*: a quad mesh representing the hotel facade, with the constraint that the vertices of each face lie on a common plane. This constrained mesh represents a layout of planar quadrilateral panels on the facade

faces planar, using affine transformations of mesh faces. The computation reduces to solving a linear system for mesh vertex positions, allowing real-time deformation. The method only works for polyhedral meshes (meshes with planar faces). Moreover, since only affine transformations are allowed, only a subset of the feasible deformations are considered, which limits the degree of freedom for shape control. Poranne et al. [6] provided an optimization approach to deform polyhedral meshes, not limited to affine transformations of faces. The deformation is computed through an alternating least-squares approach similar to [3]. However, only face planarity constraints are considered by the method. Deng et al. [7] proposed a framework to deform meshes under hard constraints, with a focus on computing local deformations. But their framework does not consider soft constraints. On the contrary, the deformation method in this chapter considers general shape constraints for meshes and allows both soft and hard constraints, providing more flexibility in shape manipulation.

Recently, computational design shape exploration tools have also been proposed for other types of architecture, such as reciprocal frame structures [9] and building layouts [10]. As these problems require other representations than polygonal meshes, they cannot be handled by our method.

2.1.2 Overview

The rest of the chapter is organized as follows. Section 2.2 briefly presents the method in [8]. Section 2.3 gives an overview of the implementation of our system. Section 2.4 provides more details about the CUDA implementation. Finally, results are presented in Sect. 2.5, followed by a discussion about limitation and future work in Sect. 2.6. The final section concludes this chapter.

2.2 Overview of the Method

In this section, we give a brief overview of the problem formulation in [8], as well as its numerical solution. Interested readers are referred to [8] for more details.

2.2.1 Problem Formulation

We consider polygonal meshes as a representation of panel layouts for free-form architectural surfaces. The mesh is deformed by changing its vertex positions while fixing its topology. During deformation, the vertex positions are subject to certain soft constraints and/or hard constraints. To control the deformation, a user specifies target positions for some vertices using handles that are freely movable. When the handles are moved, the mesh vertex positions are updated such that:

- The new mesh satisfies the soft constraints as much as possible and satisfies the hard constraints strictly.
- The handle vertices are close to their target positions.
- The non-handle vertices stay close to their original positions.
- The vertex deformation field is smooth across the mesh.

With a given topology, the shape of a mesh is determined by its vertex positions $\mathbf{p}_1, \mathbf{p}_2, \ldots, \mathbf{p}_N \in \mathbb{R}^3$, where N is the number of vertices. A shape constraint involving m vertex positions $\mathbf{p}_{i_1}, \mathbf{p}_{i_2}, \ldots, \mathbf{p}_{i_m}$ can be represented by the condition $(\mathbf{p}_{i_1}, \ldots, \mathbf{p}_{i_m}) \in \mathcal{C}$, where $\mathcal{C} \subset \mathbb{R}^{3m}$ is the feasible set. We assume that the constraint is translation invariant, meaning that applying a common translation to all involved vertices does not change the status of constraint satisfaction (which is the case for most shape constraints relevant to free-form architecture). To facilitate numerical solution, we introduce auxiliary variables $\mathbf{y}_{i_1}, \mathbf{y}_{i_2}, \ldots, \mathbf{y}_{i_m} \in \mathbb{R}^3$ and rewrite the constraint as

$$\begin{cases} \left(\mathbf{y}_{i_1} \cdots \mathbf{y}_{i_m}\right) \in \mathcal{C}, \\ \mathbf{p}_j - \mathbf{mean}\left(\mathbf{p}_{i_1}, \ldots, \mathbf{p}_{i_m}\right) = \mathbf{y}_j, \quad \text{for } j = i_1, \ldots, i_m, \end{cases} \tag{2.1}$$

where $\mathbf{mean}(\mathbf{p}_{i_1}, \ldots, \mathbf{p}_{i_m}) = (\mathbf{p}_{i_1} + \cdots + \mathbf{p}_{i_m})/m$ is the barycenter of $\mathbf{p}_{i_1}, \ldots, \mathbf{p}_{i_m}$. Note that the second constraint in (2.1) is a linear condition which can be written in matrix form $\mathbf{A}_{\mathcal{C}}\mathbf{p} = \mathbf{y}_{\mathcal{C}}$, where vector $\mathbf{p} \in \mathbb{R}^{3N}$ packs all vertex positions, vector $\mathbf{y}_{\mathcal{C}} \in \mathbb{R}^{3m}$ packs the auxiliary variables, and matrix $\mathbf{A}_{\mathcal{C}} \in \mathbb{R}^{3m \times 3N}$. For each soft constraint with feasible set $\mathcal{S}$, we introduce auxiliary variables $\mathbf{y}_{\mathcal{S}} \in \mathcal{S}$ to derive an equivalent condition $\mathbf{A}_{\mathcal{S}}\mathbf{p} = \mathbf{y}_{\mathcal{S}}$. Then the constraint violation can be measured with a function $F_{\mathcal{S}} = \|\mathbf{A}_{\mathcal{S}}\mathbf{p} - \mathbf{y}_{\mathcal{S}}\|_2^2$. Similarly for each hard constraint with feasible set $\mathcal{H}$, we introduce auxiliary variables $\mathbf{y}_{\mathcal{H}} \in \mathcal{H}$ to derive its equivalent condition $\mathbf{A}_{\mathcal{H}}\mathbf{p} = \mathbf{y}_{\mathcal{H}}$. Given N_s soft constraints and N_h hard constraints with feasible sets $\{\mathcal{S}_j | j = 1, \ldots, N_s\}$ and $\{\mathcal{H}_k | k = 1, \ldots, N_h\}$, respectively, the vertex positions $\mathbf{p}$ are computed by the following optimization:

$$\begin{aligned} \min_{\mathbf{p}, \mathbf{y}} \quad & w_h F_{\text{handle}} + w_c F_{\text{close}} + w_f F_{\text{fair}} + \sum_{j=1}^{N_s} w_j^s F_{\mathcal{S}_j} + \sum_{j=1}^{N_s} \sigma_{\mathcal{S}_j}\left(\mathbf{y}_{\mathcal{S}_j}\right) + \sum_{k=1}^{N_h} \sigma_{\mathcal{H}_k}\left(\mathbf{y}_{\mathcal{H}_k}\right) \\ \text{s.t.} \quad & \mathbf{B}\mathbf{p} = \mathbf{y}_H. \end{aligned}$$

Here $\mathbf{y} = \left[\mathbf{y}_{\mathcal{S}_1}, \ldots, \mathbf{y}_{\mathcal{S}_{N_s}}, \mathbf{y}_{\mathcal{H}_1}, \ldots, \mathbf{y}_{\mathcal{H}_{N_h}}\right]$ packs all auxiliary variables for soft constraints and hard constraints, $F_{\mathcal{S}_j}$ is the soft constraint violation function introduced above, and side condition $\mathbf{B}\mathbf{p} = \mathbf{y}_H$ collects all linear relations from the equivalent conditions of hard constraints, with $\mathbf{B} = \left[\mathbf{A}_{\mathcal{H}_1}^T, \ldots, \mathbf{A}_{\mathcal{H}_{N_h}}^T\right]^T$ and $\mathbf{y}_H = \left[\mathbf{y}_{\mathcal{H}_1}^T, \ldots, \mathbf{y}_{\mathcal{H}_{N_h}}^T\right]^T$. Functions $F_{\text{handle}}, F_{\text{close}}, F_{\text{fair}}$ measure respectively the distance from handle vertices to their target positions, the distance from non-handle vertices to their original positions, and the smoothness of the vertex deformation field based on its Laplacian:

$$F_{\text{handle}} = \sum_{i \in \Gamma} \|\mathbf{p}_i - \mathbf{t}_i\|_2^2, \quad F_{\text{close}} = \sum_{j \notin \Gamma} \left\|\mathbf{p}_j - \mathbf{p}_j^0\right\|_2^2, \quad F_{\text{fair}} = \left\|\mathbf{L}\left(\mathbf{p} - \mathbf{p}^0\right)\right\|_2^2,$$

where Γ is the index set for handle vertices, $\mathbf{t}_i$ is the target position for vertex i, $\mathbf{p}_j^0$ is the original position for vertex j, $\mathbf{p}^0$ packs the original positions for all vertices, and $\mathbf{L}$ is the Laplacian matrix. The indicator function $\sigma_{\mathcal{S}_j}\left(\mathbf{y}_{\mathcal{S}_j}\right)$ makes sure $\mathbf{y}_{\mathcal{S}_j} \in \mathcal{S}_j$ in the solution, with

$$\sigma_{\mathcal{S}_j}\left(\mathbf{y}_{\mathcal{S}_j}\right) = \begin{cases} 0, & \text{if } \mathbf{y}_{\mathcal{S}_j} \in \mathcal{S}_j, \\ +\infty, & \text{otherwise.} \end{cases}$$

Indicator function $\sigma_{\mathcal{H}_k}\left(\mathbf{y}_{\mathcal{H}_k}\right)$ is defined in the same way. w_h, w_c, w_f and $w_{\mathcal{S}_j}$ are positive weights trading off different terms. The optimization problem can be written in matrix form as

$$\begin{aligned} \min_{\mathbf{p}, \mathbf{y}} \quad & \|\mathbf{D}\mathbf{p} - \mathbf{r}\|_2^2 + w_f \left\|\mathbf{L}(\mathbf{p} - \mathbf{p}^0)\right\|_2^2 + \sum_{j=1}^{N_s} w_j^s \left\|\mathbf{A}_{\mathcal{S}_j}\mathbf{p} - \mathbf{y}_{\mathcal{S}_j}\right\|_2^2 \\ & + \sum_{j=1}^{N_s} \sigma_{\mathcal{S}_j}\left(\mathbf{y}_{\mathcal{S}_j}\right) + \sum_{j=1}^{N_s} \sigma_{\mathcal{H}_k}\left(\mathbf{y}_{\mathcal{H}_k}\right), \\ \text{s.t.} \quad & \mathbf{B}\mathbf{p} = \mathbf{y}_H, \end{aligned} \tag{2.2}$$

where

$$\mathbf{D} = \begin{bmatrix} d_1\mathbf{I}_3 & & \\ & \ddots & \\ & & d_N\mathbf{I}_3 \end{bmatrix}, \quad \mathbf{r} = \begin{bmatrix} \mathbf{r}_1 \\ \vdots \\ \mathbf{r}_N \end{bmatrix},$$

with $\mathbf{I}_3$ being the 3×3 identity matrix, and

$$d_i = \begin{cases} \sqrt{w_\mathrm{h}} & \text{if } i \in \Gamma \\ \sqrt{w_\mathrm{c}} & \text{otherwise} \end{cases}, \quad \mathbf{r}_i = \begin{cases} d_i\mathbf{t}_i & \text{if } i \in \Gamma \\ d_i\mathbf{p}_i^0 & \text{otherwise} \end{cases} \quad \text{for i} = 1, \ldots, N.$$

2.2.2 Numerical Solution

2.2.2.1 Alternating Minimization

Without hard constraints, problem (2.2) reduces to minimizing quadratic terms with indicator functions. It is solved by alternating between two steps until convergence:

1. *Projection*: fix $\mathbf{p}$ and minimize over $\mathbf{y}$.
2. *Linear solve*: fix $\mathbf{y}$ and minimize over $\mathbf{p}$.

The minimization in step 2 simply amounts to solving a symmetric positive definite (SPD) sparse linear system, hence the name. For step 1, the problem is separable for auxiliary variables from different constraints and is solved in parallel. Specifically, we solve a set of independent subproblems, each of which is associated with one constraint and has the following form:

$$\min_{\mathbf{y}_C} \left\|\mathbf{y}_C - \mathbf{x}\right\|_2^2 + \sigma_C(\mathbf{y}_C),$$

where C is the feasible set and $\mathbf{y}_C$ are the auxiliary variables for the constraint. The solution is the closest *projection* from $\mathbf{x}$ onto C, which we call the *proximal operator* of C for input data $\mathbf{x}$. For many constraints, we can derive the closed-form representation of the proximal operator. For example (see [3] for details):

- *Coplanarity*. This constraint requires $n > 3$ vertices to lie on a common plane. It can be used to model planar panels, for example, by requiring the vertices of each mesh face to be coplanar (see Fig. 2.1). The proximal operator finds n coplanar points $\mathbf{y}_1, \ldots, \mathbf{y}_n \in \mathbb{R}^3$ closest to the input data $\mathbf{x}_1, \ldots, \mathbf{x}_n \in \mathbb{R}^3$. The solution is $\mathbf{y}_i = \mathbf{x}_i - \mathbf{n}[\mathbf{n} \cdot (\mathbf{x}_i - \overline{\mathbf{x}})]$ $(i = 1, \ldots, n)$, where $\overline{\mathbf{x}} = \mathbf{mean}(\mathbf{x}_1, \ldots, \mathbf{x}_n)$ and $\mathbf{n}$ is the left singular vector of matrix $[\mathbf{x}_1, \ldots, \mathbf{x}_n]$ for the smallest singular value.
- *Regular polygon*. This constraint requires a face with $n \geq 3$ vertices to be a regular n-gon. It can be used to induce shape regularity of mesh elements (see Fig. 2.2). The proximal operator finds a regular n-gon closest to a polygon with vertices $\mathbf{x}_1, \ldots, \mathbf{x}_n \in \mathbb{R}^3$. This can be done by computing the translation, rotation, and scaling of a predefined regular n-gon to fit the target polygon, using the algorithm in [11].

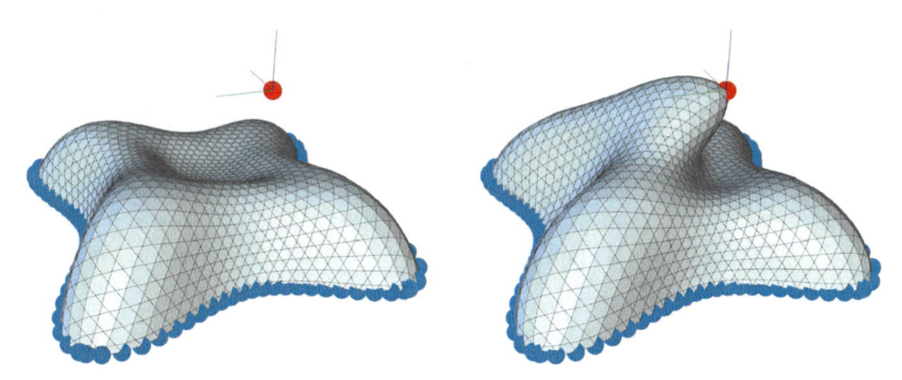

Fig. 2.2 Handle-based deformation of a constrained mesh subject to the soft constraint that each face is a regular polygon. *Left*: the initial mesh with the handles (shown in *red* and *blue*) attached to the boundary vertices and four vertices in the middle. The handles for the middle positions are moved to new target positions (shown in *red*). *Right*: the mesh deforms according to the handle positions while satisfying the soft constraints

2.2.2.2 Augmented Lagrangian Method

When dealing with hard constraints, extra care has to be taken to ensure that the linear side constraints in problem (2.2) are satisfied. This is done using the *augmented Lagrangian method* (ALM) [12], which searches for a saddle point of the following *augmented Lagrangian function*:

$$\mathcal{L}(\mathbf{p}, \mathbf{y}, \lambda; \mu) = F(\mathbf{p}, \mathbf{y}) + \lambda^T \mathbf{h}(\mathbf{p}, \mathbf{y}) + \mu \|\mathbf{h}(\mathbf{p}, \mathbf{y})\|_2^2, \tag{2.3}$$

where $F(\mathbf{p}, \mathbf{y})$ is the target function in (2.2), $\mathbf{h}(\mathbf{p}, \mathbf{y}) = \mathbf{B}\mathbf{p} - \mathbf{y}_H$ is the residual of side constraints in (2.2), λ is a vector of dual variables, and $\mu > 0$ is a penalty parameter. The solver iteratively updates $\mathbf{p}, \mathbf{y}, \lambda$ and μ until convergence. In each iteration, new values $\hat{\mathbf{p}}, \hat{\mathbf{y}}, \hat{\lambda}, \hat{\mu}$ are computed from current values $\overline{\mathbf{p}}, \overline{\mathbf{y}}, \overline{\lambda}, \overline{\mu}$ using the following steps:

1. *Primal update*: $(\hat{\mathbf{p}}, \hat{\mathbf{y}}) = \min_{\mathbf{p},\mathbf{y}} \mathcal{L}(\mathbf{p}, \mathbf{y}, \overline{\lambda}, \overline{\mu})$.
2. *Dual update*: $\hat{\lambda} = \overline{\lambda} + \overline{\mu}\,\mathbf{h}(, \hat{\mathbf{p}}, \hat{\mathbf{y}})$.
3. *Penalty update*: choose $\hat{\mu} \geq \overline{\mu}$.

The problem in step 1 has a similar structure as the one from Sect. 2.2.2.1 and is solved in the same way. Specifically, it alternates between two steps:

1. Projection step with proximal operator evaluations:

$$\min_{\mathbf{y}_{\mathcal{S}_j}} \left\| \mathbf{y}_{\mathcal{S}_j} - \mathbf{A}_{\mathcal{S}_j}\mathbf{p} \right\|_2^2 + \sigma_{\mathcal{S}_j}\left(\mathbf{y}_{\mathcal{S}_j}\right), \qquad j = 1, \ldots, N_{\mathrm{s}},$$

$$\min_{\mathbf{y}_{\mathcal{H}_k}} \left\| \mathbf{y}_{\mathcal{H}_k} - \left(\mathbf{A}_{\mathcal{H}_k}\mathbf{p} + \frac{\lambda_{\mathcal{H}_k}}{2\mu} \right) \right\|_2^2 + \sigma_{\mathcal{H}_k}\left(\mathbf{y}_{\mathcal{H}_k}\right), \qquad k = 1, \ldots, N_{\mathrm{h}},$$

where $\lambda_{\mathcal{H}_k}$ collects the components of λ in the same positions as $\mathbf{y}_{\mathcal{H}_k}$ in $\mathbf{y}_H$.

2. Solving a sparse SPD system for $\mathbf{p}$:

$$\begin{aligned} &\left(\mathbf{D}^T\mathbf{D} + w_{\mathrm{f}}\mathbf{L}^T\mathbf{L} + \mu\mathbf{B}^T\mathbf{B} + \sum_{j=1}^{N_{\mathrm{s}}} w_j^s \mathbf{A}_{\mathcal{S}_j}^T \mathbf{A}_{\mathcal{S}_j} \right) \mathbf{p} \\ &= \mathbf{D}^T\mathbf{r} + w_{\mathrm{f}}\mathbf{L}^T\mathbf{L}\mathbf{p}^0 + \mu\mathbf{B}^T\left(\mathbf{y}_H - \frac{\lambda}{2\mu} \right) + \sum_{j=1}^{N_{\mathrm{s}}} w_j^s \mathbf{A}_{\mathcal{S}_j}^T \mathbf{y}_{\mathcal{S}_j}. \end{aligned} \tag{2.4}$$

The primal update step is the most time-consuming part of the solver. We will not go into the details of steps 2 and 3, but refer the readers to [8] instead. Note that for a given problem, the linear system matrix in (2.4) only changes according to the penalty parameter μ. The penalty update scheme in [8] only generates a predefined set of values for μ, so we can precompute all linear system matrices that appear in (2.4).

2.3 General Implementation Strategies

We developed an interactive handle-based shape manipulation system for constrained meshes, based on the algorithms presented in the previous section. For an initial mesh, the user selects a set of handle vertices and specifies their target positions (which we call *handle positions*) by dragging 3D manipulators. Whenever the manipulators are moved, the system deforms the mesh according to the new handle positions, providing immediate feedback to the user (see Fig. 2.2 for an example).

Figure 2.3 shows the architecture of our system. Here we distinguish between the work of the threads from the user side (user interface, mouse and keyboard interaction, mesh display, etc.), which we gather as the *user module*, and the work done within a single thread dedicated to a GPU-based ALM solver, which we call the *optimization module*. The latter loops over three main logical steps:

1. *Input phase*: transfer current handle positions to GPU.
2. *Optimization phase*: iterate the ALM steps on GPU, until some output conditions are satisfied.
3. *Output phase*: read back updated vertex positions from GPU.

To run the ALM solver on GPU, we store on the GPU all the optimization variables, as well as other auxiliary data [such as matrices $\mathbf{A}_{\mathcal{S}_j}$, $\mathbf{B}$, $\mathbf{D}$ and vector $\mathbf{r}$ in

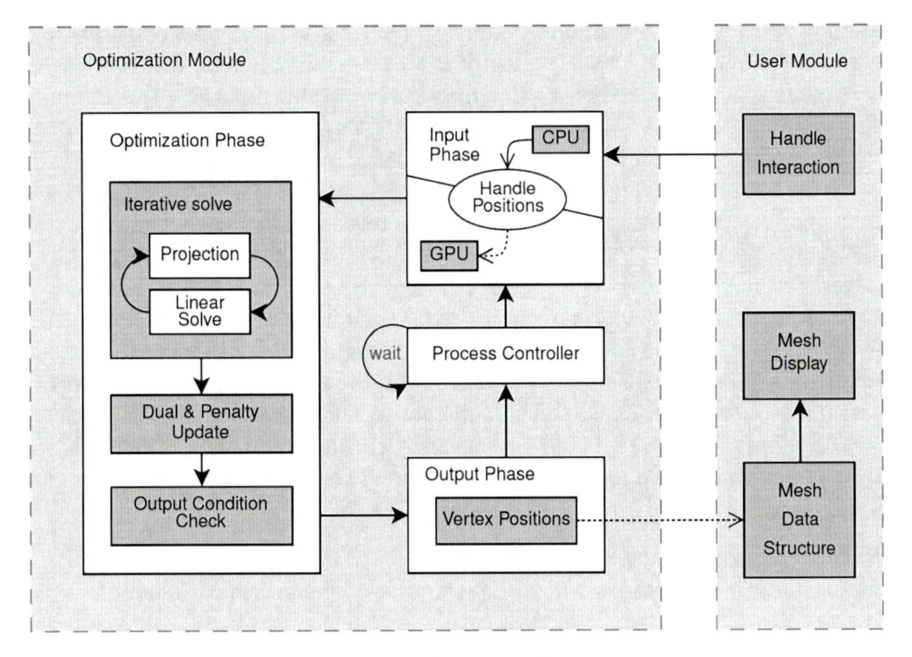

Fig. 2.3 The architecture of our GPU-based implementation

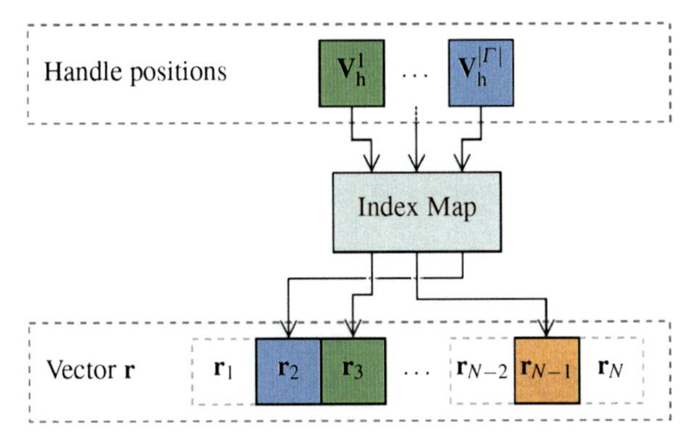

Fig. 2.4 The update of the handle data on GPU is done using a kernel that fills vector $\mathbf{r}$ according to the handle position vector (on GPU) and a precomputed index map

formulation (2.2) and the linear system matrices in problem (2.4)]. Many of these data remain constant during optimization and only need to be initialized once at the beginning. Thus, in the input phase, we only need to transfer the latest handle positions to the GPU to update the problem specification.

As an iterative solver, the optimization phase requires initial values of the variables. To initialize the current optimization phase, we always use the resulting variable values from the previous optimization phase. The motivation is that when a user drags the handles continuously, the handle positions used in two consecutive optimization phases are close to each other. Thus, their solutions will be close to each other as well, making the solution from the previous phase a good guess for the current solution.

Depending on the data, the optimization phase might take a large number of iterations to fully converge. To keep the process interactive, we allow switching from optimization phase to output phase even if it is not fully convergent yet. When the handles are dragged, they are likely to be moving at the same time as the ALM solver is running. Rather than solving the current problem to a very high accuracy, it is more important to output the current result and start a new optimization phase with the new handle positions, so that the mesh shape follows the handle positions smoothly and shows how the shape reacts to handle position changes. Even if the output mesh shape is not the exact solution, it is still a good approximation because the solver usually converges quickly to an approximate solution [13]. Therefore, we switch from optimization phase to output phase, if one of the following conditions is satisfied:

1. The optimization phase fully converges.
2. The number of iterations within the optimization phase exceeds a limit $M_{\max}$.

The output phase is responsible for reading back new vertex positions in order to update the mesh data structure in host memory, which is then used to update the mesh display. Both operations (vertex readback and mesh display update) involve data transfer between CPU and GPU. To avoid unnecessary transfer while keeping the process interactive, we only read back vertex positions if the elapsed time (in milliseconds) from the last readback is larger than a threshold ε. With such a strategy, the maximum frame rate for mesh display is $1,000/\varepsilon$ FPS.

After the output phase, depending on the availability of new handle positions and the convergence of the optimization phase, we are in one of the following cases:

- If there are new handle positions, transfer them to GPU and start a new optimization phase.
- Otherwise, if the previous optimization phase was not fully convergent, resume the optimization.
- Otherwise, wait for new handle positions.

2.4 CUDA Implementation Details

Our GPU implementation was done with CUDA. We targeted NVIDIA GeForce GTX 580 [14], which runs under the Fermi architecture [15, 16]. It has 16 streaming multiprocessors providing a total of 512 cores. Each of them has 64 kB of memory available between the L1 cache and the shared memory. The rest of this section will present the challenges and specific implementation details.

2.4.1 *Kernels*

We implemented custom kernels for two critical operations: updating the handle data and evaluating the proximal operators.

2.4.1.1 Handle Update

When starting an optimization phase with new handle positions, we need to update the GPU memory storage of vector $\mathbf{r}$ in formulation (2.2). With the number of handle vertices being usually much smaller than the number of vertices, we first transfer the handle positions onto the GPU as a contiguous vector $\mathbf{V}_{\mathrm{h}} \in \mathbb{R}^{3|\Gamma|}$. Then a custom kernel updates the entries of $\mathbf{r}$ according to $\mathbf{V}_{\mathrm{h}}$, using a precomputed index map (see Fig. 2.4). Note that the index map remains unchanged during optimization, since neither the choice of handle vertices nor the mesh topology is allowed to change.

Another strategy would be to transfer only the handle positions that are being changed by the user. This requires a *dynamic* index map for writing to vector $\mathbf{r}$, as well as checking which handles are being moved. To simplify implementation, we did not use such strategy.

2.4.1.2 Proximal Operator Evaluation

As we saw in Sect. 2.2, proximal operators are responsible for updating auxiliary variables. Each type of constraint corresponds to one proximal operator, which involves a predefined set of operations. For different constraints of the same type, their proximal operator evaluation is independent since the involved auxiliary variables do not overlap. Such characteristics make it suitable to evaluate proximal operators using custom CUDA kernels. Specifically, we implement one kernel for each type of constraint, within which each thread handles one constraint.

For high performance, we need to ensure coalesced memory access. Thus, we store the auxiliary variables $\mathbf{y}$ in formulation (2.2) into a contiguous array in global memory, where the components corresponding to the same kernel reside in a contiguous region. The input data for proximal operators are of the same dimension as $\mathbf{y}$, and we store them with an array in global memory using the same layout as $\mathbf{y}$ (see Fig. 2.6 for an example).

Another performance consideration is the grid and block sizes. We follow [17] which suggests a number of threads per block:

1. Dividing the maximum number of threads per SM, i.e., 1,536 for Fermi
2. At least 32 threads per block, i.e., the warp size
3. At most 3 blocks per SM, so as to maximize occupancy (and thus at least 1,536/ $8 = 192$ threads per block)

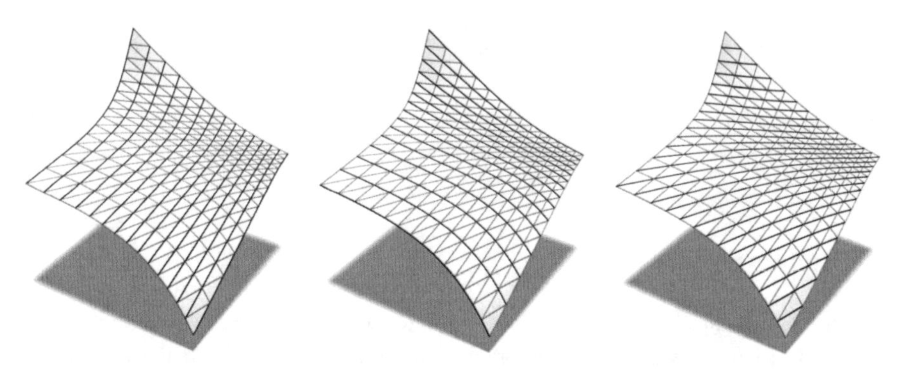

Fig. 2.5 For a regular triangle mesh (i.e., each interior vertex has valence 6, and each boundary vertex has valence no larger than 6), there exist three families of edge polylines (shown in *blue*). Being a planar web requires each polyline to be planar, namely, all vertices on the polyline lie in a common plane

Since we do not know the relation between different types of kernels, we chose to simply saturate them by using a block size of 512 threads, which proved to be sufficient for our need according to experiments.

Coplanarity Constraint

Because of specific features and limitations of GPU, additional care needs to be taken when implementing some proximal operators. Here we use the vertex coplanarity constraint as an example to show the challenges and our solutions. Coplanarity constraint is one of the most important shape constraints in free-form architecture. It can be used to model planar panels [18] (Fig. 2.1), as well as *planar webs* which consist of curve elements of planar shapes [19] (Fig. 2.5). For input data $\mathbf{x}_1, \ldots, \mathbf{x}_n \in \mathbb{R}^3$, a key step of the proximal operator is a singular value decomposition (SVD) to extract the left singular vector of $\mathbf{M} = [\mathbf{x}_1, \ldots, \mathbf{x}_n] \in \mathbb{R}^{3 \times n}$ for the smallest singular value (see Sect. 2.2.2.1).

Due to the memory layout requirement mentioned before, the global memory storage of $\mathbf{x}_1, \ldots, \mathbf{x}_n$ is already a column-major representation for matrix $\mathbf{M}$. Thus, a naïve approach is to implement an SVD solver that operates directly on the global memory storage of $\mathbf{M}$. However, this might lead to excessive access to global memory, lowering the performance significantly [20].

To reduce global memory access, we implemented the kernel as follows. First, note that the target singular vector is the same as the right singular vector of 3×3 matrix $\mathbf{M}\mathbf{M}^T = \sum_{i=1}^{n} \mathbf{x}_i \mathbf{x}_i^T$ for the smallest singular value. Thus, we create matrix $\mathbf{M}\mathbf{M}^T$ on local memory, by reading each $\mathbf{x}_i$ from global memory and summing up $\mathbf{x}_i \mathbf{x}_i^T$. Afterwards, we perform SVD on matrix $\mathbf{M}\mathbf{M}^T$. In this way, each global memory element of $\mathbf{M}$ needs to be accessed only once for computing the singular vector. Moreover, this approach only performs SVD on a 3×3 matrix. For coplanarity constraints involving a large number of vertices, this significantly reduces the

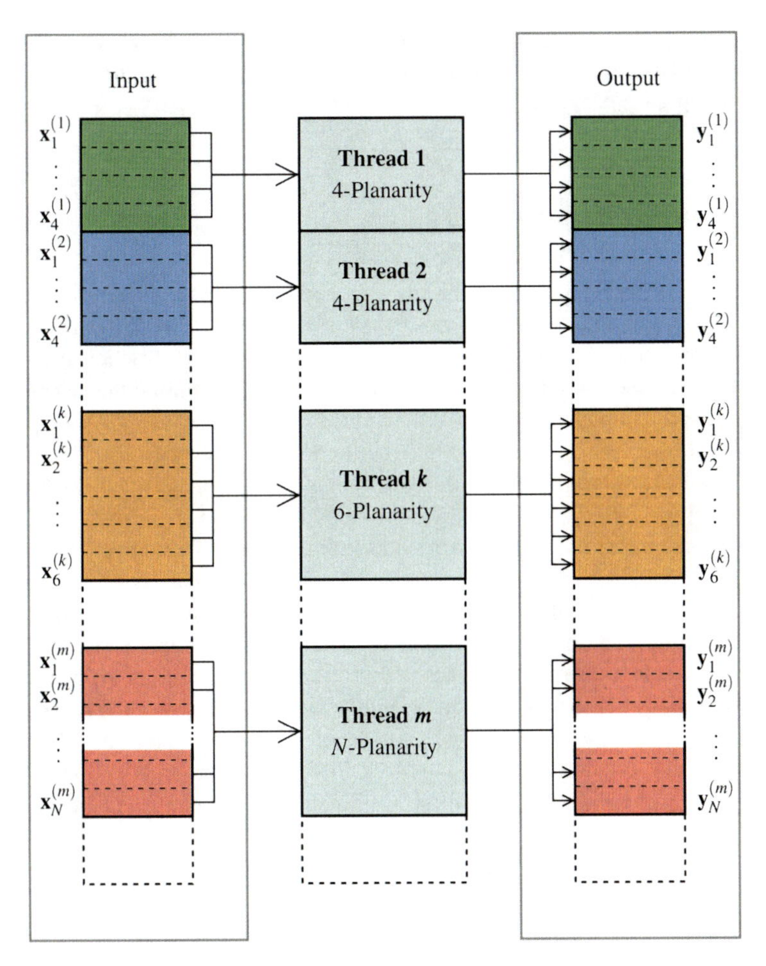

Fig. 2.6 Schematic diagram for the proximal operator kernel of coplanarity constraints. Input data $\mathbf{x}$ and output data $\mathbf{y}$ are stored in two contiguous arrays, respectively. Within each array, data associated with a thread reside in a contiguous region. Our implementation is able to handle coplanarity constraints for different number of vertices within a single kernel. Here *N-planarity* refers to a coplanarity constraint for N vertices

matrix storage on local memory compared to the original matrix $\mathbf{M}$. Such compact storage helps to reduce register spilling and L1 cache misses, which improves the performance of the kernel. Furthermore, with this approach, we are able to deal with coplanarity constraints with different number of vertices using a single kernel, by precomputing an array that stores for each coplanarity constraint the following information: (1) the number of vertices and (2) the address of input data. Using a single kernel helps to increase parallelism for the implementation, resulting in improved throughput of the system. Figure 2.6 provides a schematic diagram for the kernel of coplanarity constraints.

For 3×3 SVD, we implemented a simple SVD solver based on [21]. There exists a branch-free 3×3 SVD solver [22] that might provide higher performance, but our simple implementation turned out to be sufficient.

2.4.2 Sparse Linear Algebra

In general, all matrices in formulation (2.2) are sparse, while the vectors are all dense. Therefore, the solver requires many sparse matrix vector multiplications (SpMV). For these operations, we used the Cusp library [23] which provides an easy C++ interface for sparse linear algebra with CUDA. Among the sparse matrix formats provided by Cusp, we chose the hybrid format (ELL + COO) as it provides faster linear operations for general unstructured sparse matrices [24].

Since we are targeting large meshes, we solve the sparse linear system (2.4) using a conjugate gradient (CG) solver provided by Cusp. To warm-start the solver, we always use the previous CG solution as initial value for the current CG solving. Typically, the right-hand side of system (2.4) changes gradually within the ALM solver; thus, two consecutive solutions of problem (2.4) do not deviate significantly from each other, making this warm-starting strategy a reasonable choice. Alternatively, direct solvers based on Cholesky factorization can be more efficient. On the other hand, they often require more memory storage, because the sparsity of the linear system matrix is not preserved by its Cholesky factors. This could be an issue for GPU, since typically the amount of GPU memory is smaller than the host memory. Thus, in our implementation, we opted for a simple CG solver.

2.5 Results

In this section, we provide some performance results of our GPU-based constrained mesh deformation method and compare them against the CPU version. The CPU version follows the same optimization workflow as described in Sect. 2.3, except that all the data reside in the host memory so there is no need to transfer handle positions in input phase and read back vertex positions in output phase. For both CPU and GPU versions, the frame rate was limited to 30 FPS (i.e., the minimum elapsed time between two vertex readback operations is 33.3 ms), and the maximum number of iterations in optimization phase was set to $M_{\max} = 50$.

Both CPU and GPU versions were implemented for double-precision floating point data. We used two CPU implementations with different solvers for system (2.4): one uses CG, and the other uses a direct solver based on Cholesky factorization. Both CPU implementations reduce system (2.4) into three smaller systems for the x, y, z coordinates of the vertices, respectively, with the same system matrix [3]. This allows the three coordinates of each vertex to be solved in parallel. The CPU version utilized OpenMP for the parallelization of proximal operator evaluation and linear system solving and used the Eigen library [25] for linear algebra

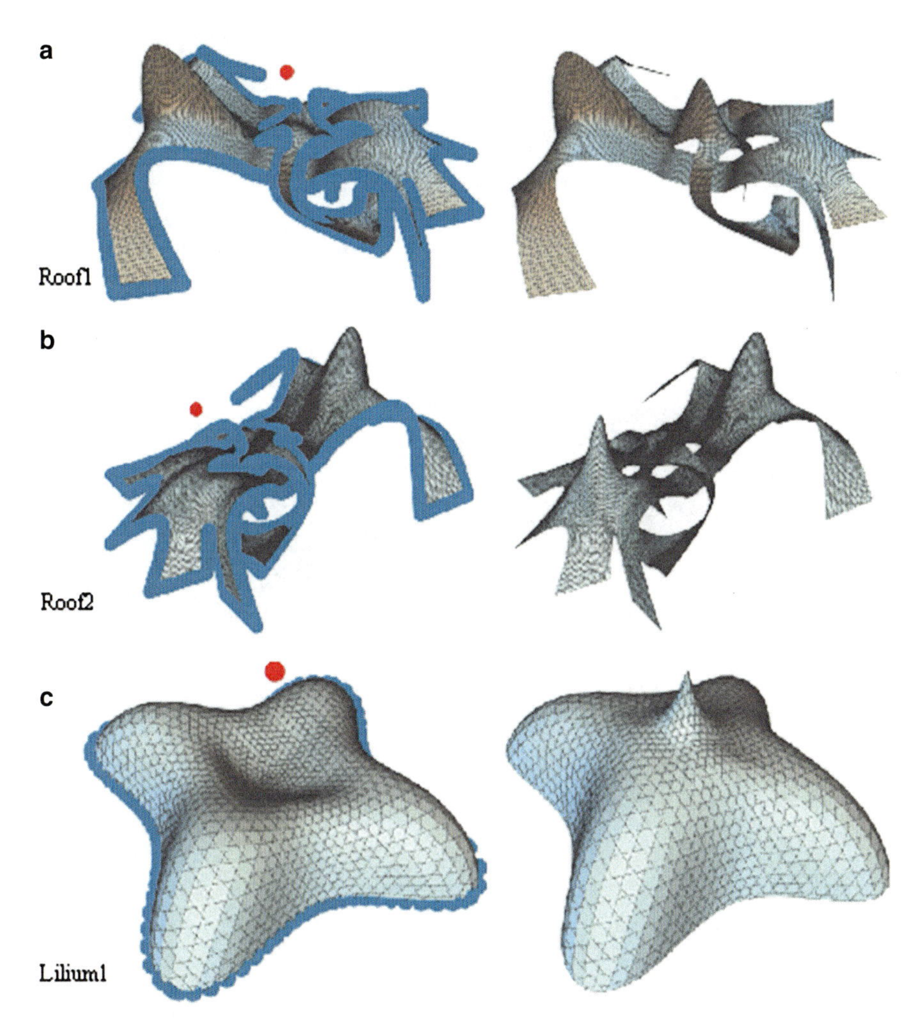

Fig. 2.7 The first sets of models with their configuration and output illustrations

operations. For the CG solver on both CPU and GPU, we set the maximum number of iterations to 100 and the tolerance for the 2-norm ratio between the residual and right-handle side to 1×10^{-6}. The CPU and GPU implementations were run on a PC with an NVIDIA GTX 580 and an Intel Core i7 870 with four cores.

For comparison, each implementation was run with the same set of meshes and constraints. Since the optimization phase spends most of the running time on proximal operator evaluation and linear system solving, we focused the performance comparison on these two steps. Thus, we only used soft constraints in our experiments, so that the optimization phase alternated between proximal operator evaluation and linear system solving. Figures 2.7 and 2.8 show the meshes used in our experiments, with the configuration of meshes and their constraints listed in

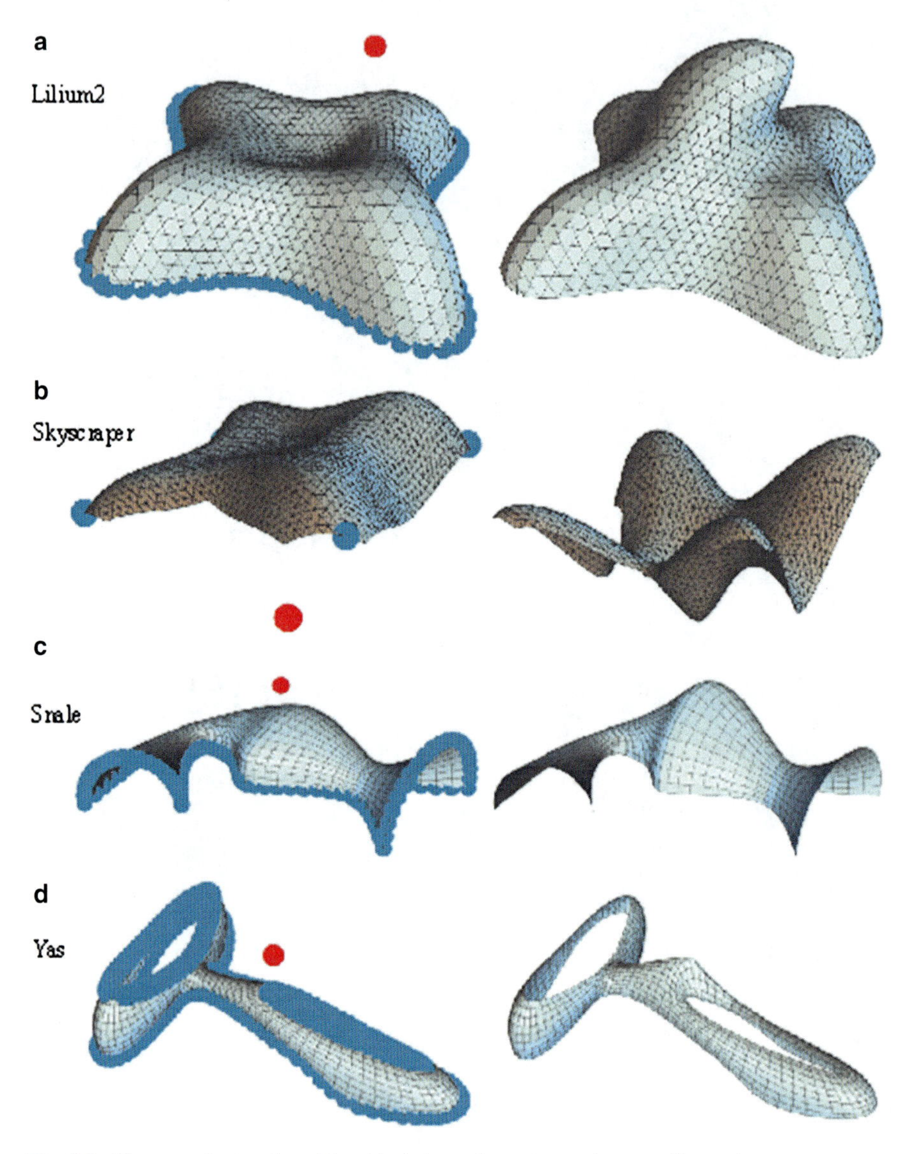

Fig. 2.8 The second sets of models with their configuration and output illustrations

Table 2.1. Here the initial mesh in Roof2 is a subdivided version of the initial mesh in Roof1, while Lilium1 and Lilium2 have the same initial mesh shape under different constraints. The coplanarity constraints (for planar faces and planar web) are applied to a face or a polyline only if it has more than three vertices, while the constraints of regular polygons are applied to all faces of a mesh.

For each mesh, some boundary vertices and interior vertices were chosen as handle vertices, with their handle positions shown in blue and red, respectively. In each experiment, the red handles were moved to trigger mesh deformation.

Table 2.1 Configurations for meshes shown in Figs. 2.7 and 2.8

Reference label	Vertices	Faces	Constraint type	Handles
Roof1	20,464	19,712	Planar faces	1,505
Roof2	80,352	78,848	Planar faces	3,012
Lilium1	3,504	3,505	Regular polygon faces	100
Lilium2	3,504	3,505	Planar faces	100
Skyscraper	1,517	2,884	Planar web	5
Snale	1,092	1,020	Planar faces	143
Yas	1,085	976	Planar faces	221

Table 2.2 Average frame time for different implementations

	Average frame time [ms]		
Mesh	CPU CG	CPU Cholesky	GPU CG
Roof1	2,159.43	791.03	29.32
Roof2	14,965.90	3,842.25	107.20
Lilium1	638.45	132.84	17.82
Lilium2	210.34	43.99	14.01
Skyscraper	119.77	279.12	3.92
Snale	115.29	63.78	23.89
Yas	94.64	52.45	3.04

Table 2.2 shows the average elapsed time between two entries to the output phase, which we refer to as *average frame time*. A system with average frame time of α milliseconds can achieve an average frame rate up to $1,000/\alpha$ FPS if the frame rate is not limited. Thus, smaller average frame time indicates more interactive result. We can see that even for a mesh with 80K vertices and 79K constraints, our GPU implementation achieves a frame rate of 9 FPS, while the frame rates for CPU implementations are much lower than 1 FPS. For a smaller model with about 1K vertices and 1K constraints, our GPU implementation can potentially achieve a frame rate of over 300 FPS, well beyond the specified upper limit. The comparison on average frame time shows that our GPU implementation gained significant speedups with respect to the CPU implementations.

The accompanying video shows the user interaction for Roof2. We can see that due to the large number of vertices and constraints, the CPU implementations failed to respond quickly to handle position changes. On the other hand, the GPU implementation remains interactive, leading to more intuitive shape manipulation.

Finally, Table 2.3 gives the timing ratio between input phase (*input*), proximal operator evaluation (*projection*), linear system solving (*CG*), and output phase (*output*) for a typical interaction session on GPU. It can be seen that linear system solving spent the largest portion of time.

Table 2.3 Ratio of running time in each part of the optimization phase on GPU

	% of the time spent in GPU optimization phase			
Mesh	Input	Projection	CG	Output
Roof1	0.34	4.27	88.29	7.10
Roof2	0.00	3.32	86.81	9.87
Lilium1	0.00	0.21	98.01	1.78
Lilium2	0.04	0.16	97.22	2.59
Skyscraper	0.00	0.85	98.43	0.72
Snale	0.01	0.04	99.89	0.07
Yas	0.28	0.74	98.71	0.28

2.6 Limitation and Future Work

In our system, the linear system solving is the bottleneck of performance. This is due to the well-known fact that SpMV involves irregular data access and thus achieves lower performance compared to dense operations on GPU. This motivates us to explore more advanced GPU SpMV techniques such as [26] to further optimize the performance. Another option is to adapt Cholesky-based direct solvers to GPU, as direct solvers outperformed CG for CPU implementations in many of our experiments.

A more ambitious improvement would be a hybrid GPU/CPU optimization. Currently, the CPU is only used for managing the GPU, and it is mostly idle during the optimization. Thus, we plan to investigate workload distribution between CPU and GPU to gain higher performance.

Our implementation requires frequent readback of vertex positions from GPU in order to update the display, which incurs some performance loss. One of our future plans is to directly update mesh display on GPU using vertex buffer object, thus totally avoiding data transfer between CPU and GPU in the output phase.

Finally, our system runs on CUDA-enabled GPUs only. We intend to develop an OpenCL-based system to make the algorithm available for a wider range of hardwares and platforms and to compare the performance between different GPUs.

Conclusion

In this chapter, we present an efficient handle-based constrained mesh manipulation system implemented on GPU. The mesh manipulation is formulated as a constrained optimization problem, which is decomposed into simple subproblems that can be solved in parallel. Utilizing the computational power of GPU, we achieve significant speedup of constrained mesh deformation compared to CPU implementations, as shown by our experiments on meshes with different sizes and constraints. On the other hand, linear system solving becomes the performance bottleneck, which provides an interesting avenue for future research.

Acknowledgments The authors thank Asymptote Architecture for providing the figure of Yas Viceroy Hotel. The mesh models are provided by Asymptote Architecture and Waagner Biro (Yas), Zaha Hadid Architects and Amir Vaxman (Lilium1 and Lilium2), and Mario Deuss (Roof1, Roof2, and Snale). This work has been supported by Swiss National Science Foundation (SNSF) grant 200021_137626.

References

1. Eigensatz, M., Kilian, M., Schiftner, A., Mitra, N.J., Pottmann, H., Pauly, M.: Paneling architectural freeform surfaces. ACM Trans. Graph. **29**(4), 1–10 (2010)
2. Yang, Y.L., Yang, Y.J., Pottmann, H., Mitra, N.J.: Shape space exploration of constrained meshes. ACM Trans. Graph. 30(6), 124:1–124:12 (2011)
3. Bouaziz, S., Deuss, M., Schwartzburg, Y., Weise, T., Pauly, M.: Shape-up: shaping discrete geometry with projections. Comput. Graph. Forum **31**(5), 1657–1667 (2012)
4. Vaxman, A.: Modeling polyhedral meshes with affine maps. In: Symposium on Geometry Processing (2011)
5. Zhao, X., Tang, C.C., Yang, Y.L., Pottmann, H., Mitra, N.J.: Intuitive design exploration of constrained meshes. In: Advances in Architectural Geometry (2012).
6. Poranne, R., Ovreiu, E., Gotsman, C.: Interactive planarization and optimization of 3D meshes. Comput. Graph. Forum **32**(1), 152–163 (2013)
7. Deng, B., Bouaziz, S., Deuss, M., Zhang, J., Schwartzburg, Y., Pauly, M.: Exploring local modifications for constrained meshes. Comput. Graph. Forum **32**(2), 11–20 (2013)
8. Deng, B., Bouaziz, S., Deuss, M., Kaspar, A., Schwartzburg, Y., Pauly, M.: Interactive design exploration for constrained meshes. Computer-Aided Design (2014)
9. Song, P. Fu, C.W., Goswami, P. Zheng, J. Mitra, N.J., Cohen-Or, D.: Reciprocal frame structures made easy. ACM Trans. Graph. **32**(4), 94:1–94:13 (2013)
10. Bao, F., Yan, D.M., Mitra, N.J. Wonka, P.: Generating and exploring good building layouts. ACM Trans. Graph. **32**(4), 122:1–122:10 (2013)
11. Umeyama, S.: Least-squares estimation of transformation parameters between two point patterns. IEEE Trans. Pattern Anal. Mach. Intell. **13**(4), 376–380 (1991)
12. Bertsekas, D.P.: Constrained Optimization and Lagrange Multiplier Methods. Athena Scientific, Belmont, MA (1996)
13. Boyd, S., Parikh, N., Chu, E., Peleato, B., Eckstein, J.: Distributed optimization and statistical learning via the alternating direction method of multipliers. Found. Trends Mach. Learn. **3**(1), 1–122 (2011)
14. NVIDIA: NVIDIA GeForce GTX 580 datasheet (2010)
15. Glaskowsky, P.N.: NVIDIA's Fermi: The First Complete GPU Computing Architecture. (2009)
16. NVIDIA: Fermi Compute Architecture Whitepaper (2009). Version 1.1
17. Torres, Y., Gonzalez-Escribano, A., Llanos, D.: Understanding the impact of CUDA tuning techniques for Fermi. In: 2011 International Conference on High Performance Computing and Simulation (HPCS), pp. 631–639 (2011)
18. Glymph, J., Shelden, D., Ceccato, C., Mussel, J., Schober, H.: A parametric strategy for free-form glass structures using quadrilateral planar facets. Automation in Construction **13**(2): 187–202 (2004), Conference of the Association for Computer Aided Design in Architecture
19. Deng, B., Pottmann, H., Wallner, J.: Functional webs for freeform architecture. Comput. Graph. Forum **30**(5), 1369–1378 (2011)
20. NVIDIA: CUDA C Programming Guide
21. Press, W.H., Teukolsky, S.A., Vetterling, W.T., Flannery, B.P.: Numerical Recipes: The Art of Scientific Computing, 3rd edn. Cambridge University Press, New york (2007)

22. McAdams, A., Selle, A., Tamstorf, R., Teran, J., Sifakis, E.: Computing the singular value decomposition of 3×3 matrices with minimal branching and elementary floating point operations. Technical Report, University of Wisconsin-Madison (2011)
23. Bell, N., Garland, M.: Cusp: generic parallel algorithms for sparse matrix and graph computations. http://cusp-library.googlecode.com (2012). Version 0.3.0
24. Bell, N., Garland, M.: Implementing sparse matrix-vector multiplication on throughput-oriented processors. In: Proceedings of the Conference on High Performance Computing Networking, Storage and Analysis, SC'09, pp. 18:1–18:11 (2009)
25. Guennebaud, G., Jacob, B., et al.: Eigen v3. http://eigen.tuxfamily.org (2010)
26. Baskaran, M.M., Bordawekar, R.: Optimizing sparse matrix-vector multiplication on GPUs. Technical Report RC24704, IBM (2008)

Chapter 3
GPU-Based Real-Time Volume Interaction for Scientific Visualization Education

Yanlin Luo, Zhongke Wu, Zuying Luo, and Yanhong Luo

Abstract In this chapter, we introduce the interaction methods of our self-developed VisEdu as a visual teaching system to teach scientific visualization courses at Beijing Normal University. VisEdu provides real-time visualization and interaction of midsize CT datasets at interactive frame rates via CUDA-based volume rendering. We describe various rendering methods through plane, superquadric, and virtual lenses tools which offer different views of the same dataset. It aids the students to better understand the feature of virtual contents and the core algorithms of the scientific visualization course such as volume rendering, volume interaction, etc.

Keywords Scientific visualization • Visual teaching system • CUDA-based volume rendering • Transfer function • Volume interaction

3.1 Introduction

Now scientific visualization courses have been taught at universities around the world. It is becoming an important part of the curriculum in a number of disciplines. It mainly studies the computational methods for converting and exploiting visual information that is easy to be understood from scientific data. At Beijing Normal University, we offer the scientific visualization course for several years. This course teaches the fundamentals of scientific visualization, the main concepts and algorithms of visualization techniques, which covers some topics including the visual programming techniques, volume rendering and isosurfacing, volume illustration, volume interaction, vector and tensor fields' visualization, etc.

Y. Luo (✉) • Z. Wu • Z. Luo
The College of Information Science and Technology, Beijing Normal University, Beijing, China
e-mail: luoyl@bnu.edu.cn; zwu@bnu.edu.cn; luoz@bnu.edu.cn

Y. Luo
The College of Electrical Engineering, Northwest University for Nationalities, Lanzhou, China

The School of Nuclear Science and Technology, Lanzhou University, Lanzhou, China
e-mail: luoyh12@lzu.edu.cn

Y. Cai, S. See (eds.), *GPU Computing and Applications*,
DOI 10.1007/978-981-287-134-3_3

Among those course contents, volume rendering-related topics take the most of time. At first the Visualization Toolkit (VTK) consisting of a C++ class library was used in assignments or teaching, but it is not visual system for easy study. The students began to use our self-developed visual teaching system VisEdu to build their visualizations since 2011. VisEdu integrates our new research results especially on volume rendering and volume interaction. Because of the limited space, we only describe its interaction method in this chapter.

Volume rendering has been an active area of research for few years. Much work concentrate on the direct rendering of complete volumes, but that process has traditionally been very slow and is thus used to obtain final images, not to perform interaction.

Volume interaction has instead usually been a process of finding clever ways to abstract something meaningful or simplify the datasets into something more rapidly displayable. Recent developments exploit the programmability features and speed of the specialized GPU, allowing the system to achieve interactive speed on an ordinary desktop.

The motion of virtual content via interaction can cause our eye reflex so as to draw our attention [1]. And during the interaction, the abstraction makes our visual system to reduce presented information and improve our understanding of complex datasets based on our structure recognition [2]. For visualization education, the real-time volume interaction aids the student to better understand the feature of virtual contents and the core algorithms of the scientific visualization course such as volume rendering, volume interaction, etc.

This chapter is structured as follows. Section 3.2 gives an overview of related work concerning GPU-based visualization and interaction. Section 3.3 follows on describing the proposed method including the design of transfer function and interaction tools. Section 3.4 shows implementation and results. The final section gives some conclusions and ideas for future work.

3.2 Related Work

Nowadays volumetric datasets are growing in terms of number and size, resulting in two challenges: maintaining performance and extracting meaningful information.

3.2.1 GPU-Accelerated Volume Rendering

Direct volume rendering (DVR) using ray casting is the most widely used and accepted technique to produce high-quality images. The volume rendering is based on the classical ray casting, which implements a simple physical light emission and absorption model without scattering effects [3]. The general approach is to shoot rays through the pixels into the field volume and to accumulate the color and opacity contributions at discrete locations to produce the final pixel color.

In the last few years, many sophisticated techniques for real-time volume rendering have been proposed on desktop platforms [4]. Current high-quality solutions, based on ray casters fully achieved in GPU, have demonstrated the ability to deliver real-time frame rates for moderate-size data, but they typically require the entire dataset to be contained in GPU memory. Long data transfer times and GPU memory size limitations are often the main limiting factors, especially for massive, time-varying, or multivolume visualization. In this issue, compression, multi-resolution schemes, and out-of-core techniques are proposed. For example, an out-of-core approach, which is based on the management of a hierarchical multi-resolution structure, is proposed by Gobbetti and Crassin et al. [5, 6]. And Balsa gives a variety of level-of-detail (LOD) data representations and compression techniques [7].

It may result in the occlusion of interesting structures when using ray-casting. Many overlapping structures may not embody the important structural details such that the cluttered images are quite difficult to understand. Occlusion is a view-dependent problem and cannot be solved easily by transfer function design. With regard to this, Haidacher introduced importance-driven rendering to create feature-emphasized visualization by defining importance via transfer function [8]. Bruckner et al. proposed the maximum intensity difference accumulation (MIDA) [9].

3.2.2 Volume Interaction

Real-time volume interaction is very important to offer different views of the same dataset and enhances its understanding. Recently, different techniques and strategies have been proposed. A clipping plane can be moved through the volume to reveal internal structures, permitting a close inspection of the entire range of data values [10, 11]. However, it can also remove important context information that leads to confusing and somewhat misleading. Bruckner et al. suggest cutaway views to focus the attention on the intersection region and ghosting views which give a better impression of the spatial location of the object [12]. Ropinski et al. propose volumetric lenses to interactively focus ROI, rendering the parts of the volume intersecting the lens defined by a convex 3D shape [13]. Monclús et al. present the Virtual Magic Lantern (VML), an interaction tool which behaves like a lantern providing the focal region visualized using a secondary transfer function or different rendering styles [14, 15]. Bruckner et al. propose style transfer functions (TF) which enables flexible data-driven illumination for inspection [16]. D'ıaz et al. develop illustrative visualization techniques to improve perception of depth and shapes of the models being inspected [17]. Most of them provide a view of the feature of interest without occlusions of other neighbor structures to prevent visual overload.

3.2.3 Volume Illustration

Illustrative techniques are generally applied to produce stylized renderings. Various illustrative styles have been applied to volumetric datasets, depicting complex structures or shapes in an easily comprehensible way such as silhouette or contour enhancement, pen and ink, stippling, hatching, etc. The early work done by Ebert et al. combines some physics-based illumination model with non-photorealistic techniques to enhance the perception of structure, shape, orientation, and depth relationships in a volume model [18]. Bruckner et al. present an illustrative volume rendering technique inspired by ghosted views [19]. Svakhine et al. utilize outlining techniques and selective depth enhancement to provide perceptual cues of object importance as well as spatial relationships [20]. Recently, these techniques are focused on to interactively modify the illustration styles and effects based on GPU for large datasets. For example, Pelt et al. adopt user-configurable particle systems to produce stylized renderings based on GPGPU paradigm [21]. Ruiz et al. propose a simple and interactive technique by the difference between the original intensity values and a low-pass filtered copy with a CUDA implementation [22].

In this chapter, we introduce the interaction methods of our VisEdu system on CUDA-based volume rendering with layered control mechanism via our transfer function design different from the above methods. We give various rendering methods through plane, superquadric, and virtual lenses tools offering different views of the same dataset and enhancing its understanding. The framework exploits the CUDA framework and the hierarchical structures such as octree for both compression of volume data and speed optimization of ray-casting process, allowing the system to achieve interactive speed.

3.3 The Proposed Method

For facilitating understanding the interior and exterior structures, we define plane, superquadric tools and virtual lenses based on our transfer function with layered control mechanism.

3.3.1 Transfer Function Design

Usually the density or grayscale varies in a certain range for different organizations in a CT dataset. Taking human tissue, for example, we describe it in Fig. 3.1 and identify it with gray value in [0,255]. According to this, we design our transfer function below.

We design transfer function which defines a mapping from data properties to optical properties by specifying the RGBA (for red, green, blue, alpha) value for

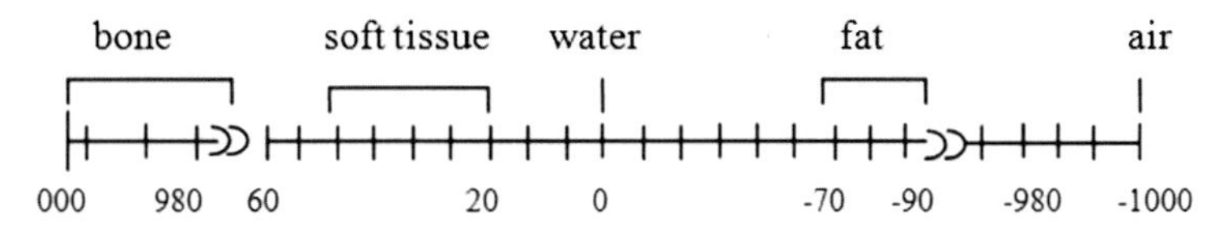

Fig. 3.1 Distribution of CT value for human tissue

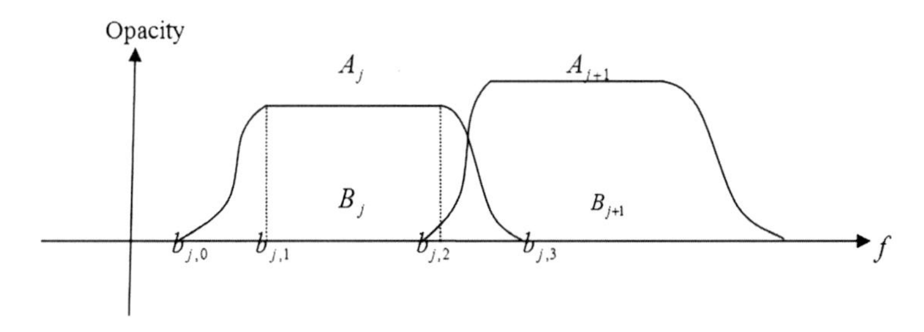

Fig. 3.2 Illustration B_j and B_{j+1} $(j=1,2,\ldots,k-1)$ of two bands. B_j is defined by a trapezoid-shaped function with a highest opacity A_j in the middle, increasing opacity on the left side, and decreasing on the right side. B_{j+1} is its adjacent band with an overlap region in scalar domain

every possible voxel value. We illustrated in Fig. 3.2 in which one trapezoid expresses one band or layer of CT datasets. The depicted transfer function is defined by all trapezoid functions.

Let k be the number of bands. Each band is parameterized by four ordered values. Let the jth $(j=1,2,\ldots,k)$ band be defined by $B_j=[b_{j,0},b_{j,1},b_{j,2},b_{j,3}]$ with color T_j and its maximum opacity A_j. Because the actual two adjacent ranges often overlap, B_j and B_{j+1} $(j=1,2,\ldots,k-1)$ have an overlapping region.

Let f_i be the scalar value of P_i,and $f_i \in B_m$ $(m=1,2,k-1)$. The weight w_i of P_i is defined by the following equation:

$$w_i = \begin{cases} g\left(\dfrac{f_i - b_{m,0}}{b_{m,1} - b_{m,0}}\right), & f_i \in [b_{m,0}, b_{m,1}] \\ 1, & f_i \in [b_{m,1}, b_{m,2}] \\ 1 - g\left(\dfrac{f_i - b_{m,2}}{b_{m,3} - b_{m,2}}\right), & f_i \in [b_{m,2}, b_{m,3}] \end{cases} \tag{3.1}$$

and $g(t)$ is a cubic function which value is in [0,1] for smooth transition. The opacity α_i of P_i is defined by the following equation:

$$\alpha_{\mathrm{i}} = \begin{cases} g\left(\dfrac{f_i - b_{m,0}}{b_{m,1} - b_{m,0}}\right) A_m & , f_i \in [b_{m,0}, b_{m,1}] \\ A_m & , f_i \in [b_{m,1}, b_{m,2}] \\ \dfrac{\left[1 - g\left(\dfrac{f_i - b_{m,2}}{b_{m,3} - b_{m,2}}\right)\right] A_m + g\left(\dfrac{f_i - b_{m+1,0}}{b_{m+1,1} - b_{m+1,0}}\right) A_{m+1}}{A_m + A_{m+1}} & , f_i \in [b_{m,2}, b_{m,3}] \end{cases} \quad (3.2)$$

where for the last overlap region $[b_{m,2}, b_{m,3}]$, the opacity is decided by their weights as 3.1.

Suppose the color c_i of P_{i} can be obtained from the color of B_m, $c_i = \alpha_{\mathrm{i}} T_m$. Accumulating α_{i} and c_i according to conventional ray-casting algorithm,

$$\begin{cases} c_i^* = c_{i-1}^* + \left(1 - \alpha_{i-1}^*\right) \alpha_i c_i \\ \alpha_i^* = \alpha_{i-1}^* + \left(1 - \alpha_{i-1}^*\right) \alpha_i \end{cases} \quad (3.3)$$

where α_i^*, c_i^* are the accumulated opacity and color corresponding to one viewing ray through the pixels into the field volume implemented by one thread in CUDA.

3.3.2 Plane Tools

Suppose a reference plane is defined by $Q.\hat{n} + D = 0$, and the view-dependent side where the normal vector toward the observer viewpoint, i.e., $Q.\hat{n} + D > 0$, is used to be clipped or keep the context within a fixed distance to the plane. We propose the following methods:

(a) Doubled-Sided Clipping

In this case, we keep information of the reference plane and take Phong shading for enhancement. For this purpose, we set

$$h = 1.0 + d_i / t \quad (3.4)$$

where t is the thickness, and if $h = 1$ is satisfied, it is on the plane.

Before accumulating, generally we use the traditional Phong model to perform the shading effect. Let k_a be an ambient reflection constant, k_d diffuse reflection constant, and k_s a specular reflection constant. We make it highlighted and more opaque on the reference plane, otherwise the original shading and less transparency off the plane. For doing so, modifying the color c_i derived from transfer function to be the shaded color $c_{i,\mathrm{shaded}}$ on the side where $d_i \geq -t$ is as follows:

$$c_{i,\text{shaded}} = \phi_i(h)c_i \tag{3.5}$$

where

$$\phi_i(h) = k_{a,\text{new}} + k_{d,\text{new}}\left(\hat{L}.\hat{g}_i\right) + k_{s,\text{new}}\left(\hat{H}.\hat{g}_i\right)^{k_e} \tag{3.6}$$

$$\begin{aligned} k_{a,\text{new}} &= k_a + (k_d + k_e)h \\ k_{d,\text{new}} &= k_d(1 - h) \\ k_{s,\text{new}} &= k_s(1 - h) \end{aligned} \tag{3.7}$$

and

Substitute c_i with $c_{i,\text{shaded}}$ into Eq. 3.3 to perform the normal accumulation process.

(b) One-Sided Clipping

In this case, the clipping plane cut away parts of volume datasets, i.e., eliminated its contribution on the side where $Q.\hat{n} + D > 0$ is satisfied. On the other side where $d_i \geq -t$ is satisfied, we use the enhancement algorithm similar to (a).

(c) Clipping Keeping Context

The view-dependent side is where the normal vector toward the observer viewpoint, i.e., $Q.\hat{n} + D > 0$, is used to keep the context within a fixed distance to the reference plane. We use the following distance-based methods to remedy the deficiencies of simple clipping plane, which removes certain parts of the volume including fine interest structures.

Let $d_{\max}$ be the maximum Euclidean distance to the reference plane for keeping the context. The distance d_i from P_i to the reference plane is defined by

$$d_i = P_i.\hat{n} + D. \tag{3.8}$$

If $d_i < d_{\max}$ is satisfied, the opacity and color are modified as follows:

$$\alpha_{i,\text{new}} = \alpha_i * \kappa_i(d_i) \tag{3.9}$$

$$c_{i,\text{new}} = \alpha_{i,\text{new}} * c_i \tag{3.10}$$

where

$$\kappa_i(d_i) = s_a\left(1 - \left|v\hat{g}_{P_i}\right|\right)^{s_e} \cdot \sin\left(\frac{\pi.d_i}{2d_{\max}}\right) \tag{3.11}$$

Substitute c_i with $c_{i,\text{new}}$ and α_i with $\alpha_{i,\text{new}}$ into Eq. 3.3 to perform the normal accumulation process. On the other side where $d_i \geq -t$ is satisfied, we take enhancement algorithm similar to (a) in Sect. 3.3.2.

3.3.3 Superquadric Tools

We set the shapes of superquadric tools including a family of geometric one defined by the following equation [23]:

$$f(x,y,z) = \left[\left(\frac{x}{r_x}\right)^{\frac{2}{e_2}} + \left(\frac{y}{r_y}\right)^{\frac{2}{e_2}}\right]^{\frac{e_2}{e_1}} + \left[\frac{z}{r_z}\right]^{\frac{2}{e_1}} = 1 \tag{3.12}$$

where r_x, r_y, and r_z are scale factors on x-, y-, and z-axis, respectively, and e_1, e_2, and e_3 are positive real numbers that determine the main features. It includes different basic shapes with rounded edges and corners. If $e_1 \approx 1, e_2 \approx 1$, it is an ellipsoid. If $e_1 << 1, e_2 << 1$, it is a cube, and if $e_1 << 1, e_2 \approx 1$, it is a cylinder. The region within the superquadric defines the ROI as shown in Fig. 3.3. We give the following superquadric tools with different functionalities.

(a) Superquadric Clipping

In this case, ROI is a clipping region. We eliminate the voxels contribution within ROI when accumulating as shown in Eq. 3.3. It entails the problem of possible occlusion because of the voxels situated between the view point O and ROI, which impede the correct visualization of the interest region inside the ROI. We cancel the accumulated color just before the first intersection M of the viewing ray and ROI.

(b) Band Picker

In ROI, we control the layer by our transfer function design as described in Sect. 3.3.1 where we set the opacity of visual band as (0, 1] and the others are 0. Outside ROI the conventional CUDA-based volume rendering is used as shown in Eq. 3.3. Also it entails the problem of possible occlusion similar to (a) in Sect. 3.3.3; the resolution is the same.

(c) Focus+Context Exploration

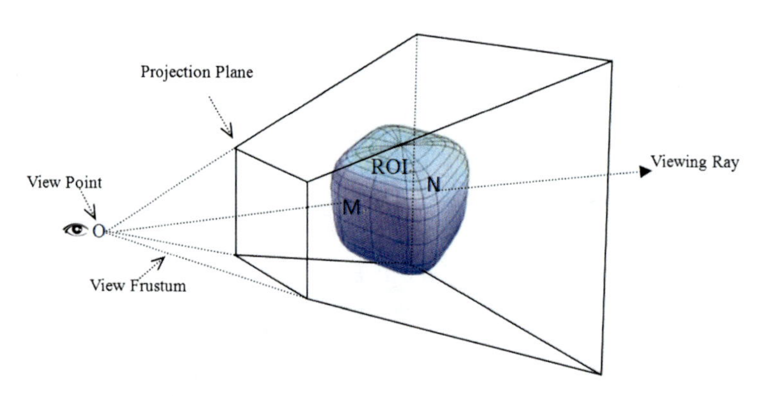

Fig. 3.3 Diagram of user-driven ROI

We divide the ROI into focus and context region by the following pq-norm [24]. The superquadrics are used to get consistent focus shapes based on pq-norm defined as follows:

$$\|P_i\|^{pq} = \|(\|(x_i, y_i)\|^p, z_i)\|^q \tag{3.13}$$

which is a generalized Euclidean distance ($p = 2, q = 2$) and

$$\|(x_i, y_i)\|^p = (|x_i|^p + |y_i|^p)^{\frac{1}{p}} \tag{3.14}$$

is the p-norm. This norm is always 1 on the surface of the superquadric.

Let $P = (x, y, z)$, and define

$$\|P\|^{pq} = \rho \qquad (\rho \leq 1\ ). \tag{3.15}$$

The focus region satisfies $\|P\|^{pq} \leq \rho \quad (\rho \leq 1\ )$, and the context region is the difference between the ROI and the focus region where $\rho < \|P\|^{pq} \leq 1$ is satisfied. In focus region, we use MImP (maximum importance projection) borrowing the idea of importance-driven volume rendering [8]. Take $\beta_{\mathrm{i}} = 1 - \delta_i$,and define

$$\delta_{\mathrm{i}} = \begin{cases} 1 & \text{if } I_i > I_{\max} \\ 0 & \text{else} \end{cases} \tag{3.16}$$

where I_{i} denotes the importance value at location P_{i} and $I_{\max}$ is the current maximum importance along the ray. The accumulated color and opacity is defined as follows:

$$\begin{cases} c_i^* = c_{i-1}^*\beta_i + (1 - \beta_i\alpha_{i-1}^*)\alpha_i c_i \\ \alpha_i^* = \alpha_{i-1}^*\beta_i + (1 - \beta_i\alpha_{i-1}^*)\alpha_i \end{cases} \tag{3.17}$$

If setting higher importance M_i of the ith band, we can focus it.

In the context region, we use the following context-preserving model. Let the gradient of P_{i} be $g_{P_i} = \nabla f(P_i)$ and $\hat{g}_{P_i}$ be its normalized gradient. We set global view with reduced detail by modifying the distance-based silhouette factor in the context area as follows:

$$\text{silhouette\_factor} = s_a\left(1 - \left|v\hat{g}_{P_i}\right|\right)^{s_e} \cdot d(P_i) \tag{3.18}$$

where s_a controls the amount of silhouette enhancement, s_e controls the sharpness of the silhouette curve, and $d(P_i)$ is defined as follows:

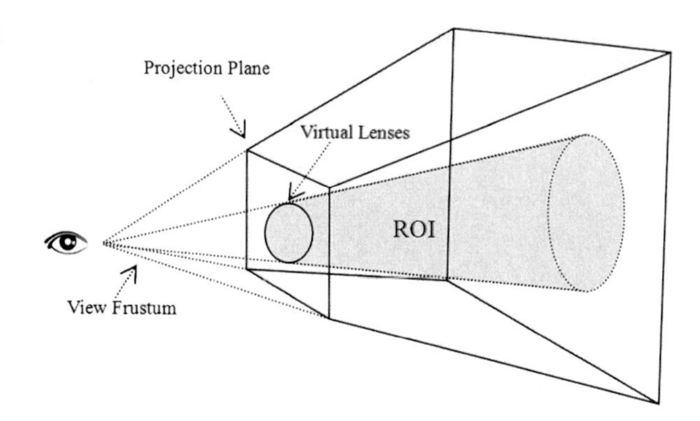

Fig. 3.4 Diagram of virtual lenses

$$d(P_i) = g_\beta\left(\frac{\|P_i\|^{pq} - \rho}{1 - \rho}\right) \tag{3.19}$$

where $g_\beta(t)$ is the Schllick function defined as follows [25].

$$g_\beta(t) = \frac{t}{e^\beta(1 - t) + t} \tag{3.20}$$

which controls more opacity far from the center due to the property of its monotone increasing.

3.3.4 *Virtual Lenses*

The main function of magnifying glass is to amplify and highlight local details in some focus region. Virtual magic lamp is an interaction tool to obtain a lantern-based inspection using a virtual cone [14], which divides the model into two regions whose illumination cone determines the region of interest. Combining magnifying glass with virtual magic lamp, this section gives a simplified model similar to real lenses, constructing 2D local region for the display of different layers according to 3D viewing frustum and viewing focus region as described in Fig. 3.4. Comparing to band picker of above superquadric tools, it eliminates the inner side view for clipping effect.

The user defines virtual lenses by the cursor position and a specified radius. According to the above simplified model, the volume dataset is divided into two parts including ROI which is the intersection of a cone with the window of virtual lenses as base and the viewpoint, and the rest of volume datasets. In ROI, we can get separated layer by modifying its opacity in (0, 1] and 0 for others. Outside ROI the conventional CUDA-based volume rendering is used.

3.3.5 *User Interaction and Implementation*

All operations which affect viewing parameters and optical properties of inner structures are performed interactively by the following transfer function widget and the user interface for interaction tools.

3.3.5.1 Transfer Function Widget

The 2D view for editing the transfer function is shown in Fig. 3.5, where the x-axis represents the voxel value and y-axis represents the opacity. All bands of trapezoid give the volume a classification, and each one has its color and maximum opacity. b0, b1, b2, b3 of the current selected one are called control points supporting dragging and moving by mouse.

3.3.5.2 GUI of Interaction

The interaction page controller is shown in Fig. 3.6. The shapes of superquadric tools include sphere, smooth cube, and cylinder. We can drag and drop our defined tools by mouse button with different combinations to mimic 6DOF manipulation, for example, the left button for rotation about z-axis, right button for translation in xy-plane, and scroll button for translation along z-axis after choosing plane or superquadric tools. For the virtual lenses, we can manipulate the mouse by dragging and dropping the left button for translation of xy-direction on the projection plane.

But the above methods are not intuitive in projection-based virtual learning environments. One of our goals is to enlarge the continuity between visualization and interaction by adopting a 3D interaction technique. We choose 3D input device such as Phantom Omni or InterSense Tracker for 6-DOF manipulation. They provide SDK and Tookit by which we can get the affine transform matrix to take 6DOF manipulation. Because it is concerned with different operation, we take the following two ways:

(a) If only interaction with the volume datasets, 2D mouse or 3D input device is independently used.

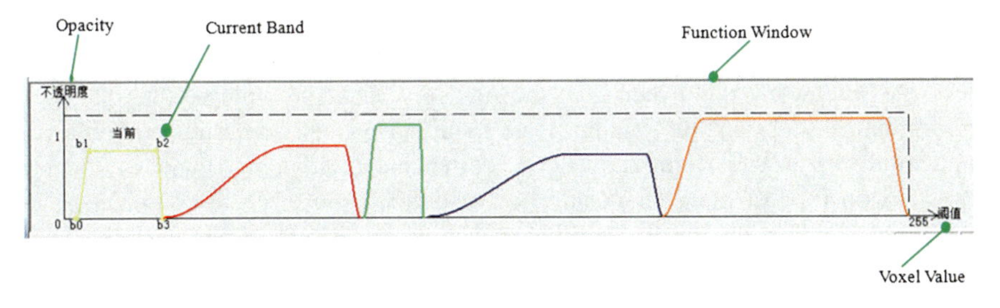

Fig. 3.5 Opacity transfer function window

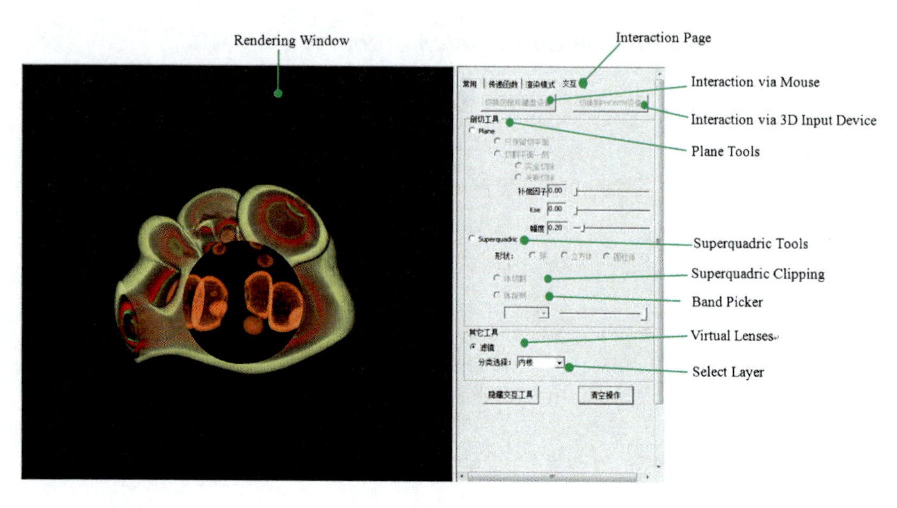

Fig. 3.6 The user interface of interaction module for VisEdu

(b) If the defined plane or superquadric tool is activated, the 2D mouse is used for volume datasets and 3D input device is used for the defined tools.

3.3.5.3 CUDA Implementation

The framework exploits the CUDA framework and the hierarchical structures such as octree for both compression of volume data and speed optimization of ray-casting process as shown in Fig. 3.7. The original volumetric datasets are organized into a coarse octree structure for the implementation of out-of-core. About the out-of-core techniques, we based on the work of Enrico Gobbetti and refer the interested reader to their paper [5, 26] for a survey. The octree contains the original data at the leaves and a filtered representation of children at inner nodes. Each node also stores the range of values and pre-computed gradients. At runtime, an adaptive loader updates a view- and transfer function-dependent working set of bricks incrementally maintained on CPU and GPU memory by asynchronously fetching data from the out-of-core octree.

Each ray is assigned to a thread by using an appropriate "kernel" function. In the beginning of ray-casting, we need to determine the thread index in CUDA and then test the intersection between the viewing ray and the volume. If intersect, resampling is done in the volume. Data mapping means from data properties to optical properties specifying the RGBA. For each sample point, it needs to decide the relationship between the voxel and the defined tools before compositing as Eq. 3.3. We adjust the opacity and color according to the tool specification if it is inside the ROI; else take data mapping from the transfer function as Sect. 3.3.1. Phong illumination model is used for shading before accumulating the color.

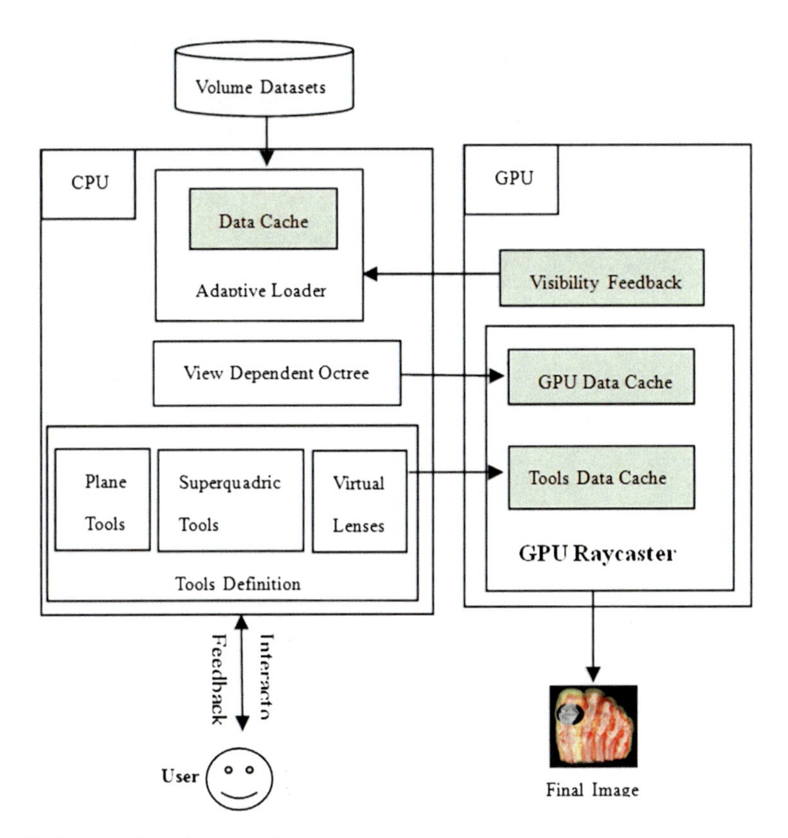

Fig. 3.7 Scheme of method overview

Because all threads run at the same time and process the data, the performance is greatly increased and real-time rendering is achieved.

3.4 Results

We have implemented our algorithm on an Intel Core 2 Quad CPU Q9400 at 2.66GHz equipped with 4 GB of RAM on Windows 7 using OpenGL and CUDA 3.0. The graphics card is a NVIDIA GeForce GTX 570 with 1GB of RAM. The framework exploits the CUDA framework and the hierarchical structures such as octree for both compression of volume data and speed optimization of ray-casting process, allowing the system to maintain an interactive frame rate with an average of 30 fps with CT $256 \times 256 \times 256$ volumetric datasets. We test the proposed techniques with CT datasets such as the angiography datasets of human head from our lab, neghip and human foot downloaded from the volume library

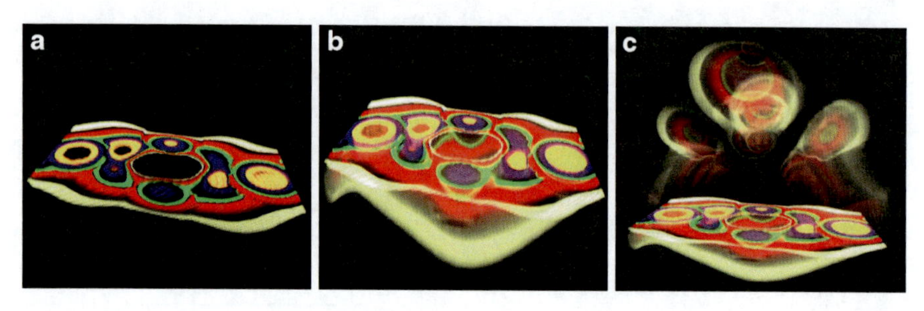

Fig. 3.8 Plane tools, (**a**) double-sided clipping, (**b**) one-sided clipping, (**c**) keeping context clipping

http://www.volren.org/, and human body from the radiology department of Geneva University.

Figure 3.8 shows the results of plane tools for neghip datasets including double-sided clipping in (a), one-sided clipping in (b), and keeping context clipping in (c). Figure 3.9 presents the plane clipping in (a) in comparison with superquadric ones in (b), (c), and (d) for sphere, smooth cube, and cylinder clipping separately. Figure 3.10 shows superquadric tools with different functionalities for human body datasets in (a) for superquadric clipping, in (b) for band picker, and in (c) for focus+context exploration. Figure 3.11 shows results of virtual lenses for different models in (a) for angiography datasets of a human head where the focus layer is blood vessels and in (b) for human foot with focus layer of bone. Figure 3.12 is snapshot of 3D interaction in virtual learning environment by Phantom Omni where it can be switched from the volume datasets to the defined plane or superquadric tools for 6DOF manipulation.

Conclusions

In this chapter, we describe the various rendering methods for scientific visualization education system, VisEdu. By specifying plane, superquadric, and virtual lenses tools, rendering offers different views of the same dataset and enhances its understanding. The implementation uses the CUDA framework to achieve real-time visualization from user inputs. For the layered control mechanism of band picker and virtual lenses, the results strongly depend on the transfer function design by which non-segmented CT datasets are automatically decomposed using a simple, threshold-based method. But the accuracy for this coarse classification still needs to be investigated in the near future.

We also hope to extend our method to support multiuser cooperative interaction by 3D device in projection-based virtual learning environments in the future, by which different users can manipulate different tools at the same time.

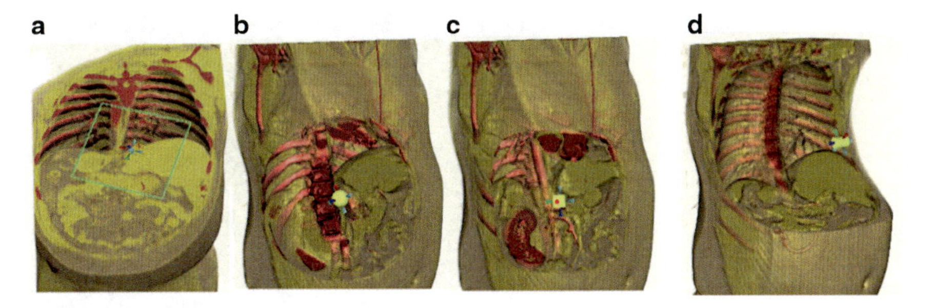

Fig. 3.9 Comparison between the plane and superquadric clipping, (**a**) clipping plane (**b**) sphere clipping, (**c**)smooth cube clipping, (**d**) cylinder clipping

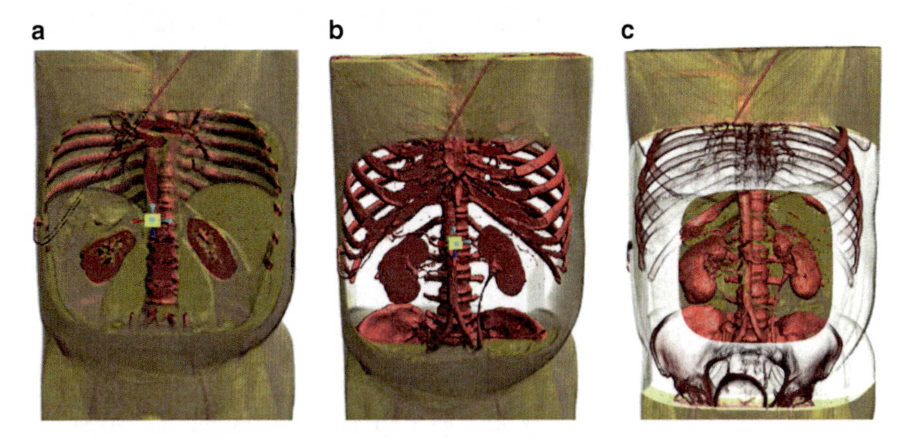

Fig. 3.10 Superquadric tools with different functionality, (**a**) superquadric clipping, (**b**) band picker, (**c**) focus+context exploration

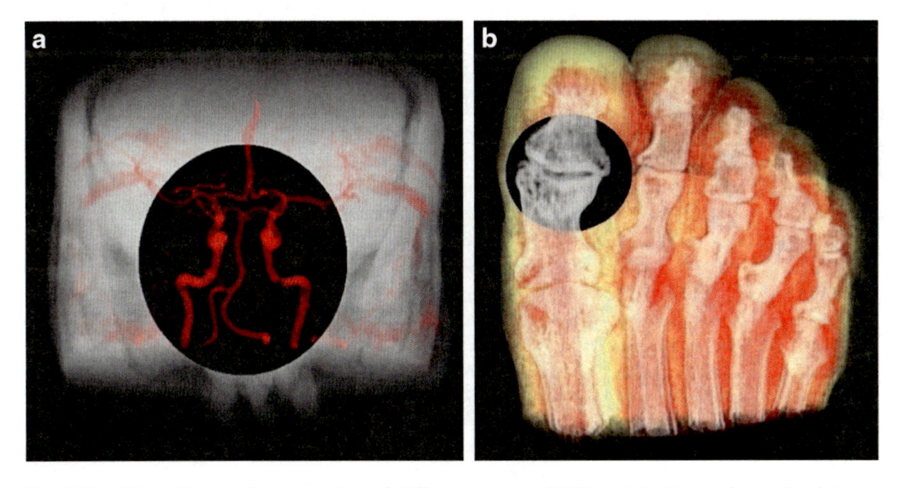

Fig. 3.11 Virtual lenses for inspection of different layers of CT models, (**a**) angiography datasets of a human head, (**b**) human foot

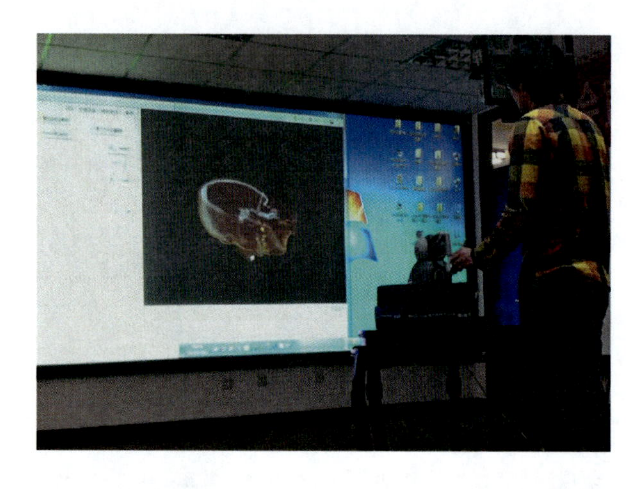

Fig. 3.12 3D volume interaction by Phantom in virtual learning environment

Acknowledgments We thank the reviewers for their useful comments. The National Natural Science Foundation of China supported this work under grants No.61170170, No.61274033, and No.61271198. Also we acknowledge the help from Visual Computing Group of CRS4 in the frame of EU Marie Curie Programme under the 3DANATOMICALHUMAN project.

References

1. Thomas, R.J., Strothotte T.: Motion enhanced visualization in support of information fusion. In: Proceedings of International Conference on Imaging Science, Systems, and Technology, CSREA Press, Las Vegas, 492–497 (2001)
2. William, R.H., Peter, N.W.: The perception of visual information. Springer-Verlag New York, Inc., Secaucus, NJ (1997)
3. Levoy, M.: Display of surfaces from volume data. IEEE Comput. Graph. Appl. **8**(3), 29–37 (1988)
4. Engel, K., Hadwiger, M., Kniss, J., Rezk-Salama, C., Weiskopf, D.: Real-time Volume Graphics. AK-Peters Publisher, Natick (2006)
5. Gobbetti, E., Marton, F., Iglesias-Guitián, J.A.: A single-pass GPU ray casting framework for interactive out-of-core rendering of massive volumetric datasets. Vis. Comput. **24**(7–9), 797–806 (2008)
6. Crassin, C., Neyret, F., Lefebvre, S., Eisemann, E.: GigaVoxels: ray-guided streaming for efficient and detailed voxel rendering. In: Proceedings of the 2009 symposium on interactive 3D graphics and games, d (212), 15–22 (2009)
7. Balsa, M., Gobbetti, E., Iglesias-Guitián, J.A.: A survey of compressed GPU-based direct volume rendering. In: Eurographics (2013)
8. Haidacher, M.: Importance-driven rendering in interventional imaging, thesis (2007), http://www.cg.tuwien.ac.at/research/publications/2007/haidacher-2007-idr/
9. Bruckner, S., Gröller, M.E.: Instant volume visualization using maximum intensity difference accumulation. Comput. Graph. Forum **28**(3), 775–782 (2009)
10. McInerney, T., Broughton, S.: Hingeslicer: Interactive exploration of volume images using extended 3D slice plane widgets. In: Proceedings of Graphics Interface, Canadian Information Processing Society, 171–178 (2006)

11. Zhang, Q., Eagleson, R., Peters, T.M.: Rapid scalar value classification and volume clipping for interactive 3D medical image visualization. Vis. Comput. **27**(1), 3–19 (2011)
12. Bruckner, S., Groller, M.E.: VolumeShop: An interactive system for direct volume illustration. In: Proceedings of the IEEE Visualization, pp. 671–678 (2005)
13. Ropinski, T., Steinicke, F., Hinrichs, K.H.: Tentative results in focus-based medical volume visualization. In: Proceedings of the 5th International Symposium on Smart Graphics. Volume 3638 of Lecture Notes in Computer Science, Springer, 218–221 (2006)
14. Monclús, E., Díaz, J., Navazo, I., Vázquez, P.P.: The virtual magic lantern: an interaction metaphor for enhanced medical data inspection. VRST **2009**, 119–122 (2009)
15. Andújar, C., Navazo, I., V'azquez, P.P.: The ViRVIG Institute. SBC Journal on 3D Interactive Systems, **2**(2), 2–5 (2011)
16. Bruckner, S., Groller, M.E.: Style transfer functions for illustrative volume rendering. Comput. Graph. Forum **26**(3), 715–724 (2007)
17. D'ıaz J., V'azquez, P.-P.: Depth-enhanced maximum intensity projection. In: IEEE/EG Volume Graphics, pp. 93–100 (2010)
18. Ebert, D., Rheingans, P.: Volume illustration: non-photorealistic rendering of volume models. In: Proceedings of IEEE Visualization, 195–202(2000)
19. Bruckner, S., Grimm, S., Kanitsar, A., Gröller, M.E.: Illustrative context-preserving exploration of volume data. IEEE Trans. Vis. Comput. Graph. **12**(6), 1559–1569 (2006)
20. Svakhine, N.A., Ebert, D.S., Andrews, and W.M.: Illustration-inspired depth enhanced volumetric medical visualization. IEEE Trans. Vis. Comput. Graph. **15**(1), 77–86 (2009)
21. Van Pelt, R.F.P., Vilanova Bartroli A. and van de Wetering, H.M.M.: GPU-based particle systems for illustrative volume rendering. In: IEEE/EG International Symposium on Volume and Point-Based Graphics, pp. 89–96 (2008)
22. Ruiz, M., Boada, I., Feixas, M., Sbert, M.: Interactive volume illustration using intensity filtering. Eurographics **2010**, 51–58 (2010)
23. Barr, A.: Superquadrics and angle-preserving transformations. IEEE Comput. Graph. Appl. **1** (1), 11–23 (1981)
24. Jaklic, A., Leonardis, A., Solina, F.: Segmentation and recovery of superquadrics. Volume 20 of Computational imaging and vision. Kluwer, Dordrecth (2000)
25. Schlick, C.: A fast alternative to Phong's specular model. In: Heckbert, P.S. (ed.) Graphics Gems IV. AP Professional, Boston, pp. 385–387 (1994)
26. Gobbetti, E., Iglesias Guitián, J.A., Marton, F.: COVRA: A compression-domain output-sensitive volume rendering architecture based on a sparse representation of voxel blocks. Comput. Graph. Forum **31**, 1315–1324 (2012)

Chapter 4
Real-Time Separable Subsurface Scattering for Animated Virtual Characters

P. Papanikolaou and G. Papagiannakis

Abstract In this chapter, we present our real-time, GPU-accelerated separable subsurface scattering method for interactive, skeletal-based deformable animated virtual characters. Our screen space implementation is based on state-of-the-art algorithms, and we propose specific algorithmic and implementation extensions so that these algorithms can be employed in real-time virtual characters. We have created a physically principled real-time rendering framework, which features a series of rendering effects based on widely available open-source tools such as Open Scene Graph, C++, and GLSL so that it can be easily integrated in modern rendering engines and scene graphs via commodity graphics h/w.

Keywords Real-time rendering • Separable subsurface scattering • Dynamic surfaces • Virtual character skin simulation

4.1 Introduction

In order to enhance the user's experience, the graphics community is continuously working on enriching these environments with realistic depiction of light interaction with objects. One of the most controversial decisions for a graphics developer is to either increase the accuracy of the scene or the rendering speed.

In order to produce high-fidelity images in an environment, where virtual characters exist, we have to depict realistically the way light interacts with human skin. To do that, we have implemented a set of physically principled effects that take place on skin rendering: real-time separable subsurface scattering including specular surface reflectance.

We have thus created a physically principled real-time rendering framework, which features a series of effects that can be applied on skeletal-based deformable animated virtual human characters. The implementation of our rendering framework is based on widely available open-source tools such as Open Scene Graph,

---

P. Papanikolaou (✉) • G. Papagiannakis
Computer Science Department, University of Crete, Heraklion, Greece

Foundation for Research and Technology Hellas, Heraklion, Greece
e-mail: ppapanik@csd.uoc.gr; papagian@csd.uoc.gr

Y. Cai, S. See (eds.), *GPU Computing and Applications*,
DOI 10.1007/978-981-287-134-3_4

C++, and GLSL so that it can be easily integrated in modern rendering engines and scene graphs via commodity graphics h/w.

4.2 Previous Work

A lot of research has been published based on how the light interacts with human skin [1, 2], real-time virtual characters [2] rendered in both VR and AR environments [3–14], and also how the BSSRDF can be alternatively modeled to run in real time [4–6]. Two major effects are among the most common that describe the light's interaction with virtual character skin. The first one is taking place on skin surface and it is called specular reflection, while the second one models light's behavior after entering the inner layers of the skin. Even though for most objects, the Phong approach suffices for real-time specular reflection implementation, it does not approximate realistically this effect for human skin. The Phong model fails to depict accurately the specularity at grazing angles. The Kelemen/Szirmay-Kalos model is a common approach that is used to implement the specular reflection on human skin, because not only it depicts the effect with high quality, but it also uses a Fresnel reflectance factor that handles accurately the specularity at grazing angles.

Subsurface scattering (SSS) is the second effect that takes place during interaction between light and skin. After the light hits the surface, only a low percent (6 %) of the incident radiance is reflected directly, while the rest enters the skin. Subsurface scattering is the phenomenon which describes the light's behavior after its entrance to inner layers. While it travels through the skin, it is either absorbed partially or scattered many times before it exits to a neighboring area. The initial rendering equation formulation of light's subsurface propagation is referred to [7], which is extended in [1] to model the human multilayered skin.

To implement the subsurface scattering effect, we have two options. The first one is the texture space diffusion method [8], which, even though it approximates this effect with high accuracy, needs a lot of adjustments in order to make it run efficiently in real time. The other method is the screen space one [9], which not only offers great rendering speed but also scales better in large environments. The main difference is that it applies a 2D convolution on screen instead of object's texture.

In order for the subsurface scattering effect to be simulated accurately, we need to take into account the light's absorption while it travels through human skin. Knowing that light consists of three colors (red, green, and blue), we have to calculate the attenuation for each one of them. We will refer to each color's exhaustion as diffusion profile. A fast and accurate approach of diffusion profile is explained in [8], which is based on [2].

In the last few years, a lot of systems that support the subsurface scattering effect for human skin have been published, but none of them support skeletal-based deformable animated characters. Jimenez's demo applies the SSS effect only on static human head along with a lot of post-processing effects. Also, NVidia™ recently has presented a demo, which applies a wide range of effects on animated virtual human heads, whose animation is not only implemented via morphing, instead of skeletal based, but also in order to run properly, it requires specific and expensive high-end graphics cards (Titan). Our rendering framework applies subsurface scattering along with other effects on skeletal-based deformable animated virtual characters. It also achieves high visual accuracy along with efficient real-time execution in modern commodity graphics h/w.

4.3 Separable Subsurface Scattering for Dynamic Surfaces

There are mainly two kinds of reflectance that take place when a light ray interacts with the human skin. The first one is the surface reflectance, which causes a small percentage (6 %) of the light to be reflected directly without being colored. It occurs due to the topmost oily layer of skin and it can be modeled using a specular reflection function. In our rendering framework, we have implemented this behavior using Kelemen/Szirmay-Kalos [10]. The other kind of interaction between the light and the skin is the subsurface reflectance (subsurface scattering). This effect occurs due to the fact that the human skin is a translucent material. Translucent objects allow the light to pass through, but with a high degree of absorption. The subsurface scattering effect is described in terms of Bidirectional Surface Scattering Reflectance Distribution Function (BSSRDF), which relates the outgoing radiance L_0 at the point x_0 in direction ω_0 to incident radiance at the point x_i from direction ω_i:

$$S_d(x_i, \omega_i, x_0, \omega_0) = \frac{1}{\pi} F_t(x_i, \omega_i) R\left(\|x_i - x_0\|_2\right) F_t(x_0, \omega_0). \tag{4.1}$$

where F_t is the Fresnel transmittance function and R is the diffusion profile. We can convert the BSSRDF into BRDF if we consider $x_0 = x_i$ The outgoing radiance can be computed by using the equation

$$L_0(x, \upsilon) = \int_A \int_\Omega S_d(x_i, \omega_i, x_0, \omega_0) L_{in}(x, I)(I.n) dI \tag{4.2}$$

where A is the area affected by subsurface scattering.

Subsurface scattering describes the way light behaves after it enters the skin. While it travels through the skin, it is either absorbed partially or scattered many times before it exits a neighboring area. In order for virtual human faces to be rendered realistically, we have to simulate this effect. In this chapter, we will discuss about our implementation of subsurface scattering (SSS), which can be supported by high-detailed skeletal-based animated models. We have used the [9] implementation of approximating light's scattering underneath the surface of a translucent material. To simulate the subsurface scattering effect for thin parts of human face (ears, nostrils), we have employed a different technique, which first introduced again by [11] and it is based on [12].

What makes virtual skin rendering difficult to simulate is the fact that it consists of multiple translucent layers. Each one of them has a large variety of properties and interacts with the light in a different manner. The more layers we take into account, the more realistic results our system will produce. In our system, we assume that skin consists of three layers of translucent material [2]. In this section, we will discuss only the real-time implementation of the subsurface scattering effect.

4.4 Implementation

The main use of subsurface scattering (SSS) algorithm in a system is to blur the high-frequency details in human skin. To implement it in our system, we employ the [9] method, which is based on the idea of performing the diffusion approximation in screen space. The reason why we prefer this method to the texture space is because it eliminates a lot of problems that affect the real-time performance. These problems are explained extensively in Jimenez's paper.

In screen space method, diffusion profile is applied directly to the image with the face. Two passes are needed in order to apply a horizontal and a vertical convolution. We do not have to sample all the texels in a straight line. Instead we use only 17 jittered samples retrieved from Jimenez's method and generated based on [13]. At this point, we have to mention that the number of samples is dependent to the resolution. The higher the resolution of our output image, the more samples we have to use in order to keep the final result undistorted. In our system, we render with 1,024 × 768 resolution; thus, 17 samples are sufficient. Each sample has its own red, green, and blue weights, which describe the attenuation for each of the light's colors. They also contain information of how far they are from the main pixel, which is the first sample. Eight samples are used for each side of the current direction (horizontal or vertical).

We implement the subsurface scattering (SSS) in two passes in fragment shader. The first pass takes as input the depth buffer of the main camera and a rendered image of the face without SSS. It applies the horizontal convolution using 17 jittered samples. Then, it renders the result to a texture, which will be used as input in the second pass, along with the depth buffer. The second pass will convolute the image vertically. Its result will be the final in the whole procedure and it will render to screen. At this point, we have to mention the fact that the subsurface scattering

Fig 4.1 Skin texture of character's head

effect has to be applied only on skin surfaces, leaving hair and cloths unaffected. That is why we use a 1-bit per color texture with white color indicating the skin (Fig 4.1).

An important factor that has to be considered is that the convolution has to be applied only on points that are close to each other in 3D space. This condition is not always true for adjacent points in a 2D image. That is why we need the depth buffer in both passes. By comparing the depths between each sample and the main pixel, we can reduce the error of using nonadjacent points in convolution. We will take into account only those samples that the depth difference does not exceed a threshold. In case the depth difference is greater than the threshold, we will use the main pixel's color along with the sample weights for the convolution. At this point, we also have to add the specular reflection calculated in the main pass and rendered in a separate texture. The reason why we had to do this is because specular reflection must be applied after the 2D convolution of subsurface scattering on the surface color. By doing so, we avoid artifacts created by the specular highlight on virtual character's surface (Fig 4.2).

Fig 4.2 Specular reflection applied on surface before (*left*) and after (*right*) subsurface scattering

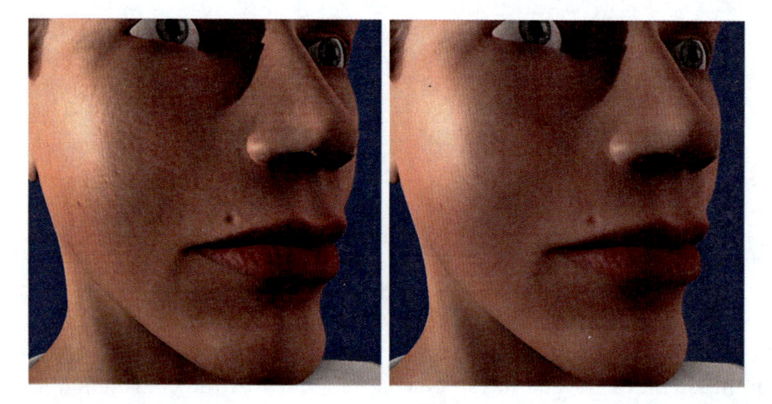

Fig 4.3 Character's head as rendered without subsurface scattering (*left*) and with subsurface scattering (*right*)

The following code shows the fragment shader implementation of subsurface scattering effect in our rendering framework. The variable scale_separable in the code segment below, modifies the area that is affected by the light striking at any point and the user through a GUI widget can adjust it. Increasing its value means the light travels farther under the skin surface. In Fig 4.3, we compare the rendering results of the character's head without and with subsurface scattering. Notice how the high-frequency details in the human skin are blurred.

Subsurface Scattering Convolution Passes

```
void main(){
float depth_threshold = 0.3;
float kernel_range = 2; //kernels range from -2 to 2.
vec2 texcoords = vec2(
(gl_FragCoord.x - 0.5)/(window_width-1.0),
(gl_FragCoord.y - 0.5)/(window_height-1.0));

float depth = texture2D(camera_depth_texture, texcoords).r;
vec4 color = texture2D(color_texture, texcoords);
vec3 final_color = color.rgb;
bool is_skin=texture2D(skin_texture,
gl_TexCoord[1].xy).r == 1.0;
vec2 final_step = vec2(0.0,0.0);
if (is_skin){
final_color = final_color * sample_kernel[0].rgb;
final_step = scale_separable * blur_dir
* 0.0025 * 1.0/depth * 1.0/kernel_range;
float eye_depth = -(depth * (far-near) + near);
for (int i = 1; i < 17; i++) {
vec2 offset = texcoords + sample_kernel[i].a * final_step;
vec4 sample_color = texture2D(color_texture,
                offset);
float sample_depth =
texture2D(camera_depth_texture, offset).r;
float sample_eye_depth = -(sample_depth *
                (far-near) + near);
if (abs(eye_depth - sample_eye_depth) < depth_threshold)
final_color.rgb += sample_color.rgb *
                sample_kernel[i].rgb ;
else
final_color.rgb += color.rgb *
                sample_kernel[i].rgb;
}
}
if (blur_dir.y == 1)
        final_color +=
texture2D(specular_texture,texcoords).rgb;
gl_FragData[0] = vec4(final_color, 1);
}
```

4.4.1 Light's Transmission Through Thin Skin

The simulation of subsurface scattering effect with convolution gives great results.

The light travels underneath the skin surface affecting only the neighboring areas, because it is fully attenuated after some distance. Although this method simulates the effect realistically for thick surfaces, it does not suffice for thin surfaces. When a light ray hits a thin surface, it might pass through the skin and be emitted from the other side. In real life, this effect can be observed in ears, nostrils, or any other thin part of the skin.

In our rendering framework, we are based on the screen space algorithm from Jimenez et al. (2010), and we extend it to support real-time, deformable characters in commodity graphics h/w. The greatest problem of screen space algorithms is the lack of information about geometry other than what the user can see. In this case, we do not know how the thin surfaces are shaped from the other side. What we need to know is the irradiance of a surface that is not observable by the user. In order to calculate the irradiance at the back of a surface, we need the surface's normal. The key concept to Jimenez's method is to assume that the normal at the back of an object is the reverse of the current pixel normal. Certainly, this is just an approximation, which reduces the accuracy of the effect, but this assumption solves our problem and now we can proceed to irradiance computation.

Another matter that we have to deal with is the calculation of the distance the light travels before it exits from the other side. For that purpose, we have used Green's method as described in [12]. The reason why we need this is to know how attenuated the light will appear. Based on the diffusion profile [8], we can compute its emittance color as implemented in [11]. In order for this effect to be independent of the near and far planes, the distance is calculated in view space.

The depth buffer that is generated in a previous pass-by light causes here some issues that we have to solve in order to improve the rendering results. The problem is caused to due to the low resolution of the depth buffer. Specifically, some artifacts appear around the projection's edges, where the depth distance between two adjacent pixels can be huge. A prompt fix is to use shadow maps with higher resolution, as we already do to fix shadow-mapping issues. But even with four times higher resolution than screen's resolution, we still get the annoying artifacts. There are two ways to deal with this. The first one is mentioned in Green's SSS approach [12], in which he grows the vertices towards normal when the shadow map is rendered. Jimenez on the other hand prefers to shrink the vertices in normal direction when creating the shadow coordinates for a vertex. We use Jimenez's approach, but instead of the fragment shader, we implement it in the vertex shader. Figure 4.4 shows the difference between shrinked and non-shrinked vertices. The two following code snippets show how we have implemented the vertex shrinking in GLSL vertex shader and light's transmission in GLSL fragment shader. The

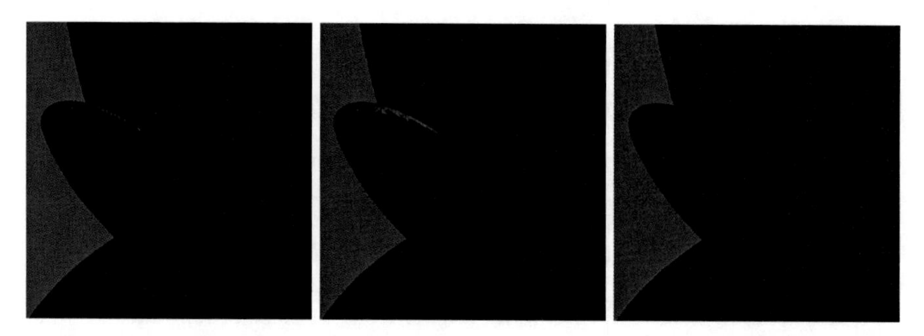

Fig 4.4 Light transmission through character's ear with low-resolution depth map and no vertex shrinking (*left*), with high-resolution depth map and no vertex shrinking (*middle*), and with high-resolution depth map and vertex shrinking (*right*)

Fig 4.5 Character's head rendered with SSS. Light is placed behind the head

variable shrinked_bias is configurable by the user and it handles the amount of vertex. Figures 4.5 and 4.6 show the subsurface scattering results along with light's transmission through thin skin in our rendering framework.

Our Vertex Shrinking Algorithm in a GLSL Vertex Shader

```
vec4 shrinked_vertex =
      vec4(position_attribute.xyz - shrinked_bias *
      normalize(normal_attribute.xyz), 1);

shadow_coords  =
      light_projection * light_view * shrinked_vertex;
shadow_coords = shadow_coords / shadow_coords.w;
shadow_coords.xy = shadow_coords.xy / 2.0 +
      0.5 * shadow_coords.w;
shadow_coords.z = (light_view * shrinked_vertex).z;
```

Our Light Transmission Algorithm in a GLSL Fragment Shader

```
vec3 getTransmission(){
  vec4 shadowmap_depth =
    texture2D(light_depth_texture, shadow_coords.xy).r;
  float eye_depth = -(shadowmap_depth *
      (light_far-light_near) + light_near);
  float depth_diff = scale_transmittance *
      abs(shadow_coords.z - eye_depth);
  vec3 profile = getProfile( -depth_diff * depth_diff
);
  float irrandiance =
    clamp(0.3 + dot(light_dir, world_normal) , 0.0,
1.0);
  vec3  transmission = profile * irrandiance;
  return transmission;
}
```

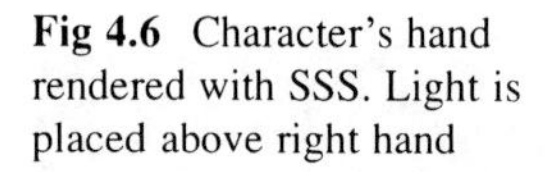

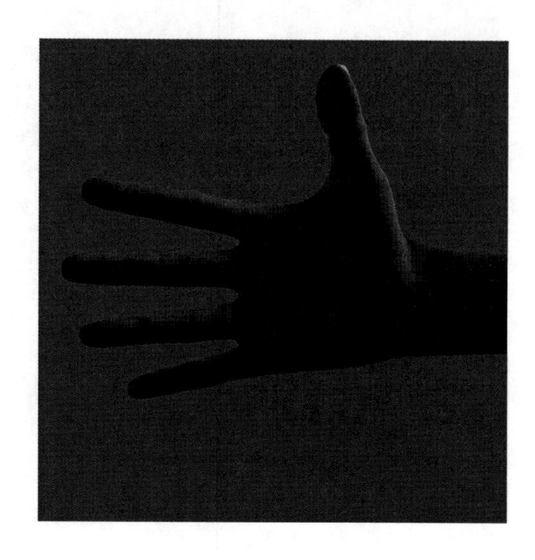

Fig 4.6 Character's hand rendered with SSS. Light is placed above right hand

4.4.2 Subsurface Scattering Implementation Novelties

As we have mentioned in this section, we based on Jimenez's screen space method to simulate the subsurface scattering effect in our rendering framework. But even though this implementation is highly efficient, we had to significantly extend this approach for real-time virtual characters. The first one appears in the vertex shrinking part, which is used to avoid projection errors on the edges of the dynamic surfaces. In Jimenez's demo, it is implemented in the fragment shader of the main pass, while in our system, it is implemented in the vertex shader of the same pass to allow for performance boost with minimal artifacts. We made this decision due to the fact the vertex shader is executed with far more minimal cost than the fragment shader for our virtual character, in each rendering pass of our framework.

Our next major improvement in Jimenez's method appears again in the separable passes. If the retrieved jittered sample has high depth difference with the main pixel, Jimenez interpolates the sample's color with the main pixel's color based on that depth difference. In our implementation, we have defined a depth difference threshold. Every sample that exceeds that distance does not contribute to the 2D convolution of the subsurface scattering. That way we avoid the interpolation in the fragment shaders of the two separable passes.

Apart from the above major in real-time performance extensions, we also have modified the way user adjusts the width that SSS affects the virtual skin area. In Jimenez's method, a common modifier is used for both 2D screen convolution in separable pass and light's transmittance through thin skin in main pass. In our system, we use two separate modifiable parameters, which allow the user to handle the way light passes through skin in a different manner than the way light enters and exits from a neighboring area. This technique can also be used to ameliorate visual errors caused by multi-geometries in body parts that normally should allow light to pass through them, but they do not without affecting the subsurface scattering 2D convolution on screen space.

4.5 Comparison with Ground Truth

Our real-time separable subsurface scattering effect is implemented in screen space. Screen space techniques run faster than the respective ray tracing ones, making them able to be used in real-time rendering systems. But how well can they match the realism produced by ray tracing? No matter how fast a rendering technique is, it is of no use if the visual results are not correct. In order to test the visual accuracy of our SSS effect, we have to compare it with ground truth images. For this purpose, we use 3D Studio Max to create our scene with the same camera and light position. To render the scene in 3ds max, we use the offline mental ray renderer, because it supports ray tracing and the generated images are depicted with great realism. In this section, we compare the images produced by our effect with those produced by the mental ray renderer of 3ds max. We consider the image generated by 3ds max to be the ground truth. That way we will be able to evaluate our results based on rendering time and accuracy criteria.

Subsurface scattering is the effect that describes the light's behavior after it enters the human skin. This phenomenon is taking place because the skin is a translucent material and allows the light to travel through. While it propagates underneath the skin surface, it is either absorbed or scattered many times before it exits a neighboring area. In this section, two experiments will take place. In the first one, we will compare the images produced by our system and 3ds max for SSS effect in thick surfaces, where the light is either absorbed completely or it exits from a neighboring area. In the second experiment, we will study this effect for thin surfaces where the light passes through and is emitted from the other side.

To implement the subsurface scattering effect in our system, we need five passes. In the first two we store information in textures that are going to be used in the main pass, which is the third one. In the fourth pass we apply a horizontal convolution on image in order to blur the high frequencies, and in the fifth pass we blur the image vertically. This effect can be produced in 3ds max by using a mental ray material: subsurface scattering fast skin, according to which the object is considered to consist of multiple layers. In the figures below, we compare the SSS in our system and in with the Autodesk 3ds max offline renderer. To understand how the SSS affects the skin, we have to also present how the model was. We can see that the visual results between our method and 3ds max are very similar. Both methods are blurring the image with a red tone color. Our system produces this result in 6 ms (160 fps) and 3ds max in 2 s (Figs 4.7 and 4.8).

In the second experiment, we will compare the results for light's transmission through thin body parts. To conduct this experiment, we have to change the position

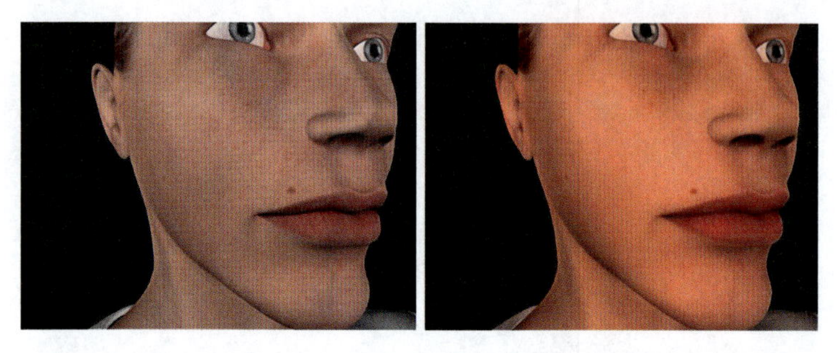

Fig 4.7 Character rendered by 3ds max without SSS (*left*) and with SSS (*right*)

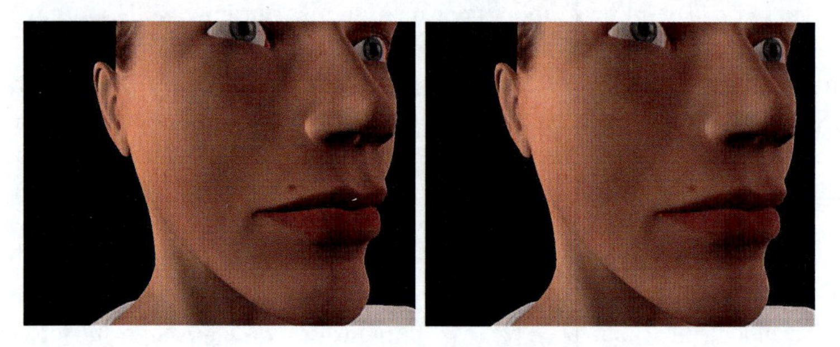

Fig 4.8 Character rendered by our system without SSS (*left*) and with SSS (*right*)

Fig 4.9 Light's transmission through thin skin as rendered by our method (*left*) and by 3ds max using ray tracing (*right*)

of the light source and place it behind the head to observe how light passes through the ear. The light will not be able to be absorbed completely, and it will be emitted from the other side with different colors due to its attenuation. In the figure below, we compare this behavior between our system and 3ds max. It is noticeable that the results look alike. The red color is the result of light's attenuation, which is approximated by the diffusion profile that we use. Skin tends to absorb blue and green colors more than red. This image is rendered by 3ds max in 8 s, while we need only 4 ms (250 fps) (Fig. 4.9).

Conclusions

The main purpose of this work was to prove that the subsurface scattering along with other physically principled lighting and shading effects can be applied on a skeletal-based deformable animated virtual human character and executed with high accuracy in real time. The reason why such a project is necessary is because even though there are similar systems, they either apply only a subset of the effects we have implemented or they are not supported for animated full virtual human characters. Jimenez in 2012 presented at SIGGRAPH12 a demo set of extremely realistic effects that run in real time but on a static virtual human head model. In 2013 NVidia presented a similar set of algorithms, and they used an animated virtual human head instead of a static one, utilizing their latest cluster of high-end dual GPU configuration. Unlike these two examples, our system not only applies a set of dynamically calculated realistic effects in real time using modern commodity hardware, but also it supports fully animated virtual human characters as shown in the figure below. To the best knowledge of the authors, there is currently no similar rendering framework in the bibliography (Fig. 4.10).

Even though our visual results are very close to the ones produced by ray tracing algorithms, there are still several aspects that need to be improved. First of all, we could use more than one point light sources for sharp shadows or area lights for soft ones. Although the number of passes would be increased, by combining effects such as shadow mapping from multiple light sources, we can create more realistic scenes. Apart from that, we could also use area lights for the SSS instead of point lights. The benefit would be the generation of low-frequency shadows instead of high-frequency shadows. We could also use different specular reflection functions based on each different virtual character geometry material.

Fig 4.10 Our real-time, skinned virtual character along with spot-based lighting, image-based lighting, ambient occlusion, shadow mapping, subsurface scattering, and environment mapping

Acknowledgments The research leading to these results has received funding from the European Union Seventh Framework Programme (FP7/2007–2013) under grant agreement no 274669 and the HIFI-PRINTER Marie Curie IEF project.

References

1. Donner, C., Jensen, H.W.: A spectral BSSRDF for shading human skin. In: Proceedings of the Eurographics Symposium on Rendering Techniques, pp. 409–417 (2006)
2. Donner, C., Jensen, H.W.: Light diffusion in multi-layered translucent materials. In: Proceedings of SIGGRAPH 2005, ACM Transactions on Graphics (2005)
3. Magnenat-Thalmann, N., Papagiannakis, G.: Virtual worlds and augmented reality in cultural heritage applications. In: Baltsavias et al (ed.) Recording, modeling and visualization of cultural heritage, pp. 419–430, ISBN-10: 041539208X, Taylor & Francis Group, 1–11 (2006)
4. Jiaping, W., Shuang, Z., Xin, T., Stephen, L., Zhouchen, L., Yue, D., Baining, G., Shum, H.-Y.: Modeling and rendering heterogeneous translucent materials using diffusion equation. ACM Trans. Graph. **27**(1)
5. Yan, L.-Q., Zhou, Y., Kun, X., Wang, R.: Accurate translucent material rendering under spherical gaussian lights. Comput. Graph. Forum **31**(7), 2267–2276 (2012)
6. Kun, X., Gao, Y., Li, Y., Tao, J., Shi-Min, H.: Real-time homogeneous translucent material editing. Comput. Graph. Forum **26**(3), 545–552 (2007)
7. Jensen, H.W., Marschner, S.R., Levoy, M., Hanrahan, P.: A practical model for subsurface light transport. In Proceedings of SIGGRAPH 2001 (2001)
8. d'Eon, E.Luebke, D.: Advanced Techniques for Realistic Real-Time Skin Rendering. In: Nguyen, H. (ed.) *GPU Gems 3*, pp. 293–347. Addison-Wesley, (2007)

9. Jimenez, J., Sundstedt, V., Gutierrez, D.: Screen-space perceptual rendering of human skin. ACM Trans. Appl. Perception **6**(4), 1–15 (2009)
10. Kelemen, C., László, S.-K.: A microfacet based coupled specular-matte BRDF model with importance sampling. In: Presentation at Euro-graphics (2001)
11. Jimenez, J., Whelan, D., Sundstedt, V., Gutierrez, D.: Real-time realistic skin translucency. Comput. Graph. Appl. IEEE **30**(4), 32–41 (2010)
12. Simon, G.: Real-time approximations to subsurface scattering. In: Randima, F. (ed.) GPU Gems, pp. 263–278. Wesley, Addison (2004)
13. Hable, J., Borshukov, G., Hejl, J. Fast skin shading. In: W. Engel, (ed.) Shader X7. Charles river media, Chap. 2.4, 161–173 (2009)
14. Egges, A., Papagiannakis, G., Magnenat-Thalmann, N.: Presence and interaction in mixed reality environments. Vis. Comput. **23**(5), 317–333 (2007)

Chapter 5
Adaptive NURBS Tessellation on GPU

Yusha Li, Xingjiang Lu, Wenjing Zhang, and Guozhao Wang

Abstract This chapter presents a method for adaptively tessellating NURBS surfaces on GPU. The method involves tessellation interval estimation, conversion from NURBS to rational Bézier patches, and gap-free tessellation of rational Bézier patches. All the computations are performed on GPU. The main contributions of the chapter lie in two aspects: (1) we improve Zheng and Sederberg's tessellation interval estimation for rational curves and surfaces to give larger tessellation interval and thus to produce fewer triangles, and (2) we propose an adaptive tessellation strategy that allows to tessellate each rational Bézier patch on GPU independently and meanwhile avoid gaps between rational Bézier patches. By using GPU, complicated NURBS models can be easily rendered in real time.

Keywords NURBS • GPU • Gap-free • Real-time tessellation

5.1 Introduction

NURBS technique is the most frequently used design tool in CAD/CAM industry. It provides high-quality shape descriptions with limited data set. Compared to the tremendous data set of polygon models, NURBS models have advantages in storing, transmitting, and editing, which make them useful in animation. NURBS models easily control the shape using a few control points rather than editing a group of polygons.

To render a NURBS surface, in current rendering pipeline, we have to convert it to some primitives such as triangles or quads, which the pipeline can process. Tessellation is such an operation which maps a regular grid in parameter domain onto the surface. Then the surface is only evaluated at these grid points and rendered as polygons. The approximation error is computed as the maximal distance between the original surface and the approximation polygons. The first step

Y. Li (✉) • W. Zhang
School of Computer Engineering, Nanyang Technological University, Singapore, Singapore
e-mail: yli1@e.ntu.edu.sg; wzhang1@e.ntu.edu.sg

X. Lu • G. Wang
Department of Mathematics, Zhejiang University, Hangzhou 310013, China
e-mail: xjlu@zju.edu.cn; wanggz@zju.edu.cn

Y. Cai, S. See (eds.), *GPU Computing and Applications*,
DOI 10.1007/978-981-287-134-3_5

for tessellation is estimating the tessellation intervals to ensure a specified approximation error over the whole surface. The density of the sampling grid is usually selected uniformly or adaptively according to the variance of the surface.

Good tessellation intervals for curves/surfaces should be as large as possible while making the approximation error within the provided error tolerance. Plenty of methods have been proposed to estimate the tessellation intervals of curves/surfaces while ensuring the approximation error within a bound [1–6, 11, 12]. Since the approximation error and tessellation interval are closely related to the shape of the surface, generally, the step size is determined according to the variance of the surface. A most widely adopted criterion is using the upper bounds of the second derivatives over the whole surface [7]. Applying this criterion on nonrational surface is pretty easy. For rational surfaces, researchers must spend much more effort, either computing the maxima of second-order derivatives very costly [7–10] or using other strategies to avoid this computation [11, 12]. Cheng [11] and Zheng and Sederberg [12] proposed such kind of method. They evaluated the upper bounds of the second derivatives of the polynomial curves/surfaces instead of rational curves/surfaces. On the other hand, since the polynomial curves/surfaces are obtained on the lower dimension space under a standard perspective projection, the key technique of their methods is to estimate the effect of perspective transformation. By contrast, Zheng and Sederberg's approach [12] provided a more precise bound and a larger tessellation interval which is more effective. However, their results depend on the affine coordinate system. More intuitively, the result is related to the furthest point on the curve $r = \sup_t \|r(t)\|$ with respect to the origin. The further this point is, the smaller the tessellation interval size is. In their paper, they found the min-max bounding box of the curve/surface and took the center of the bounding box as the new origin. This can decrease the value r and improve the results sometimes. However, when the curve/surface is translated, other items in the formula are also changed, which makes the final results unpredictable.

In this chapter, we make some improvements to Zheng and Sederberg's method. We take the formula as a whole, finding a relatively optimal value of the whole formula rather than only one item in it. Obviously this is more reasonable than the original approach. Besides, we subdivide the curve/surface once and use the new control points to compute r, which is more compact than using the original control points.

To evaluate a NURBS surface, the general approach would probably be to first convert it to primitives that are more real-time friendly, such as Bézier patches. Then a group of Bézier patches can be tessellated in parallel into triangles. However, since the tessellation intervals for each Bézier patch are determined independently, gaps may be introduced between patches, as shown in Fig. 5.1. To solve this problem, in previous literatures, they either used shaded fat lines to fill the gaps [13, 14] or redesigned the connectivity between patches [15–17]. In this chapter, we adopt the later one. After estimating the tessellation intervals for all patches, we create a transition region between adjacent patches with inconsistent tessellation intervals to stitch them together seamlessly while ensuring that the approximation error is still under the given tolerance.

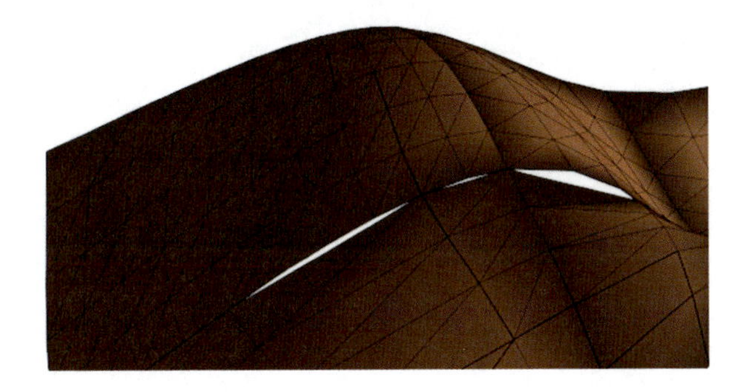

Fig. 5.1 Gaps between adjacent patches with different tessellation intervals

In many applications such as animation and dynamic tessellation, the parametric surface is often required to be tessellated in real time. In this chapter, we use GPU to accelerate the computations. In recent years, GPU becomes more powerful and easy to make use of. CUDA is an API for NVIDIA's GPUs. Using CUDA, the memory writes are more flexible. We can first copy the data from CPU to GPU, and each thread in the GPU can read the data required from the GPU memory accordingly. In our work, all the main stages are carried out on the GPU in a patch-parallel way. Since all the patches are handled independently, before creating transition regions, we need to extract and record their neighbor information first and then redesign the topology and output triangles.

The rest of the chapter is organized as follows. First the tessellation interval estimation method for rational Bézier curves is introduced. Then the method is extended to rational Bézier surface cases. Afterwards, a gap filling algorithm is proposed for patches with different tessellation intervals. In addition, taking each rational Bézier patch as a unit, we perform all the main stages of the program on GPU to accelerate the computations. Finally, the experiment results are presented.

5.2 Estimating the Tessellation Intervals

This section describes how we estimate the tessellation intervals for rational Bézier surfaces. For simplicity, we explain our idea and approach using rational Bézier curves first. Then the extension to rational Bézier surfaces becomes straightforward.

5.2.1 Tessellation Intervals for Rational Bézier Curves

Given a rational curve over the domain $[\alpha, \beta]$,

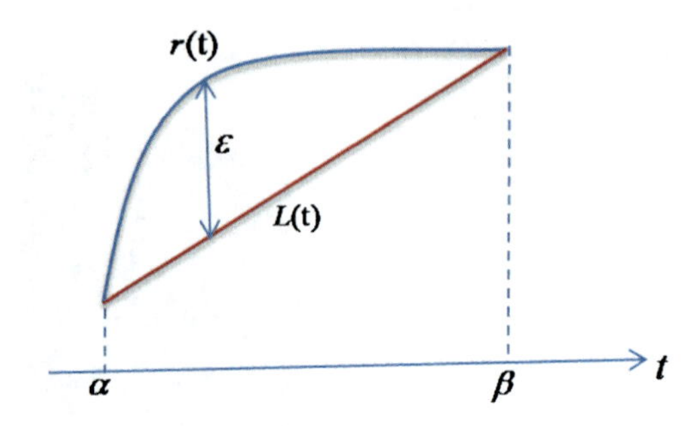

Fig. 5.2 Error between rational curve and its approximation line segment

$$r(t) = \frac{R(t)}{w(t)}, t \in [\alpha, \beta] \tag{5.1}$$

let $L(t)$ be a fractional-linearly parameterized line segment that connects $r(\alpha)$ and $r(\beta)$. Zheng and Sederberg showed that the approximation error (as shown in Fig. 5.2) can be guaranteed not higher than ε if the step size δ satisfies [12]

$$\delta = \beta - \alpha \leq \begin{cases} \sqrt{\dfrac{8\inf_t\{w(t)\}\varepsilon}{\sup_t(\|R''(t)\| + (r-\varepsilon)|w''(t)|)}}, & \varepsilon < r \\ \sqrt{\dfrac{8\inf_t\{w(t)\}\varepsilon}{\sup_t\|R''(t)\|}}, & r \leq \varepsilon < 2r \\ 1, & 2r < \varepsilon \end{cases} \tag{5.2}$$

where r is defined as the furthest point on the curve $r = \sup_t \|r(t)\|$ with respect to the origin and $R''(t)$ is the second-order derivative of the nonrational curve. Applying the above results to a rational Bézier curve of degree n

$$r(t) = \frac{R(t)}{w(t)} = \frac{\sum_{i=0}^{n} P_i w_i B_i^n(t)}{\sum_{i=0}^{n} w_i B_i^n(t)}, t \in [0, 1] \tag{5.3}$$

the bounds of the second-order derivatives of the nonrational Bézier curve can be estimated by

$$\|R''(t)\| \leq n(n-1) \max_i \|\Delta^2(w_i P_i)\| \tag{5.4}$$

$$\|R''(t)\| + (r-\varepsilon)|w''(t)| \leq n(n-1) \max_i \left(\|\Delta^2(w_i P_i)\| + (r-\varepsilon)|\Delta^2 w_i|\right) \tag{5.5}$$

The qualified step size for the Bézier curve can be obtained by substituting the above results to formula (5.2). In Zheng and Sederberg's approach, they estimated the effect of perspective transformation from rational space to nonrational space first and then define the approximation line segment L(t) also in a rational form. After that, the approximation error between r(t) and L(t) can be determined in nonrational space which is less computational intensive. Theoretically, using this method, the results are related to the position of the origin of the perspective projection.

In their work, they improved the results by translating the curve and made r be minimal with respect to the new origin of the coordinate system. From formula (5.2), it can be learnt that the smaller formula (5.4) and formula (5.5) are, the larger the step size is. One disadvantage of this method is that the other parts in the formula (5.5), $\|\Delta^2(w_iP_i)\|$, also change with the translation, which makes the final result unpredictable. Here instead we compute the minimal of the whole $\|\Delta^2(w_iP_i)\| + (r - \varepsilon)|\Delta^2 w_i|$. Assume O is the new origin, after translation, the right parts of formula (5.5) becomes

$$
\begin{aligned}
& n(n-1)\max_i \left(\left\|\Delta^2(w_i(P_i - O))\right\| + (r_o - \varepsilon)\left|\Delta^2 w_i\right|\right) \\
&= n(n-1)\max_i \left(\left\|\Delta^2 w_i P_i - \Delta^2 w_i O\right\| + (r_o - \varepsilon)\left|\Delta^2 w_i\right|\right) \\
&= n(n-1)\max_i \left|\Delta^2 w_i\right|\left(\left\|\frac{\Delta^2 w_i P_i}{\Delta^2 w_i} - O\right\| + (r_o - \varepsilon)\right)
\end{aligned}
\tag{5.6}
$$

Moreover, since ε is a constant given by the user and r can be roughly defined as the distance from the origin to the furthest control point $r = \max_j \|P_j\|$ rather than the furthest point on the curve, the above formula changes to

$$
\begin{aligned}
& n(n-1)\max_i \left|\Delta^2 w_i\right|\left(\left\|\frac{\Delta^2 w_i P_i}{\Delta^2 w_i} - O\right\| + \|P_j - O\|\right) - \left|\Delta^2 w_i\right|\varepsilon \\
& \qquad i = 0 \ldots n-2; j = 0 \ldots n
\end{aligned}
\tag{5.7}
$$

Given a point O and two groups of discrete points, the main part of above formula $\left\|\frac{\Delta^2 w_i P_i}{\Delta^2 w_i} - O\right\| + \|P_j - O\|$ indicates that the sum of the distance from O to the furthest point in each group. However, according to formula (5.2), our goal is to choose a O to make the sum of the distances be minimized. However, one additional element in the formula $|\Delta^2 w_i|$ varying with different points will affect the final results unpredictably. Therefore, when estimating O, the geometry positions of these points are not the only determinant factors. For the discrete points with high weights, there is a high probability that one of them leads to the maximal value of the distance. Hence, the O should be closer to them, to balance the effects of their weights. Based on the above analysis, we set the following objective function to determine O:

$$\min_{O}\left(\sum_{j=0}^{n}(P_j-O)^2+\sum_{i=0}^{n-2}|\Delta^2 w_i|\left(\frac{\Delta^2 w_i P_i}{\Delta^2 w_i}-O\right)^2\right) \tag{5.8}$$

Let

$$F=\sum_{j=0}^{n}(P_j-O)^2+\sum_{i=0}^{n-2}|\Delta^2 w_i|\left(\frac{\Delta^2 w_i P_i}{\Delta^2 w_i}-O\right)^2 \tag{5.9}$$

Then O can be obtained by letting the first-order derivative of the function equal to zero:

$$\frac{\partial F}{\partial O}=2\sum_{j=0}^{n}(O-P_j)+2\sum_{i=0}^{n-2}|\Delta^2 w_i|\left(O-\frac{\Delta^2 w_i P_i}{\Delta^2 w_i}\right)=0 \tag{5.10}$$

$$O=\frac{\sum_{j=0}^{n}P_j+\sum_{i=0}^{n-2}|\Delta^2 w_i|\frac{\Delta^2 w_i P_i}{\Delta^2 w_i}}{n+1+\sum_{i=0}^{n-2}|\Delta^2 w_i|} \tag{5.11}$$

To further improve the results, we subdivide the rational Bézier curve once to obtain a tighter bounding box of the curve. This will decrease r with a high probability. We represent the subdivision control points as Ps.

$$O=\frac{\sum_{j=0}^{k}Ps_k+\sum_{i=0}^{n-2}|\Delta^2 w_i|\frac{\Delta^2 w_i P_i}{\Delta^2 w_i}}{k+1+\sum_{i=0}^{n-2}|\Delta^2 w_i|} \tag{5.12}$$

To summarize, we have the following algorithm:

1. Subdivide the Bézier curve once. Get Ps.
2. Estimate the new origin O using equation (5.12).
3. Find the furthest control point Ps_k with respect to O. Compute $r=\max_k\|Ps_k-O\|$.
4. Find the max value of formula (5.4) or (5.5).
5. Compute the step size using formula (5.2).

If $|\Delta^2 w_i|=0$, the discrete point $\frac{\Delta^2 w_i P_i}{\Delta^2 w_i}$ becomes infinity; we just ignore this kind of points and apply the same algorithm to all the left points. Besides, when the user changes the error tolerance, there is no need to recompute O. New step sizes can be obtained very quickly by only changing some constant values. We also used weighted points to give the curve a tighter convex hull [18]. In this case, all the forward difference points $\frac{\Delta^2 w_i P_i}{\Delta^2 w_i}$ except the first and last one will be replaced by $\left\| \frac{\Delta^2(w_{i-1}P_{i-1})+\Delta^2(w_i P_i)}{\Delta^2 w_{i-1}+\Delta^2 w_i} \right\|$. Other procedures of the algorithm remain the same.

5.2.2 *Tessellation Intervals for Rational Bézier Surfaces*

The approach described in Sect. 5.2.1 can be easily extended to the surface case. According to Zheng and Sederberg's method [12], given a C^2 rational surface defined on a domain T, the approximation error between the original surface and the approximate triangular mesh is smaller than ε if

$$D_{ss}\delta_s^2 + 2D_{st}\delta_s\delta_t + D_{tt}\delta_t^2 \le 8\varepsilon \inf_{(s,t)} w(s,t) \tag{5.13}$$

where

$$r \ge \sup_T \|r(s,t)\|$$
$$D_{ss} = \begin{cases} \sup_T \left(\left\| R_{ss}''(s,t) \right\| + (r-\varepsilon)\left| w_{ss}''(s,t) \right| \right), & \varepsilon < r \\ \sup_T \left\| R_{ss}''(s,t) \right\|, & r \le \varepsilon \le 2r \\ 0, & 2r \le \varepsilon \end{cases} \tag{5.14}$$

The other two elements D_{st}, D_{tt} are defined in the same way as $D_{ss.}$ δ_s and δ_t are the tessellation step sizes. Under a specified error tolerance, according to the formula (5.13), in order to maximize the step sizes, we should make the three values, D_{ss}, D_{st}, and D_{tt}, as small as possible. For a rational Bézier surface defined as follow:

$$r(s,t) = \frac{R(s,t)}{w(s,t)} = \frac{\sum_{i=0}^{n}\sum_{j=0}^{m} w_{i,j} P_{i,j} B_i^n(s) B_j^m(t)}{\sum_{i=0}^{n}\sum_{j=0}^{m} w_{i,j} B_i^n(s) B_j^m(t)} s,t \in [0,1], \tag{5.15}$$

the bounds of D_{ss}, D_{st}, D_{tt} of this rational Bézier surface can be computed

$$D_{ss} = n(n-1) \max_{\substack{0 \le i \le n-2 \\ 0 \le j \le m}} \left\{ \left\| w_{i+2,j}P_{i+2,j} - 2w_{i+1,j}P_{i+1,j} + w_{i,j}P_{i,j} \right\| + (r-\varepsilon)\left| w_{i+2,j} - 2w_{i+1,j} + w_{i,j} \right| \right\}$$

$$D_{st} = nm \max_{\substack{0 \le i \le n-1 \\ 0 \le j \le m-1}} \left\{ \left\| w_{i+1,j+1} P_{i+1,j+1} - w_{i+1,j} P_{i+1,j} - w_{i,j+1} P_{i,j+1} + w_{i,j} P_{i,j} \right\| \right.$$

$$\left. + (r - \varepsilon) \left| w_{i+1,j+1} - w_{i+1,j} - w_{i,j+1} + w_{i,j} \right| \right\} \tag{5.16}$$

$$D_{tt} = m(m-1) \max_{\substack{0 \le j \le m-2 \\ 0 \le i \le n}} \left\{ \left\| w_{i,j+2} P_{i,j+2} - 2 w_{i,j+1} P_{i,j+1} + w_{i,j} P_{i,j} \right\| \right.$$

$$\left. + (r - \varepsilon) \left| w_{i,j+2} - 2 w_{i,j+1} + w_{i,j} \right| \right\}$$

Let $r = \max_{\substack{0 \le i \le n \\ 0 \le j \le m}} \left\| P_{ij} \right\|$. We can derive the objective function in the same manner as curve case:

$$\min_{O} \sum_{k,h} (P_{k,h} - O)^2 + \sum_{i,j} \left| \Delta^2 W_{i,j} \right| (Q_{i,j} - O)^2 \tag{5.17}$$

In this situation, $Q_{i,j}$ is the combination of

$$\begin{aligned}
&\frac{w_{i+2,j} P_{i+2,j} - 2 w_{i+1,j} P_{i+1,j} + w_{i,j} P_{i,j}}{w_{i+2,j} - 2 w_{i+1,j} + w_{i,j}} \quad 0 \le i \le n-2, 0 \le j \le m \\
&\frac{w_{i+1,j+1} P_{i+1,j+1} - w_{i+1,j} P_{i+1,j} - w_{i,j+1} P_{i,j+1} + w_{i,j} P_{i,j}}{w_{i+1,j+1} - w_{i+1,j} - w_{i,j+1} + w_{i,j}} \quad 0 \le i \le n-1, 0 \le j \le m-1 \\
&\frac{w_{i,j+2} P_{i,j+2} - 2 w_{i,j+1} P_{i,j+1} + w_{i,j} P_{i,j}}{w_{i,j+2} - 2 w_{i,j+1} + w_{i,j}} \quad 0 \le i \le n, 0 \le j \le m-2
\end{aligned} \tag{5.18}$$

and their corresponding weights are the denominators. The results can also be further improved by subdividing the surface once to obtain a smaller r. Besides, the weighted point method is used to find a smaller convex hull for the surface. Using these points to compute the discrete forward difference points $Q_{i,j}$ will be better.

5.3 Creating Transition Regions

5.3.1 Extracting Bézier Patches

Given a NURBS surface, we first convert it to a set of rational Bézier patches. Then the uniform tessellation intervals for each Bézier patch are determined independently. For a NURBS surface with an m by n control grid, only (m−3)*(n−3)

Bezier patches are extracted. The details of conversion can be found in Piegl and Tiller's book [19]. We will not explain it here.

5.3.2 *Filling the Gaps*

For adjacent patches with different tessellation intervals, gaps unavoidably occur when rendering. We need to change the boundary connections to stitch them together. On the coarse patch, we choose the column/row next to the common boundary to create a transition region using existing vertices, as shown in Fig. 5.3. This can ensure that the approximation error in this region is still below the given tolerance after topology is changed. Except this row/column, the topology of other parts on the patch keeps unchanged.

We reconnect the vertices on the boundary of this region according to some criteria:

1. For each point on the coarse side, find its nearest point on the opposite edge of this region and connect them as an edge.
2. For the rest unconnected points on the dense edge, find its nearest point on the coarse edge and connect them as an edge.
3. If there still are quadrangles in this region, connect its shorter diagonal.

These criteria can guarantee that there is no cross of all edges. Figure 5.4 is a surface composed of nine rational Bézier patches with different tessellation intervals only in the vertical direction. It is obvious that there are many T-junctions along their boundaries which will lead to gaps. We can see that there are no T-junctions anymore and the edges are connected in a very decent manner after applying the algorithm. Highlights are two inconsistent patches and their corresponding transition regions.

The criteria can also handle the situation that a patch needs more than one transition region along its boundaries, as the left-top patch in Fig. 5.5.

5.4 Implementations on GPU

We implement all the main stages in the algorithm: including extraction of rational Bézier patches, estimation of tessellation intervals, creation of transition regions, and tessellation on a dual-core NVIDIA graphic card GTX 590 (one core is used) with memory bandwidth 163 GB/s. Taking a single Bézier patch as a unit, the program is run in a patch-parallel level. Each patch is assigned a thread to run all the stages successively. The work flow is shown in Fig. 5.6.

Given a NURBS surface, firstly all the control points are sorted in a row-wise order and passed to the GPU memory. Then the control points' indices that are needed for computing each rational Bézier patch are extracted. In each block, there

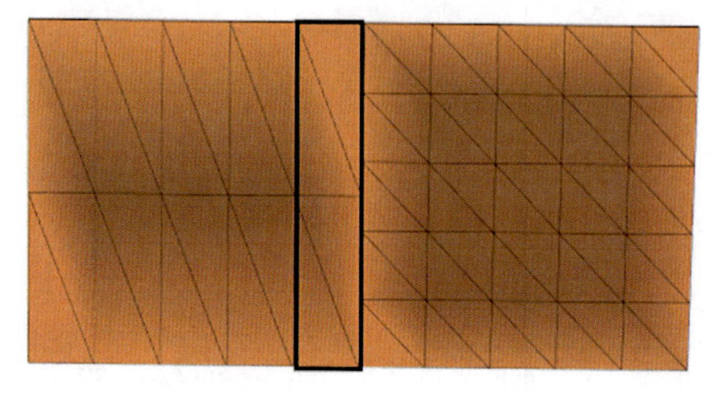

Fig. 5.3 The domain to create transition region between two adjacent patches

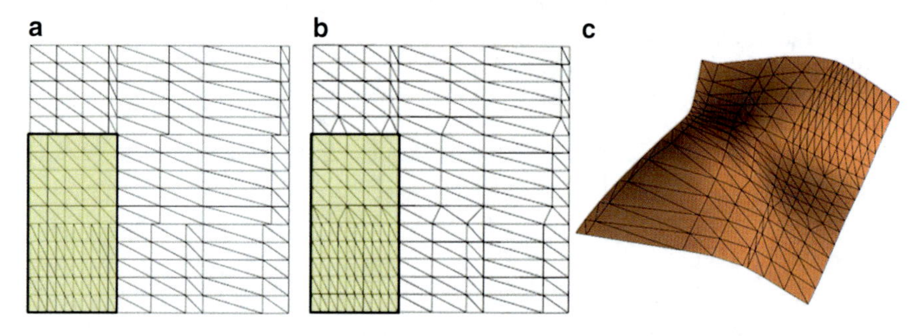

Fig. 5.4 (**a**) Nine Bézier patches with different tessellation intervals in vertical direction, (**b**) transition regions, (**c**) corresponding gap-free surface

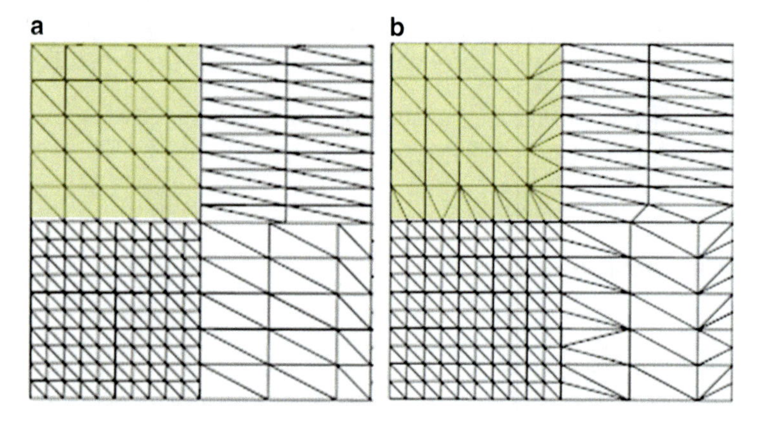

Fig. 5.5 (**a**) Bézier patch with different tessellation intervals in both directions, (**b**) after creating transition regions

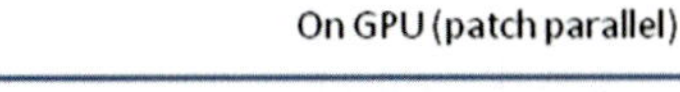

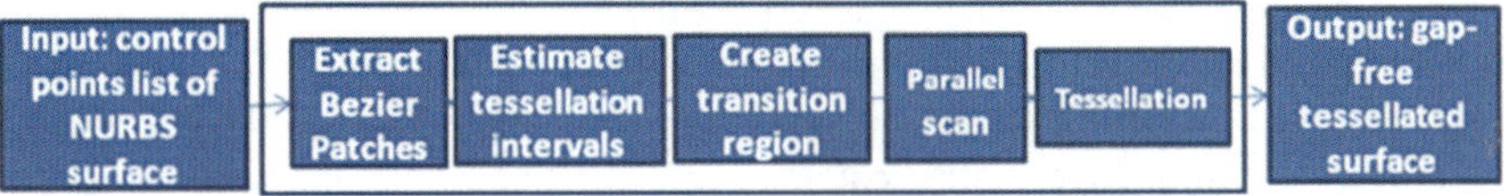

Fig. 5.6 Program workflow

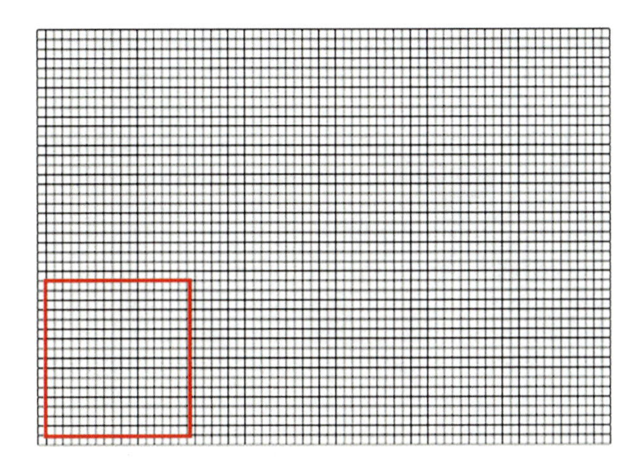

Fig. 5.7 Data read by one block

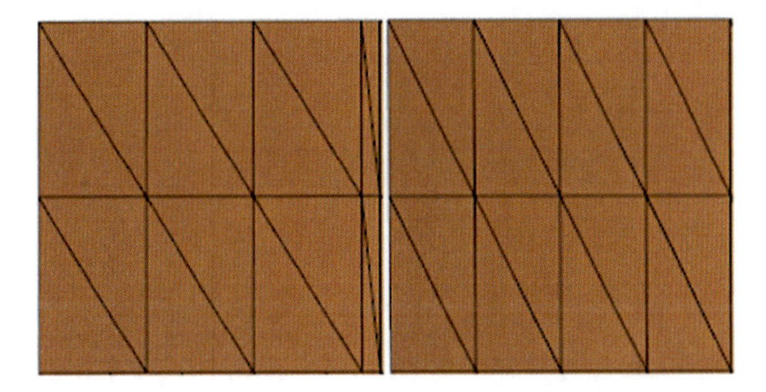

Fig. 5.8 Adjust tessellation pattern to reduce inconsistency and avoid slender triangles

are 16 by 16 threads. Each thread deals with a patch. Take degree 3 NURBS, for example: each patch needs 16 control points to compute rational Bézier control points. Reading the control points for patches in sequent row/column order is not an efficient way. Since a control point is needed by four patches, we read a block of 19 by 19 control points for 16 by 16 patches into the shared memory first, as shown in Fig. 5.7. This reduces the memory transfer greatly.

After estimating tessellation intervals and how many vertices are on each direction, we evenly distribute the tessellation vertices while keeping their number unchanged, as shown in Fig. 5.8. This may decrease the approximation error but not reduce the tessellation efficiency. Using this approach, it can reduce the number of inconsistent patches. For instance, tessellation step sizes between 0.25 and 0.3 are all considered as consistent factors since they will evenly generate five vertices on one boundary. Therefore, we just need to search the numbers rather than tessellation interval to determine inconsistent patches. This also helps to avoid slender triangles which are caused by dividing step size with small remainder in the parameter domain.

For each patch, we store the number of vertices in its two parameter directions. These will be used to search adjacent patch pairs with different tessellation intervals and determine how many vertices on each boundary of each patch. Here, according to these tessellation intervals, we evaluate all the vertices on each patch without considering the transition regions. One benefit of the regular tessellation pattern is that the blending functions of one parameter direction for all the vertices in one row/column are the same. These values can be pre-computed only once and stored for multiple uses in direct evaluation of NURBS surface. All the vertices in a patch are computed sequentially using direct evaluation method in our implementation.

Then triangles are generated for both interior region and transition region. For those with more than one inconsistent boundary with neighbors, we create transition regions first in one parameter direction and then in the other direction. The patches need transition regions in two directions which cannot run at once since conflicts may occur at corner using our algorithm. Since the NURBS surface is tessellated adaptively, each patch may have different numbers of vertices and triangles. We cannot decide the offset of a certain vertex or triangle in the mesh. Parallel scan [20] should be applied to obtain these offset once the tessellation intervals for each patch is obtained. Afterwards, the vertex buffer and index buffer with corresponding sizes are created on the global memory.

The final data in vertex buffer and index buffer can be directly used for rendering.

5.5 Experiment Results

5.5.1 Comparisons to Zheng and Sederberg's Method

To compare the results with Zheng and Sederberg's method [12], we simply use the same configuration with their experiments and choose their best results to compare with our method. First, we estimate the step sizes for some single curves whose degree is from 1 to 8. They are listed in Table 5.1. All of them are planar curves with coordinates (x, y, w) which is **not** homogenous coordinate.

In their paper, they run all the examples using translating control point method, weighted point method, and the combination of these two methods. We choose their best results among these methods as shown in Table 5.2.

Table 5.1 List of Bézier curves

Degree	Control points
$C_1(t)$	(2. 5, 5.6), (1. 8, 0.7)
$C_2(t)$	(−3,−10, 0.96), (6, 8, 2.3), (2, 4, 0.63)
$C_3(t)$	(19, 61, 0.08), (−61, 52, 0.5), (17, 55, 1), (49,−20, 0.4)
$C_4(t)$	(1, 5, 6.1), (7, 7, 0.39), (−8,−10, 18.4), (−1,−10, 1.1), (−6,−3, 0.03)
$C_5(t)$	(53,−6, 0.7), (−7, 66, 1.8), (−64,−46, 147), (−71, 43, 6.6), (97,−68, 4), (−66, 57, 0.7)
$C_6(t)$	(36,−23, 1.7), (48, 54, 0.8), (14,−13, 0.2), (64, 13, 1), (−68, 54, 1.4), (43,−1, 0.4), (34, 92, 0.2)
$C_7(t)$	(9, 9, 1.5), (−4, 0, 3.1), (5, 0, 3.3), (−7, 0, 2.7), (10,−10, 2.6), (2,−3, 0.7), (−4, 9, 1.1), (1,−5, 1.3)
$C_8(t)$	(3, 2, 0.2), (−5, 8, 1.6), (1, 4, 0.8), (7, 10, 0.8), (1,−8, 1.1), (−7, 5, 1.3), (0, 10, 0.4), (4, 1, 0.7), (6,−1, 2.3)

Table 5.2 The step size for some single curves with error tolerance 0.1

δ	$C_1(t)$	$C_2(t)$	$C_3(t)$	$C_4(t)$	$C_5(t)$	$C_6(t)$	$C_7(t)$	$C_8(t)$
Zheng's best results	1.0	0.0551	0.0081	0.0021	0.001	0.0052	0.0132	0.0075
Our method	1.0	0.0551	0.0090	0.0021	0.001	0.0051	0.0135	0.0085

From the above results, we can see that in some cases our method is very effective. However, sometimes the weights of the discrete points negatively affect the results a little. That is because we did not assign a weight to the control points; in some cases, the new origin is pulled close to the discrete forward difference points by their high weights, but it becomes far away from the control point. After multiplying the weight, $|\Delta^2 w_i|r$ increases more than the decrease of $\left|\Delta^2 w_i\right|\left\|\frac{\Delta^2 w_i P_i}{\Delta^2 w_i}\right\|$.

To get the statistics of the overall performance, we also ran the algorithm on 300 randomly generated degree 3 rational Bézier curves. The x, y coordinates are randomly distributed in three intervals, [−10, 10], [−100, 100], [−1000, 1000], each interval containing 1,000 curves. The weight of each control point is the ratio of two random numbers between 1 and 10,000. In most cases, the improved method is better. For all intervals, the average step size increases (Table 5.3).

5.5.2 *Run Time on CPU and GPU*

We run the three models in Fig. 5.9 both on CPU and GPU. The run time on CPU is taken as a reference compared to the GPU program. The three models contain hundreds of or more than ten thousands of patches. We firstly use our method to estimate the tessellation intervals and then tessellate them on the GPU. We can see that even generating millions of triangles, it still can achieve real-time performance (Table 5.4).

Table 5.3 Average step size of a group of curves with error tolerance 0.1

	Zheng and Sederberg's method	Our method	No. of curves (our method is better)
[−10, 10]	0.02389	0.02810	849
[−100, 100]	0.00705	0.00835	842
[−1000, 1000]	0.00240	0.00278	839

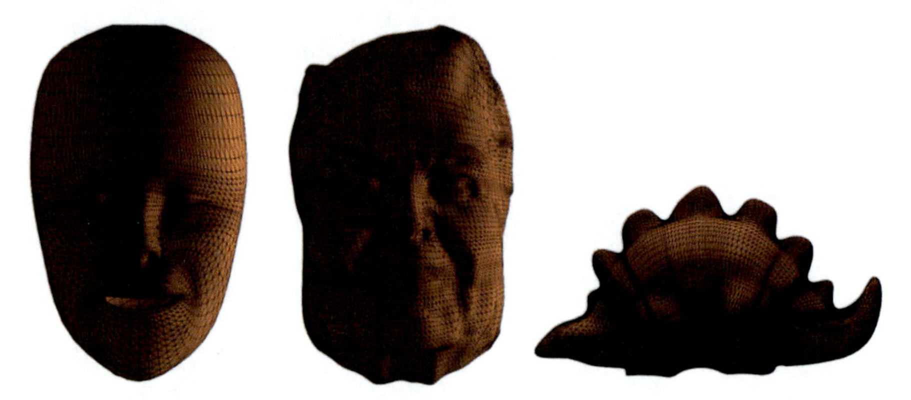

Fig. 5.9 Models after adaptive tessellation

Table 5.4 Run time on CPU and GPU (adaptive tessellation)

	Patch no.	Triangle no.	Vertex no.	Run time on CPU	Run time on GPU
Head	360	2298	2850	1.76 ms	0.416 ms
Old man face	3828	10216	17775	17.91 ms	1.58 ms
Stegosaurus	12312	25530	50150	41.25 ms	0.66 ms

The run time is not proportional to the number of total vertices/triangles since it is determined by the most time-consuming thread. It can be seen that the old man model is most unsmooth. It generates more vertices on each patch compared to the stegosaurus model. Therefore, though containing less vertices and triangles, its run time is higher. Besides, the head model generates more vertices on each patch on average than "old man," but run time is much less. That is mainly because the run time on GPU is determined by the most time-consuming thread in each block. Some of the patches on the "old man" may need more time to process. One solution to this problem is that firstly rank the patches according to the number of vertices and number of transition triangles and then let the patches with similar complexity be processed in the same block.

Since most of the composed Bézier patches are very flat in the three models, the generated vertices are very few on each patch. We also do a uniform tessellation. Each Bézier patch is tessellated to 10 by 10 vertices to test the performance of our program. Compared to adaptive tessellation, there is a sharp growth of the data size

Table 5.5 Run time on CPU and GPU (uniform tessellation)

	Patch no.	Triangle No.	Vertex no.	Run time on CPU	Run time on GPU
Head	360	29160	29503	14.01 ms	2.83 ms
Old man face	3828	310068	331633	161.3 ms	3.95 ms
Stegosaurus	12312	997272	1039186	500 ms	10.1 ms

but run time only increases a little. That is mainly because all the patches are tessellated in parallel. The consuming of more time only happens when the vertex data and index data that need to be computed on each patch increase (Table 5.5).

Conclusions

In this chapter, we improved the tessellation interval estimation algorithm for rational Bézier curves/surfaces, which makes the tessellation more effective. Besides, after estimating tessellation intervals, we adjust the vertices to make them evenly distributed. This effectively reduces the number of inconsistent patches and slender triangles. We also create transition regions between inconsistent patches to guarantee a gap-free polygon surface. Using CUDA, all the programs are executed on GPU in a patch-parallel way. The performance is significantly improved compared to CPU. Even very complicated models still can be rendered in real time.

Our improvement for the tessellation interval estimation is not theoretically optimized. The weighted method only excels in some cases. The final result may be further improved by assigning a suitable weight for each control point.

This method can be applied to T-spline surface, but more complicated. Since there are T-junctions, there may be more than one tessellation interval along a patch boundary. We need to develop other methods to handle the connections between patches while making them suitable for GPU implementation. The algorithm should be explored at a different level of parallelism.

Moreover, direct evaluation is not the best way for GPU evaluation. We can use vertex-parallel manner to compute the vertex data or use forward differencing method which is more effective than direct evaluation for sequential vertex evaluation.

Acknowledgment This research is supported by Multi-plAtform Game Innovation Centre (MAGIC), funded by the Singapore National Research Foundation under its IDM Futures Funding Initiative and administered by the Interactive & Digital Media Programme Office, Media Development Authority, and the National Science Foundation of China (Grant no. 60933008, 61272300).

References

1. Abi-Ezzi, S.S., Shirman, L.A.: Tessellation of curved surfaces under highly varying transformations. In: Proceedings of Eurographics, vol. 91, pp. 385–397 (1991)
2. Sheng, X., Hirsch, B.E.: Triangulation of trimmed surfaces in parametric space. Comput. Aided Des. **24**, 437–44 (1992)
3. Piegl, L.A., Richard, A.M.: Tessellating trimmed NURBS surfaces. Comput. Aided Des. **27**, 16–26 (1995)
4. Elber, G.: Error bounded piecewise linear approximation of freeform surfaces. CAD Comput. Aided Des. **28**, 51–7 (1996)
5. Tookey, R., Cripps, R.: Improved surface bounds based on derivatives. Comput. Aided Geom. Des. **14**, 787–91 (1997)
6. Yeo, Y.I., Bin, L., Peters, J.: Efficient pixel-accurate rendering of curved surfaces. In: ACM SIGGRAPH Symposium on Interactive 3D Graphics and Games, pp. 165–173. New York, NY (2012)
7. Filip, D., Magedson, R., Markot, R.: Surface algorithms using bounds on derivatives. Comput. Aided Geom. Des. **3**, 295–311 (1986)
8. Floater, M.S.: Derivatives of rational Bézier curves. Comput. Aided Geom. Des. **9**, 161–74 (1992)
9. Wang, G.J., Sederberg, T.W., Saito, T.: Partial derivatives of rational Bézier surfaces. Comput. Aided Geom. Des. **14**, 377–81 (1997)
10. Wang, G.Z. The subdivision method for finding the intersection between two Bezier curves or surfaces. Zhejiang Univ. J. (in Chinese). 108–119 (1984)
11. Cheng, F.: Estimating subdivision depths for rational curves and surfaces. ACM Trans. Graph. **11**, 140–51 (1992)
12. Zheng, J., Sederberg, T.W.: Estimating tessellation parameter intervals for rational curves and surfaces. ACM Trans. Graph. **19**, 56–77 (2000)
13. Balázs, Á., Guthe, M., Klein, R.: Fat borders: Gap filling for efficient view-dependent LOD NURBS rendering. Comput. Graph. **28**, 79–85 (2004)
14. Guthe, M., Balázs, Á., Klein, R.: GPU-based trimming and tessellation of NURBS and T-Spline surfaces. In Proceedings of SIGGRAPH, 3 edn, vol. 24, pp. 1016–1023 (2005)
15. Moreton, H.: Watertight tessellation using forward differencing. In: Workshop, pp. 25–32 (2001)
16. Abi-Ezzi, S.S., Subramaniam, S.: Fast dynamic tessellation of trimmed NURBS surfaced. Computer Graphics Forum **13**, 107–26 (1994)
17. Schwarz, M., Stamminger, M.: Fast GPU-based adaptive tessellation with CUDA. Comput. Graph. Forum **28**, 365–74 (2009)
18. Farin, G.: Tighter convex hulls for rational Bézier curves. Comput. Aided Geom. Des. **10**, 123–5 (1993)
19. Piegl, L.A., Tiller, W.: The NURBS book. Springer, New York, c1997, 2nd edn. (1997)
20. Sengupta, S., Harris, M., Zhang, Y., Owens, J.D.: Scan primitives for GPU computing. In: Proceedings of the SIGGRAPH/Eurographics Workshop on Graphics Hardware, pp. 97–106 (2007)

Chapter 6
Graphics Native Approach to Identifying Surface Atoms of Macromolecules

Huagen Wan, Yunqing Guan, and Yiyu Cai

Abstract Classification of "surface atoms" or "interior atoms" of proteins or other macromolecules is significant for many biochemical tasks, particularly for molecular docking. We present a simple and easy-to-implement algorithm for identifying surface atoms of macromolecules from interior atoms. Unlike existing methods that are based on geometry computations, our approach takes the advantage of graphics hardware, and most of the computations are fulfilled with graphics processing unit (GPU). The algorithm can be easily incorporated within visualization applications for macromolecules to enable the removal of interior atoms from a macromolecular structure, thus simplifying the graphics display and manipulation.

Keywords Molecular surface • Solvent accessible surface • Surface atoms • Interior atoms • Graphics algorithm • Graphics hardware • Rendering

6.1 Introduction

The structure of proteins and other macromolecules is fundamental for the underlying biological interactions. As biological molecules interact at their surfaces, an understanding of the surface characteristics of the participating molecules would be particularly useful for studying interactions among them. Although the boundary surface of the electronic density surrounding a molecule is not well defined, the term of molecular surfaces was first introduced by Richards in 1977 to describe a molecular envelope accessible, e.g., by a solvent molecule [1]. There are several

H. Wan
State Key Lab of CAD&CG, Zhejiang University, Hangzhou, China 310027
e-mail: hgwan@cad.zju.edu.cn

Y. Guan
Institute for Media Innovation, Nanyang Technological University, Singapore, Singapore 637553
e-mail: yunqing.guan@ntu.edu.sg

Y. Cai (✉)
School of Mechanical and Aerospace Engineering, Nanyang Technological University, Singapore, Singapore 639798
e-mail: myycai@ntu.edu.sg

Y. Cai, S. See (eds.), *GPU Computing and Applications*,
DOI 10.1007/978-981-287-134-3_6

representational schemes to define the molecular surface model. These include the isovalue electronic density surface, van der Waals surface, Richards's molecular surface, and solvent accessible surface (SAS) [2].

The isovalue electronic density surface is described as the molecular envelope consisting points with the same electronic density values, generally 0.002 au, in a given volume.

The van der Waals surface is, however, defined as the molecular envelope containing the atomic spheres with van der Waals radii. It is simply constructed from overlapping van der Waals spheres of the atoms. Given the spherical representation of the atoms with van der Waals radii, the van der Waals surface is represented as the union of all portions of all atomic sphere surfaces not occluded by neighboring atomic spheres.

Richards's molecular surface is composed of two different kinds of surface patches: the contact surface and the reentrant surface [1]. Imagine the approach of a small "probe" molecule up to the van der Waals surface of a macromolecule. Depending on the size of the probe molecule (except for a probe of zero size), there will be regions of "dead space," crevices that are not accessible to the probe as it rolls about on the macromolecule. The molecular surface is traced out by the inward-facing part of the probe molecule sphere as it rolls on the van der Waals surface of the macromolecule. The contact surface is formed by the part of the van der Waals surface of each atom that is accessible to the probe sphere. The reentrant surface corresponds to the inward-facing part of the probe sphere when it is simultaneously in contact with two or three atoms forming crevices too narrow for the probe molecule to penetrate. Richards's molecular surface is usually defined using a water molecule as the probe, represented as a sphere with radius 1.4 Å. In [3], Connolly has proposed an analytical method for calculating Richards's molecular surface, with which a set of curved regions of spheres and tori, joined together at circular arcs, are used to describe the molecular surface.

The solvent accessible surface (SAS) corresponds to the molecular envelope of the surface that is traced by the center of the probe molecule sphere as it rolls on the van der Waals surface of the macromolecule [4, 5]. The center of the probe molecule can thus be placed at any point on the accessible surface and not penetrate the van der Waals spheres of any of the atoms in the macromolecule. Mathematically, it is equivalent to a van der Waals surface in which the atomic radii have been extended by the probe radius.

Figure 6.1 illustrates the last three kinds of representational schemes for the molecular surface model.

Molecular surface modeling has several applications. One direct benefit with molecular surfaces is the protein or macromolecule visualization [6–8]. Various physical chemical properties such as electrostatic potential and hydrophobicity [9] can be mapped onto the molecular surface and color coded [10–14]. Crucial in protein-protein interaction and interface study [15], molecular surfaces have been applied to the protein-protein docking problem which is the prediction of a complex between two proteins given the three-dimensional structures of the individual

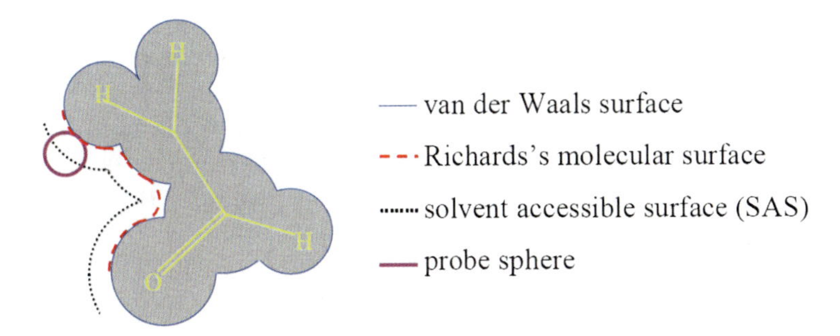

Fig. 6.1 Schematic view of van der Waals surface, Richards's surface, and SAS

proteins [16–18]. Identifying binding pockets on protein surfaces to help in rational or structure-based drug design [19–25] is another major purpose of molecular surface investigation.

For those atoms of a protein or other macromolecules, a significant number of them lie buried beneath the molecular surface of the protein or macromolecule. Interactions among these macromolecules are often dominated by interactions with the "surface atoms," although interactions with the interior atoms of the macromolecule certainly contribute to the total intermolecular interaction energy. Therefore, a classification with "surface atoms" or "interior atoms" of proteins or other macromolecules is significant for biochemical tasks, particularly for molecular docking. For such a classification, several factors should be considered, e.g., the running time of the classification algorithm, number of surface atoms correctly identified, and the numbers of surface atoms and interior atoms incorrectly identified [26].

In this chapter, we present a simple, graphics hardware-based approach to identifying surface atoms of macromolecules from interior atoms. The chapter is organized as follows. In Sect. 6.2, we review the related research works. In Sect. 6.3, we describe the overview of our algorithm as well as its implementation details. Section 6.4 presents some experimental results and discussions and the final section concludes our study.

6.2 Prior Work

Deanda et al. [26] propose a definition for surface atoms as follows: "An atom will be classified as an 'effective surface atom' if its SAS area is greater than a user specified minimum threshold value for the atomic SAS area $SA_{acc}^{\min}$." Accordingly, they develop an SAS approach to distinguishing the surface atoms of macromolecules from the interior atoms. The SAS approach is a computational one that calculates the atomic contributions to the SAS area and designating beforehand a constant value as the minimum threshold for the atomic SAS area. They adopt a surface area and volume package (SAVOL3) [27, 28] to calculate the atomic SAS area. In their paper, they also summarize several other methods for surface atom identification: (1) the NIN (number of intersecting neighbors) approach based on the intuitive notion that the number of intersecting neighbors (i.e., atomic spheres intersect one another) would be far greater for interior atoms than for surface atoms, (2) the SOV (sum of vectors) approach which is a variation of the NIN approach and uses the norm of the SOV to its neighbors as a criterion for classifying surface atoms from interior atoms, (3) the UCSF (University of California at San Francisco) approach that imbeds the macromolecule within a 3D lattice and associates the atoms with the lattice points for classifying surface atoms [29], and (4) the MDS (molecular dot surface) approach which uses the molecular cloud point representation to identify surface atoms [30].

All those algorithms are geometry based. While the NIN, SOV, UCSF, and MDS approaches suffer from ambiguities for identifying surface atoms (i.e., atoms are often misclassified) [26], the SAS approach needs geometry computations of atomic SAS areas which are often performed with specific software packages. With the rapid development of graphics processing unit (GPU), numerous applications have been developed based on graphics hardware [31–35, 39–42]. We believe that techniques developed for graphics hardware rendering will be very useful for bio-related tasks, such as the identification of surface atoms for proteins or other macromolecules.

6.3 Algorithm Overview and Implementation

The kernel idea behind the definition of surface atoms in [26] is that if an atom of a macromolecule contributes to the molecule's SAS, then the atom will be considered as a "surface atom" of the molecule. Bearing this in mind, we adjust slightly the surface atom definition as follows. Let an atom A (with van der Waals radius r) of a macromolecule M be represented as a hard sphere HS and the counterpart of HS with the radius being extended by the probe radius pr to $(r + pr)$ be denoted as an extended hard sphere (EHS), and then atom A will be classified as a "surface atom" if EHS can be seen from outside of the solvent accessible surface (SAS) of the molecule M.

6.3.1 Algorithm Overview

Our algorithm is based on the rendering of the *EHS*s with commercially available graphics hardware. Therefore, we can exploit the hardware to increase performance.

Imagine that the solvent accessible surface of a macromolecule M is surrounded by a bounding box and that each face of the box is a viewing plane. An image is

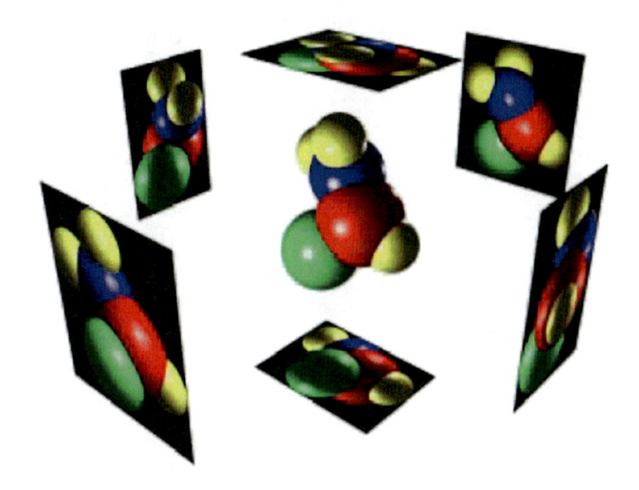

Fig. 6.2 Identifying surface atoms with color and depth buffers

generated for each face by parallel projecting onto it the *EHS*s of the macromolecule *M* with hidden surfaces removed by depth comparison (Fig. 6.2).

Therefore, if the *EHS* of an atom appears in one or more of the six images, then the atom will be classified as a surface atom. Resolutions for the faces are chosen so that there are enough pixels for classifying the surface atoms.

6.3.2 Implementation

The implementation of the algorithm takes the advantage of graphics hardware capabilities (e.g., color buffer and depth buffer), OpenGL graphics library as well as the OpenGL utility toolkit (GLUT) [36, 37]. Apart from the objects positioning and orientation in the scene, OpenGL offers facilities to define a viewing volume and to specify the way objects are projected on the screen. There are two kinds of projection: orthographic and perspective. The orthographic projection draws object without affecting their relative size. The perspective projection is similar to our vision mode: the further an object is, the smaller it appears, and two parallel straight lines seem to converge in the distance. In both cases, viewing volumes are hexahedra: a box or a truncated pyramid respectively (Fig. 6.3).

In our algorithm, the orthographic projection is used and the bounding box of the macromolecule's SAS is adopted as the viewing volume. An image is generated for each of the six faces of the viewing volume by rendering the *EHS*s of the macromolecule with hidden surfaces removed.

For graphics hardware rendering with OpenGL, the color information at each pixel can be stored either in RGBA mode or in color-index mode. In the first mode, the R, G, B, and possibly alpha values are kept for each pixel. In the second mode, however, only a single number (called the color index) is stored for each pixel. Each color index indicates an entry in a color table that defines a particular set of R, G,

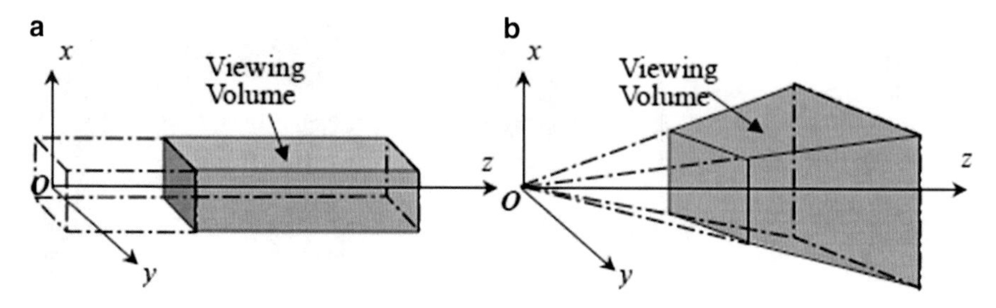

Fig. 6.3 (**a**) Orthographic and (**b**) perspective views

and B values. In either RGBA or color-index mode, a certain amount of color data is stored at each pixel. This amount is determined by the number of bitplanes in the frame buffer. A bitplane contains one bit of data for each pixel.

For most commonly available low-end graphics cards, at least 16 bitplanes are provided for color storage in RGBA mode, and at most 8 bitplanes are available in color-index mode. Considering there are often several hundreds to thousands of atoms in a typical macromolecule, we choose the RGBA mode in this implementation. It would be more straightforward with the color-index implementation, and high-end graphics workstations can be used to improve its efficiency (e.g., with 12 bitplanes on SGI Octane workstations for color-index buffers).

Each atom is firstly initialized with a unique identity, and a color table (with the number of atoms of the macromolecule in size) is created with each of its components corresponding to an atom identity, and then each atom's *EHS* is rendered with the color (in the color table) corresponding to the atom's identity. Subsequently, the color values of the rendered atoms' *EHS*s are read from the color buffer and used to determine the appearance of the *EHS*s in the images. To do so, a Boolean array is used as a flag list to indicate which atom is a surface atom and which one is not. The display list is used for rendering *EHS*s with a high performance.

It is worth noting that the same viewing matrix is used for a pair of rendering (e.g., front and back, left and right, and top and bottom). This is done by setting the depth comparison logic on one of the renderings to save the z-depth values farthest away instead of closest with glDepthFunc( ) and set the face culling logic on the same rendering to eliminate the front polygons of *EHS*s with glCullFace( ). For instance, when rendering the two images for the front and back pair of the viewing volume, firstly the viewing matrix for the front view is set, and then the first image (corresponding with the front view) is generated by culling back polygons (of *EHS*s of the molecule) which face away from the front view and setting the depth comparison logic to GL_LEQUAL to make the depth test satisfied if the incoming z value is less than or equal to the stored z value and finally the second image (corresponding to the back view) is rendered by culling front polygons which face toward the front view and setting the depth comparison logic to GL_GREATER to make the depth test passed if the incoming z value is greater than the stored z value.

Figure 6.4 lists the pseudo code of our algorithm.

```
01  proc IDENTIFYING_SURFACE_ATOMS()
02      Initialize macromolecule data;
         /* e.g. read data file, calculate the bounding box of
            the molecule's SAS, assign an identity for each of the
            atom and build the color table, and etc. */
03      Initialize graphics;
         /* e.g. set color display mode, open display window,
            enable depth test and culling test,
            compile display list for EHS, and etc. */
04      Initialize a flag array for identified surface atoms;
         /* each value of the array is initialized to false to
            indicate that all the atoms are not surface atoms */
05      for each pair of the box faces
            /*e.g. front & back, left & right, and top & bottom */
06         Set viewing matrix with the bounding box parameters;
07         for each face of the pair
08              Set the rendering attributes;
                 /* e.g. face culling logic,
                    depth comparison logic, and etc. */
09              Clear color buffer & depth buffer;
10              for each EHS of the macromolecule
11                 Set rendering color to the EHS's given color;
12                 Render the EHS;
13              end_for
14              Read color values from the color buffer;
15              Update the flag array with the color values read;
                  /* if the EHS of an atom appears in the image,
                     change the value of the array element
                     corresponding to the atom to true */
16          end_for
17      end_for
18      Output the result;
19  end_proc
```

Fig. 6.4 Pseudo code of our algorithm

6.3.3 *Improvements*

The above algorithm can quickly and successfully classify most surface atoms of any macromolecules. The main limitation of the above approach is that it may miss concavities. If some *EHS* of an atom contributes to the molecule's SAS and is not visible from any of the six faces of the viewing volume, then this atom will not be properly classified. The algorithm, however, can be easily improved by adding more viewing planes. For instance, we can sample from the four diagonals of the above bounding box to add 8 more viewing directions and construct viewing planes to render the atoms' *EHS*s (Fig. 6.5). Furthermore, we find using higher resolution of the viewing plane can also improve the classification. We will show with experiment how they help in the next section.

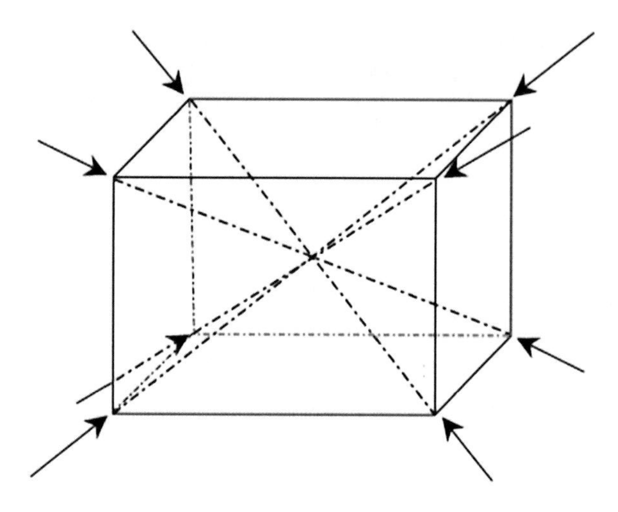

Fig. 6.5 Improving the algorithm by sampling from additional 8 viewing directions

6.4 Experimental Results and Discussions

Several macromolecules from the Protein Data Bank (PDB) [38] were tested under the resolution of 1792 * 1344 in the true color mode (32-bit mode). Figure 6.6 shows the contents of the color buffer when performing the test with a triose-phosphate isomerase (1TIM). Table 6.1 lists the testing results with a dihydrofolate reductase (1RA2), a thermolysin (7TLN), and a triose-phosphate isomerase (1TIM). The tests were performed under different resolutions of the viewing plane (e.g., 100 * 100, 400 * 400, 800 * 800, 1000 * 1000, and 1182 * 1182) and with different configurations of viewing planes (e.g., 6 viewing planes and 14 viewing planes). For comparison reason, the experimental data of the SAS approach selected from [26] were listed in Table 6.2. Their experiments were performed on an SGI Indigo with an R4400 processor.

From Table 6.1, we can clearly see that the number of classified surface atoms increases with the increment of both the viewing planes and the rendering resolution. However, while the accuracy of the classification is nearly constantly improved with more sampling view planes, the number of classified surface atoms increases nonlinearly with the increment of the rendering resolution. For the number of classified surface atoms of the 3 testing macromolecules, there is only a subtle degree of difference for the resolutions of 1000 * 1000 and 1182 * 1182.

Theoretically, there may be an "accurate" or "exact" number of surface atoms for a macromolecular structure, and there may exist a "clear" borderline between surface atoms and interior atoms. However, to our knowledge, there is yet to have a theoretical solution at present time to calculate the "accurate" or "exact" surface atom number. It is a challenging job as well to numerically find out this "accurate" or "exact" number and/or "clear" borderline. In fact, the accuracy of the SAS approach [26] is dependent upon the user-specified minimum threshold value for

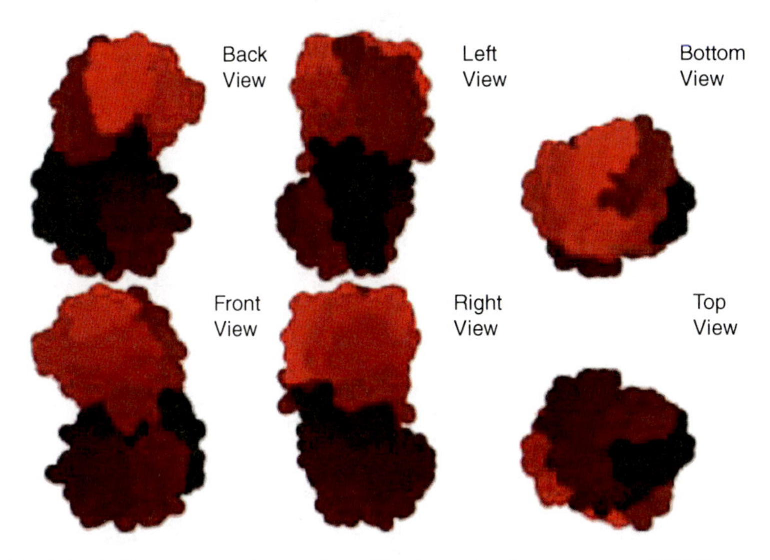

Fig. 6.6 Color buffer contents when testing with triose-phosphate isomerase (1TIM)

the atomic SAS area and the precision of the atomic SAS area calculation. On the other hand, the accuracy of our graphics hardware-based approach depends upon both the viewing plane setting and the rendering resolution. Still, we think that the numerical solutions are worth trying when "accurate" theoretical solutions are not available. Also, we believe that the numbers of classified surface atoms from our approach show kinds of tendency of convergence when the viewing directions and resolution are increased. This again turns out as an interesting yet difficult research topic.

Conclusions

This chapter presents a fast and easy-to-implement algorithm for identifying surface atoms of macromolecules from interior atoms, which is based on the color buffer and z-buffer. The algorithm can be easily incorporated within visualization applications for macromolecules as a preprocessing step to enable the removal of interior atoms from the macromolecular structure. Doing so, a simplified macromolecular structure can be generated for graphics display which can reduce the time required for display and manipulation of macromolecules.

Unlike existing methods for identifying surface atoms of macromolecules mainly based on geometry computations performed by general CPU, our approach takes the advantage of widely available graphics hardware and most of the computations are fulfilled with the graphics processing unit (GPU). As our algorithm is based on the color buffer and z-buffer, its

(continued)

Table 6.1 Experimental results with several macromolecules from Protein Data Bank

	Resolution	100 × 100		400 × 400		800 × 800		1000 × 100		1182 × 1182	
Molecule	Number of sampling viewing planes	6	14	6	14	6	14	6	14	6	14
IRA2 Total atom number: 1268	Number of surface atoms	682	703	723	752	732	756	732	762	733	763
	Running time in seconds	0.34	0.81	0.48	1.14	0.91	2.20	1.34	3.08	1.83	4.20
7TLN Total atom number: 2436	Number of surface atoms	1006	1049	1107	1155	1120	1172	1125	1174	1128	1175
	Running time in seconds	0.63	1.48	0.77	1.83	1.16	2.89	1.45	3.69	1.92	4.66
ITIM Total atom number: 3740	Number of surface atoms	1463	1594	1680	1861	1721	1829	1730	1844	1730	1852
	Running time in seconds	0.98	2.25	1.08	2.63	1.39	3.69	1.64	4.48	1.92	5.36

Table 6.2 Experimental data of the SAS approach (selected from [26])

Molecule	Threshold	$SA^{acc}_{min} = 1A^2$	$SA^{acc}_{min} = 0.01A^2$
IRA2 Total atom number: 1268	Number of surface atoms	604	730
	Running time in seconds	7.57	7.57
7TLN Total atom number: 2436	Number of time in seconds	891	1181
	Running time in seconds	15.73	15.73
ITIM Total atom number: 3740	Number of surface atoms	1565	2216
	Running time in seconds	23.45	23.45

complexity is independent of the molecule complexity but dependent on the rendering resolution and its viewing plane setting.

With the computational power of graphics hardware outperforming that of general CPU by Moore's law [34], we believe that algorithms based on GPU for biochemical tasks will be very promising in the future.

Acknowledgments The authors would like to thank the partial funding support from Singapore MOE Tier 1 (RG 10/12).

References

1. Richards, F.M.: Areas, volumes, packing, and protein structure. Ann. Rev. Biophys. And Bioeng. **6**, 151–176 (1977)
2. Leach, A.R.: Molecular Modelling: Principles and Applications, 2nd edn. Essex, Pearson Education EMA (2001)
3. Connolly, M.L.: Analytical molecular surface calculation. J. Appl. Crystallogr. **16**, 548–558 (1983)
4. Lee, B., Richards, F.M.: Interpretation of protein structures: estimation of static accessibility. J. Mol. Biol. **55**, 379–400 (1971)
5. Hermann, R.B.: Theory of hydrophobic bonding. II. The correlation of hydrocarbon solubility in water with solvent cavity surface area. J. Phys. Chem. **76**, 2754–2759 (1972)
6. Quarendon, P.: A general approach to surface modeling applied to molecular graphics. J. Mol. Graph. **2**, 91–95 (1984)
7. Connolly, M.L.: Depth buffer algorithms for molecular modeling. J. Mol. Graph. **3**, 19–24 (1985)
8. Connolly, M.L.: Plotting protein surfaces. J. Mol. Graph. **4**, 93–96 (1986)
9. Ooi, T., Oobatake, M., Nemethy, G., Scheraga, H.A.: Accessible surface areas as measure of the thermodynamic parameters of hydration of peptides. Proc. Natl. Acad. Sci. USA **84**, 3086–3090 (1987)
10. Chapman, M.S.: Mapping the surface properties of macromolecules. Protein Sci. **2**, 459–469 (1993)
11. Nicholls, A., Bharadwaj, R., Honi, B.: GRASP: graphical representation and analysis of surface properties. Biophy. J. **64**, A166 (1993)

12. Heiden, W., Moeckel, G., Brickmann, J.: A new approach to analysis and display of local lipophilicity/hydrophilicity mapped on molecular surfaces. J. Comput. Aided Mol. Des. **7**, 503–514 (1993)
13. Duncan, B.S., Macke, T.J., Olso, A.J.: Biomolecular visualization using AVS. J. Mol. Graph. **13**(5), 271–282 (1995)
14. Altman, R.B., Hughes, C., Gerstein, M.B.: Methods for displaying macromolecular structural uncertainty: application to the globins. J. Mol. Graph. **13**, 142–152 (1995)
15. Janin, J., Chothi, C.: Surface, subunit interfaces and interior of oligomeric proteins. J. Mol. Biol. **204**, 155–164 (1988)
16. Zielenkiewicz, P., Rabczenko, A.: Protein-protein recognition: method for finding complementary surfaces of interacting proteins. J. Theor. Biol. **111**, 17–30 (1984)
17. Santavy, M., Kypr, J.: A fast computer algorithm for finding an optimum geometrical interaction of two macromolecules. J. Mol. Graph. **2**, 47–49 (1984)
18. Cherfils, J., Janin, J.: Protein docking algorithms: simulating molecular recognition. Current Opinion in Structural Biology **3**, 265–269 (1993)
19. Kuntz, I.D.: Structure-based strategies for drug design and discovery. Science **257**, 1078–1082 (1992)
20. Navia, M.A., Murcko, M.A.: Use of structural information in drug design. Curr. Opin. Struct. Biol. **2**(2), 202–210 (1992)
21. Bugg, C.E., Carson, W.M., Montgomery, J.A.: Drugs by Design. Scientific American **269**(6), 92–98 (1993)
22. Murcko, M.A., Rotstein, S.H.: GenStar: a program for de novo drug design. J. Comput. Aided Mol. Des. **7**, 23–43 (1993)
23. Verlinde, C.L.M.J., Hol, W.G.J.: Structure-based drug design: progress, results and challenges. Structure **2**, 577–587 (1994)
24. Whittle, P.J., Blundell, T.L.: Protein structure-based drug design. Ann. Rev. Biophys. Biomol. Struct. **23**, 349–375 (1994)
25. Jackson, R.M., Sternberg, M.J.E.: Protein surface-area defined. Nature **366**(6456), 638 (1993)
26. Deanda, F., Pearlman, R.S.: A novel approach for identifying the surface atoms of macromolecules. J. Mol. Graph. Model. **20**, 415–425 (2002)
27. Pearlman, R.S.: Molecular surface area and volume: their calculation and use in predicting solubilities and free energies of desolvation. In: Dunn III, W.J., Block, J.H., Pearlman, R.S. (eds.) Partition Coefficient: Determination and Estimation, pp. 3–20. Pergamon Press, New York, NY (1986)
28. Savol3: surface & volume algorithms, http://www.chem.ac.ru/Chemistry/Soft/SAVOL3.en.html. Last visit 4 Oct 2003
29. Bash, P.A., Pattabiraman, N., Huang, C., Ferrin, T.E., Langridge, R.: van der Waals surfaces in molecular modeling: implementation with real-time computer-graphics. Science **222**, 1325–1327 (1983)
30. Brusniak, M.-Y.K.: Development and application of software for CADD. Ph.D. Dissertation, The University of Texas, Austin (Chapter 2) (1996)
31. Baxter, W.V., III, Sud, A., Govindaraju, N.K., Manocha, D.: GigaWalk: interactive walkthrough of complex environment. UNC-CH Technical Report TR02-013 (2002)
32. Govindaraju, N.K., Sud, A., Yoon, S.E., Manocha, D.: Parallel occlusion culling for interactive walkthroughs using multiple GPUs. UNC Computer Science Technical Report TR02-027 (2002)
33. Karabassi, E.A., Papaioannou, G., Theoharis, T.: A fast depth-buffer-based voxelization algorithm. J. Graph. Tools **4**(4), 5–10 (1999)
34. Lin, M., Manocha, D.: Interactive geometric computations using graphics hardware. In: Siggraph'2002 course notes (2002)
35. Tomas, A.M., Eric, H.: Real-Time Rendering, 2nd edn. A.K. Peters Ltd, Natick, MA (2002)
36. McReynolds, T.: Programming with OpenGL: Advanced Rendering. In: SIGGRAPH'96 course notes (1996)

37. GLUT specification, http://www.opengl.org/developers/documentation/glut/index.html. Last visit 4 Oct 2003
38. Protein Data Bank, http://www.rcsb.org/pdb/. Last visit 4 Oct 2003
39. Colberg, P. H., Höfling, F.: Highly accelerated simulations of glassy dynamics using GPUs: Caveats on limited floating-point precision. Comp. Phys. Comm. **182** (5), 1120–1129, (2011)
40. Ufimtsev, I.S., Martinez, T.J.: Graphical processing units for quantum chemistry. Comp. Sci. Eng. **10**(6), 26–34 (2011)
41. Pronk, S., Larsson, P., Pouya, I., Bowman, G.R., Haque, I.S., Beauchamp, K., Hess, B., Pande, V.S., Kasson, P.M., Lindahl, E.: Copernicus: a new paradigm for parallel adaptive molecular dynamics. In: 2011 International Conference for High Performance Computing, Networking, Storage and Analysis, pp. 1–10, 12–18 (2011)
42. Dror, R.O., Dirks, R.M., Grossman, J.P., Xu, H., Shaw, D.E.: Biomolecular simulation: a computational microscope for molecular biology. Annu. Rev. Biophys. **41**, 429–452 (2012)

Chapter 7
A Scalable Software Framework for Stateful Stream Data Processing on Multiple GPUs and Applications

Farhoosh Alghabi, Ulrich Schipper, and Andreas Kolb

Abstract During the past few years, the increase of computational power has been realized using more processors with multiple cores and specific processing units like graphics processing units (GPUs). Also, the introduction of programming languages such as CUDA and OpenCL makes it easy, even for non-graphics programmers, to exploit the computational power of massively parallel processors available in current GPUs. Although CUDA and OpenCL relieve programmers from considering many low-level details of parallel programming on multiple cores on a single GPU, the same support at a higher level of parallelization for multiple GPUs is still under research. In particular, fundamental issues of memory management and synchronization must be dealt with directly by the programmer. In this chapter, we introduce concepts for CUDA-based frameworks which are designed for stateful stream data processing for graph-like arrangements of processing modules on two or more GPUs in a single compute node. We evaluate these concepts and further elaborate on the approach of our choice. Our approach relieves the programmer from error-prone chores of memory management and synchronization. The chapter presents detailed evaluation results which demonstrate the scalability of the proposed framework. To demonstrate the usability of our framework, we utilize it for demanding online processing in the areas of crystallographic structure detection and video decryption.

Keywords GPGPU • Software framework • Multi-GPU • Stream data processing

F. Alghabi (✉) • U. Schipper • A. Kolb
Institute for Vision and Graphics, University of Siegen, Hölderlinstr. 3, 57076 Siegen, Germany
e-mail: farhoosh.alghabi@uni-siegen.de; ulrich.schipper@uni-siegen.de; andreas.kolb@uni-siegen.de

Y. Cai, S. See (eds.), *GPU Computing and Applications*,
DOI 10.1007/978-981-287-134-3_7

7.1 Introduction

Although the idea of parallel processing has been around for some decades, the interest to this field and its applications in various scientific and engineering areas has grown significantly in the past few years. There are two reasons that have played a major role in this growth. One reason is the advancements in hardware industry which have enabled processor manufacturers to put more processing cores on a single die, thereby moving the parallel programming from expensive mainframes and clusters to desktop computers. This fact is verified by noticing the widespread use of multi-core CPUs and many-core GPUs in almost any PC around the world. The second reason is the introduction of languages, libraries, and tools that ease the task of parallel programming for these processors. Particularly, we can mention CUDA and OpenCL which both target GPUs and unleash the huge computational power even to programmers not familiar with computer graphics.

Both OpenCL and CUDA offer general-purpose application programmers with great support for parallel programming. This is accomplished by providing concepts and features that easily map to real-world problems which are parallel in nature, thus enabling efficient exploitation of computational power delivered by the numerous cores of a single GPU with minimal effort. Although these features work well for cases where there is only one GPU available in the compute node, they are not so easily extensible to the cases where multiple GPUs exist in a single node. Thus, it remains the task of programmer to take care of any details in order to provide the same degree of scalability at this new level of parallelization (***multi-GPU, single-node parallelization***) as the one available across the cores on a single GPU.

This chapter specifically addresses the concept and realization of a CUDA-based framework for multi-GPU, single-node parallelization problems, where GPU-scalability is a major concern. The framework has been designed with easy use by application programmers in mind. As a consequence, transparency is an important property of the proposed framework, mainly with regard to memory management and synchronization. Actually the most important programming task left to the application programmer is writing CUDA kernels responsible for processing of data as if they would run on a single-GPU node.

We assume the data is provided as sequences of homogeneous data sets $D^i(t_j)$ (*frames* at time t_j), where i indicates the last processing module M_i with which the data has been processed. We address a specific class of stream processing [1] problems, which can be characterized as follows (see also Fig. 7.1):

Module-based stream processing: We assume data to be loaded to the framework via one or several *source modules* and to be processed by one or several *processing modules*. The resulting data is exported via one or several *sink modules*.

A module M_i can be seen as a CUDA kernel which processes stream data, i.e., transforming input data $D^{i-1}(t_j)$ to output data $D^i(t_j)$ that is fed into subsequent module(s).

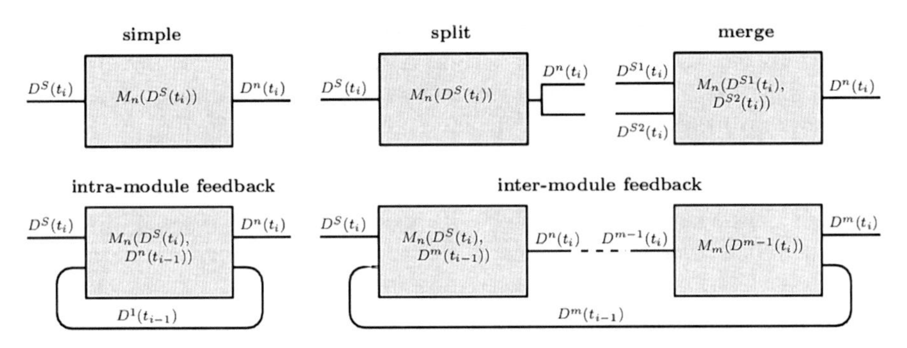

Fig. 7.1 Processing modules and their arrangement, including stream splits (*top middle*), merges (*top right*), intra-module feedback (*bottom left*), and the optional inter-module feedback (*bottom right*)

Graph-based layout: The stream data is transferred between modules, which can be arranged like a graph, including stream splits and stream merges.

Stateful processing of data: It means that previous data or processing results are required for processing newly arrived data. This is realized using *intra-module feedback*; here, the processing in module M_i of frame t_j also depends on the prior result of *the same* module, i.e., on $D^i(t_{j-1})$.

Inter-module feedback (optional): Inter-module feedback improves on the intra-module feedback by letting two distinct modules be connected via feedback.

As a result, the addressed problem class is more general than standard pipeline processing, and thus it has a wider range of applications. Stateless problems, nevertheless, can still be subject to automatic multi-GPU processing.

In order to give an impression of how useful our framework is, we briefly discuss two applications from different domains here. One application lies within the scope of information security. To protect against unauthorized access to information, various cryptographic and steganographic algorithms have been developed. Not surprisingly, videos form an important subclass of data which are required to be protected against unauthorized access. The rapid growth in size of videos (due to increasing resolution, color depth, frame rate, etc.), however, has made the task of applying complex methods computationally quite expensive. One scenario shows the online application of cryptographic and steganographic methods.

The second application lies within the field of crystallography. One common practice in the community is to study the structure of crystals by examining a sample using x-ray imaging. Here, a crystal sample is illuminated by an x-ray beam radiation and the scattered radiation pattern is detected by an energy-dispersive pn-type charge-coupled device (pnCCD) sensor. This camera generates images with 384 * 384 pixels and 2 bytes of information per pixel at currently 400 frames per second, yielding an overall data rate of beyond 112 MB/s. The overall goal of these kinds of experiments is to have near real-time data analysis in order to be able to directly detect improper adjustments of the setup or wrong experimental parameters. Furthermore, in the near future, these experimental setups should be applied for continuous analysis of large sample sets. In a separate section, we show how successfully our framework is used to address this problem.

The remainder of this chapter is organized as follows: Section 7.2 gives an overview of works done in the area of multi-GPU as well as stream data processing. Section 7.3 describes the parallelization concepts and implementation details for the framework. Section 7.4 presents some experimental evaluations. In Sect. 7.5, as mentioned, two real-world applications where our framework has been utilized are elaborated. Finally, a brief discussion is presented as conclusion.

7.2 Related Work

As stated in the introduction, we focus on multi-GPU, single-node parallelization for stream data processing. Consequently, in the following, we first mention works mainly characterized by running on multi-GPU systems and then those which deal with stream data processing.

In [2] Enmyren and Kessler propose a skeleton programming library for systems with multiple CPU cores and GPUs. This is accomplished by use of CUDA and OpenCL as the back ends for code running on GPU and OpenMP for CPU code. The operations supported by their library follow MapReduce model and are in the form of a C++ template library. [3] proposes an approach for high-performance scientific computing on single- and multi-GPU systems. An important feature of the prototype implemented in the paper is the separation of algorithm description from mapping to the hardware which is achieved through the definition of a domain-specific language. The language is defined in close collaboration with experts of the domain for which the framework is intended. In [4] Chen et al. propose a task-based queue scheme for systems with one or multiple GPUs. The main goal of the scheme is dynamic load balancing which is achieved by breaking down the computations into fine-grained tasks and then dynamically assigning them to GPUs. Note that in the case of single-GPU systems, this reduces to assignment of tasks to CUDA cores available on a GPU which is reported to outperform the CUDA scheduler in case of unbalanced workload. Chen et al. further develop on this work to support GPUs on different nodes in a cluster [5]. They also improve their scheme for dynamic load balancing on individual nodes with multiple GPUs. As an interesting application, Stuart et al. [6] have proposed a multi-GPU design for volume rendering. In their implementation, parallel volume rendering has been fit into MapReduce model and run on a cluster of nodes equipped with GPUs. As a rather innovative work, [7] presents a performance prediction model for multi-GPU systems, which gives an estimate of the expected performance improvement when moving from a single-GPU to a multi-GPU system, based on the performance results on a single-GPU system.

In summary, all of the above-mentioned approaches either focus on a problem domain that does not include the problem domain addressed in this chapter or they use a different hardware setup, e.g., CPU clusters, for which the concepts cannot be directly applied to our hardware setup.

Considering related works mainly characterized by stream data processing, [8] presents a framework for processing of multiple data streams on heterogeneous systems where both CPUs and GPUs are used as processors. The paper proposes a method for assignment of streams to CPUs and GPUs such that hard real-time constraints of stream data processing are satisfied. Yamagiwa et al. [9] elaborate on their efforts for porting an already existing framework for stream data processing on single GPU from previous GPU generations to present ones. To this end, they use CUDA. This, in addition to the use of OpenGL and DirectX for GPUs of old generations, leads to the development of a framework capable of running on different generations of GPUs. Teodoro et al. [10] introduce a stream data processing framework capable of exploiting the computational power of both CPUs and GPUs. A significant point with their framework is a mechanism for determining on which type of processor (CPU or GPU) the processing should be done (provided that the code for both types of processors is given). The framework uses CUDA as computational back end on GPUs. Houzet et al. [11] present a programming model which can be used for stream data processing on multi-GPU systems. The innovation of this work is its use of system design language SystemC which is used as a high-level language for description of the desired processing, thereby hiding many low-level details from users. Zhang and Mueller propose a scalable stream data processing framework which runs on GPU clusters and is based on CUDA [12]. It makes extensive use of template-based generic programming techniques in C++ to offer programmability and uses MPI for internode communication. As the last work in this section, Vogelgesang et al. [13] have developed a GPU-based image processing framework which supports CPU usage as well. Similar to [10], their framework chooses between CPU and GPU codes provided that both codes exist. The framework supports processing on a cluster of nodes and uses OpenCL as computational back end.

All of the mentioned stream data processing approaches lack support for either multi-GPU or the problem domain addressed in this chapter (i.e., stateful stream data processing).

7.3 The Framework

In this section, we first describe possible parallelization concepts for the addressed problem domain (Sect. 7.3.1). The evaluation of these concepts in Sect. 7.4.1 forms the basis for final implementation, which is described in Sect. 7.3.3.

Remember that our framework assumes that all or majority of processing is done on GPUs, thus a processing module can be safely considered as a user-defined CUDA kernel in most cases. The *processing graph* is a collection of modules which describe the flowchart of processing done on data, including stream source and stream sink modules (see Fig. 7.1).

7.3.1 Basic Concepts

An important design question while developing the framework is how to distribute the computational load over several GPUs and, as a consequence, how the synchronization and the data management is organized.

Since in our treatment of the framework the computational load is decomposed into modules, this question reduces well to that of how to assign different modules to GPUs. We consider two completely different approaches, i.e.:

Distributed Graph: In this first concept, the processing graph is divided into *N* subgraphs, where *N* is the number of GPUs, and modules within each subgraph are strictly assigned to a separate GPU.

Multiple Graph Instantiation: In this concept, on the other hand, one instance of each module or more precisely one instance of the whole processing graph runs on each GPU.

Table 7.1 summarizes their main characteristics. Note that there are two variants of the Distributed Graph approach (see Sect. 7.3.2).

In order to select one of the concepts for final implementation, we have implemented preliminary versions of both concepts. These preliminary versions are fully functional in terms of data management, synchronization, and process control. Based on the preliminary implementation, the performance of the concepts has been evaluated (see Sect. 7.4.1). The essence of the evaluation is that the Multiple Graph Instantiation approach outperforms the two variants of Distributed Graph in almost all test cases except when the number of intra-module feedbacks is large enough. Thus, we made the choice to fully implement the Multiple Graph Instantiation approach. Consequently, the technical description of the Distributed Graph concept is less detailed than the one for Multiple Graph Instantiation.

Table 7.1 Characteristics of the different concepts

	Multi-threaded distributed graph	Single-threaded distributed graph	Multiple graph instantiation
Architecture	One instance Modules distributed over GPUs One CUDA stream per module	One instance Modules distributed over GPUs Two CUDA streams per GPU	Multiple instances One instance per GPU One or more CUDA streams per GPU instance
Synchronization	CPU-thread synchronization	CUDA stream barrier	CPU-thread synchronization
Memory transfers	Source, sink, GPU borders	Source, sink, GPU borders	Source, sink, feedback
Load distribution	Module distribution	Module distribution	Built-in
Inter-module feedback	Not supported	Container modules	Main memory

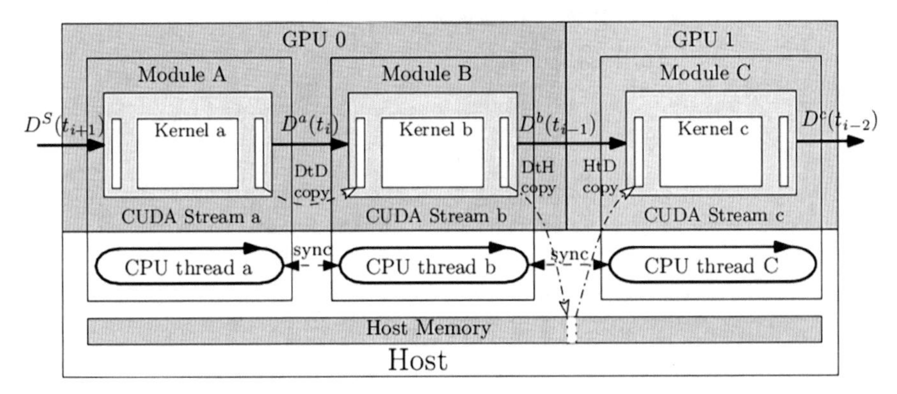

Fig. 7.2 Distributed Graph, multi-threaded variant: Each module runs in a separate CPU thread. The processing is synchronized via access to the ring buffer. Data transfers across GPU borders are managed via the main memory

7.3.2 Distributed Graph

There are two variants of the Distributed Graph approach. The major difference between these two variants is the number of CPU threads used for controlling the modules, which strongly influences the synchronization method to be applied. In multi-threaded variant, each module is controlled by a separate CPU thread (see Fig. 7.2). The module stores its result in a small output ring buffer. If a module is idle, it polls the output buffer of the predecessor for new data to process. If this is the case, new data is copied to an input buffer (DeviceToDevice copy) and processed. If no new data is available, it yields its time slice. If a module has a successor that is located on a different GPU, the output ring buffer is mirrored to the host memory (DeviceToHost copy). On the other hand, if a module has a predecessor that resides on a different GPU, it copies the data from main memory to its GPU memory (HostToDevice copy). The modules are synchronized via the access to the output ring buffer. In single-threaded variant, all modules are controlled by the same CPU thread (see Fig. 7.3), which calls all CUDA functions (kernel launches and memory transfers) asynchronously. Two CUDA streams are used for each GPU, one for data transfer and the other for kernel calls, thus partially hiding data transfer time by overlapping kernel launch and memory transfer. Before the next frame is processed, the CUDA streams are synchronized by a barrier.

This approach requires a manual decomposition of the complete processing graph into N subgraphs to be distributed to the N GPUs. The load distribution is a direct result of this decomposition and thus a difficult task left to the user.

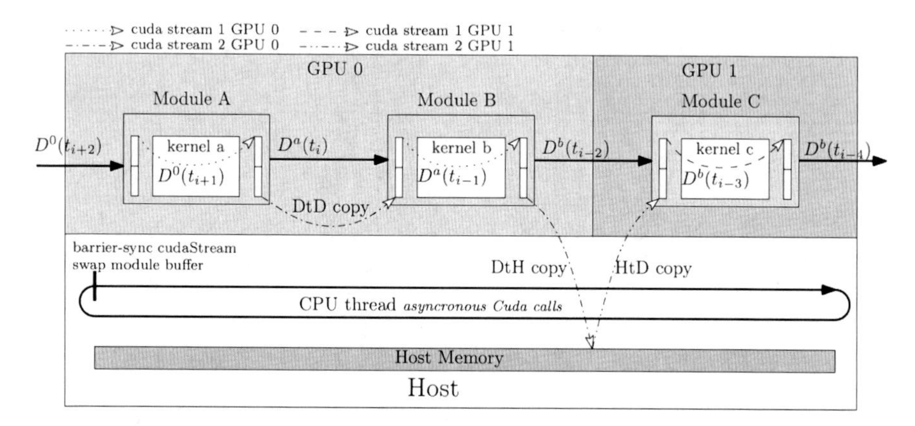

Fig. 7.3 Distributed Graph, single-threaded variant: In this concept, all modules are triggered within a single CPU thread using asynchronous CUDA calls. Two CUDA streams for each GPU are used to partially hide data transfer time. A CUDA stream barrier is used to synchronize after each process iteration

7.3.3 Multiple Graph Instantiation

At the very heart of the proposed framework lies a simple idea: processing all the input stream(s) data at a specific time step t_i by a single GPU (see Fig. 7.4). Precisely speaking, for $N > 1$ GPUs, numbered from 0 to N - 1, the data from all input streams at time step $t \geq 0$ is processed by GPU $t \bmod N$. This has an immediate consequence of nearly perfect load distribution over GPUs in case of data-independent processing.

Although the basic idea behind the proposed framework is quite simple, there are still a few other considerations which affect the framework design in a significant way. The two most important considerations are synchronization and main memory management which are largely influenced by the stateful processing requirement of the framework, i.e., the realization of the intra- and inter-module feedback functionalities. For the Multiple Graph Instantiation approach, feedback data is transferred first from the memory of one GPU to the main memory of the system and then from there to the memory of another GPU. This leads to two memory transfer operations between host and device with additional synchronization requirements, whereas in the Distributed Graph concept, this data remains on the same GPU.

Besides the two aforementioned considerations, there are still a few less important ones which are specifically taken care of to exploit useful features of GPUs offered by CUDA. Notably, GPU memory management and concurrent CUDA kernel launches and memory copies are among these. These last two points together with synchronization and main memory management are separately considered in the following four subsections.

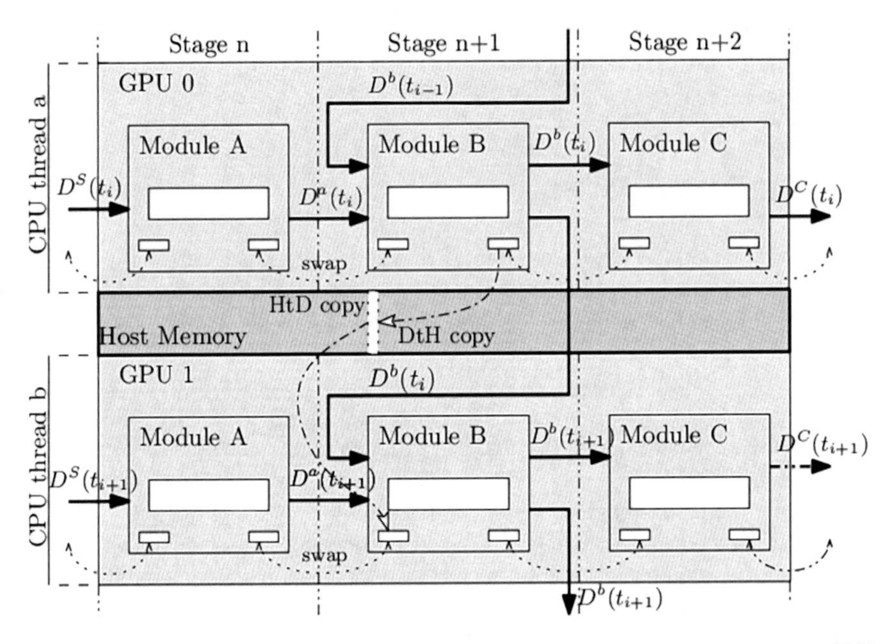

Fig. 7.4 Multiple Graph Instantiation: The whole processing graph is executed on each GPU (here, only 2 GPUs are shown). Data transfers for inter- and intra-module feedbacks are handled via main memory. The input and output buffers are swapped during stage changes to save GPU memory (see Sect. 7.3.3)

7.3.3.1 Synchronization

Considering the basic idea of the framework, there should be a mechanism which ensures us that the GPUs both read the inputs from sources and write the outputs into the sinks in correct order. In order to realize this behavior, the framework launches as many CPU threads as GPUs where each CPU thread is in full charge of a GPU. This, in turn, lets the framework control the order of accesses to input as well as output streams by different GPUs through the use of synchronization objects defined at CPU-thread level. The same mechanism is used to let each GPU access the processing results of input(s) at previous time step, thereby enabling the stateful processing property of the framework.

7.3.3.2 Main Memory Management

Main memory can be regarded as the major gateway of the framework for communication with the outside world. Actually, it is the place where inputs represented by sources are read from by GPUs, and also it is the place where outputs represented by sinks are written into by GPUs. In addition to these two functionalities, the main memory also serves another important purpose: providing a place for exchange of data between GPUs. This latter point combined with previously mentioned synchronization mechanism which is used to synchronize accesses to common main memory areas between two GPUs realizes the stateful processing capability of the framework.

7.3.3.3 GPU Memory Management

Although a straightforward way for GPU memory management is to allocate memory for inputs and outputs of all modules in the processing graph, the framework employs another strategy for this. The motivation for this has been better utilization of precious GPU memory. To implement this strategy, the framework introduces the concept of stage. A stage is defined as composed of modules whose inputs are produced in previous stage(s). Such a definition is a recursive one and the only requirement is to define the first stage. To complete our definition, the first stage is considered to be composed of only sources.

Now that we have organized all the modules in the processing graph into stages, GPU memory management can be described as allocation of two separate areas on GPU memory. From one of the GPU memory areas, the inputs for all modules in the current stage are read and into the other the outputs of all modules of the current stage are written. The roles of the two GPU memory areas are swapped when finishing current stage and starting a new one. This way the output area of current stage becomes the input area of the new stage, thus ensuring the desired behavior. This swap process is repeated whenever a stage is complete and a new one begins. Note that this GPU memory management strategy is done for each GPU separately and the two GPU memory areas are allocated on global memory of GPUs. This latter point ensures the data are persistent between two consecutive stages.

7.3.3.4 Concurrent Kernel Launches and Memory Copies

A useful concept introduced in CUDA is that of CUDA streams. An immediate consequence of this concept is the possibility of concurrent kernel launches as well as concurrent kernel execution and copies between main and GPU memories. With the aim of increasing performance, the framework is designed to exploit this valuable feature as well. For this purpose, the framework provides the user with some CUDA streams on which to launch kernels.

7.4 Experimental Evaluation

In accordance with how the effort for development of the framework is divided into two main phases (see Sect. 7.3), the evaluations carried out are well categorized into two major groups, i.e., those aimed at the selection of a concept for final implementation (Sect. 7.4.1) and those to depict the scalability of the final implementation (Sect. 7.4.2). Note that the system used for running all the experiments in this chapter is equipped with 4 Tesla C2050 GPUs each having 448 CUDA cores and connected via a separate PCI Express 2.0 × 16 interface. The system also has two Intel Xeon E5630 2.53 GHz Quad-Core CPUs with 24GB of RAM. Finally the system runs Windows Server 2008 R2 as the operating system.

7.4.1 Comparison of Preliminary Implementations

The evaluation of the preliminary implementations is based on three different processing graphs. The stream data for all experiments consists of 10.000 data frames of 384 * 384 2-byte data elements, adding up to some 2.75 GB. Furthermore, we vary the amount of computation performed in each module. Therefore, we use two different CUDA kernels, one *light kernel*, inducing relatively little computational effort, and one *heavy kernel* with high computational costs. Then the average time measurement is reported. As the last point, in Distributed Graph experiments the distribution of modules among GPUs is done manually in order to get the best load balance for each processing graph.

The first processing graph examined is a *serial processing graph*, in which the processing modules are connected sequentially and their number varies from 1 to 10. Figure 7.5 shows the result for this experiment. This experiment is ideal for parallelization, since the least amount of data transfer is required, i.e., no feedback, splitting, or merging. The Multiple Graph Instantiation completely outperforms the two variants of Distributed Graph in both light and heavy kernels. There is, however, an interesting observation: for the heavy kernel, the Multiple Graph Instantiation implementation performs almost linear, whereas this is almost constant in light kernel version. This effect is due to the fact that the computation done in heavy kernel is large enough to constitute most of the measured time whereas in light kernel version, other operations such as host (CPU)-to-device (GPU) and device-to-host memory transfer times dominate the computation time in kernels, leading to an almost constant performance. Note that these two types of memory transfer operations are performed exactly the same number of times regardless of the number of processing modules in the serial processing graph.

The next experiment is conducted using a *parallel processing graph*, where the processing modules are arranged in a purely parallel fashion and their count varies between 1 and 10. The results of experiments are shown in Fig. 7.6. The Multiple Graph Instantiation concept again outperforms the two variants of Distributed Graph. Once again, the same effect as the one in Fig. 7.5 can be seen for light and heavy kernel modules used in the Multiple Graph Instantiation. This can well be explained by the same line of reasoning as the one stated for serial processing graph.

In the last processing graph, we use a more complex arrangement consisting of 23 processing modules (see Fig. 7.7). In this processing graph, some of the processing modules have an intra-module feedback, the number of which ranges between 0 and 23. As can be seen in Fig. 7.8, the Multiple Graph Instantiation

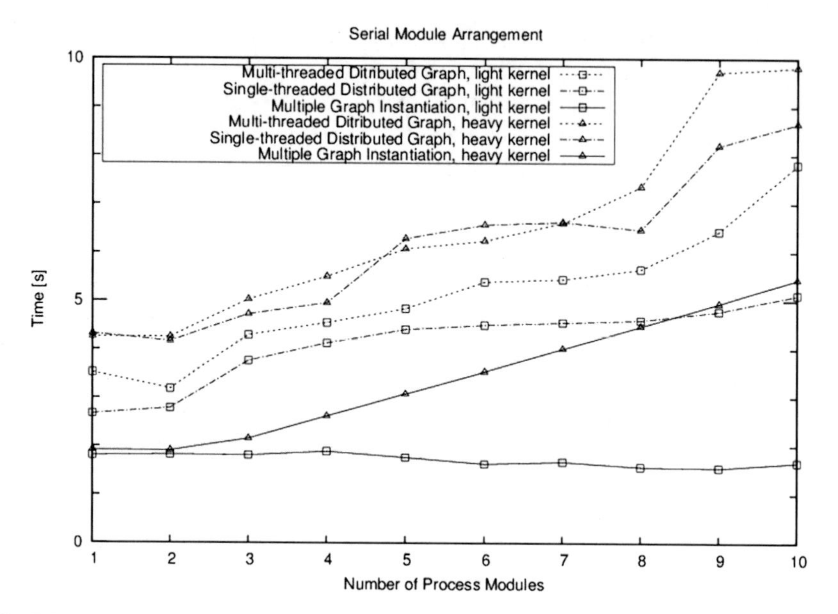

Fig. 7.5 Serial processing graph experiment performed with 1–10 processing modules consisting of either light or heavy kernels for all three concepts

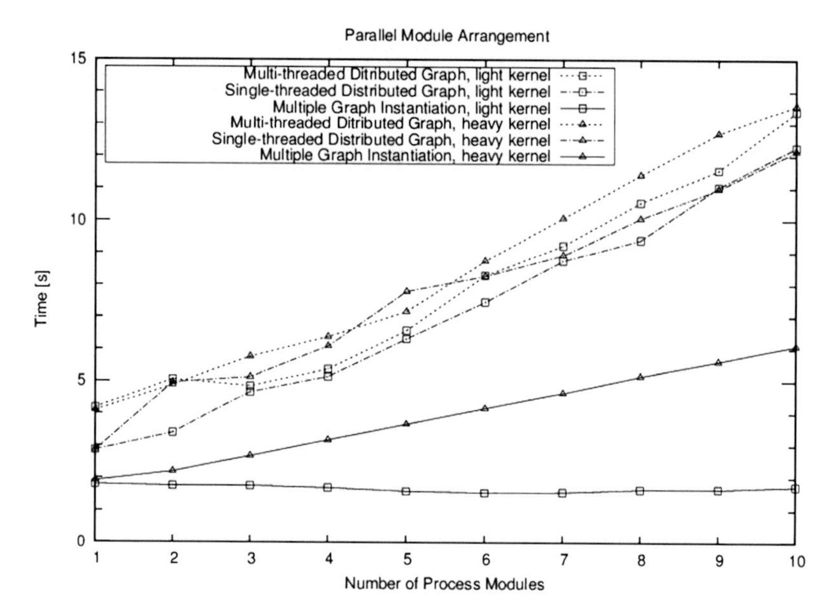

Fig. 7.6 Parallel processing graph experiment performed with 1–10 modules consisting of either light or heavy kernels for all three concepts

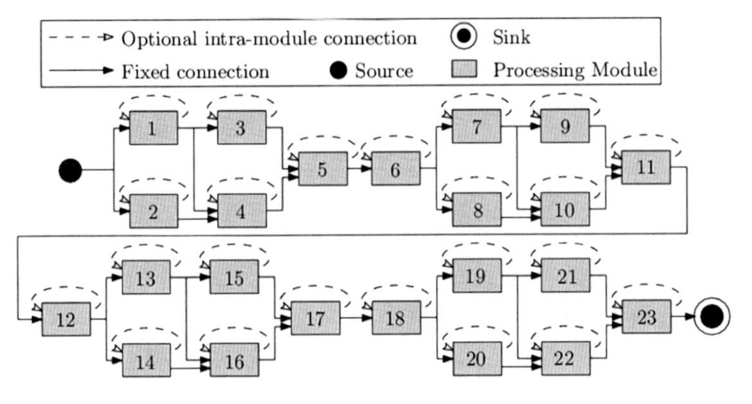

Fig. 7.7 Complex processing graph used in Sect. 7.4.1

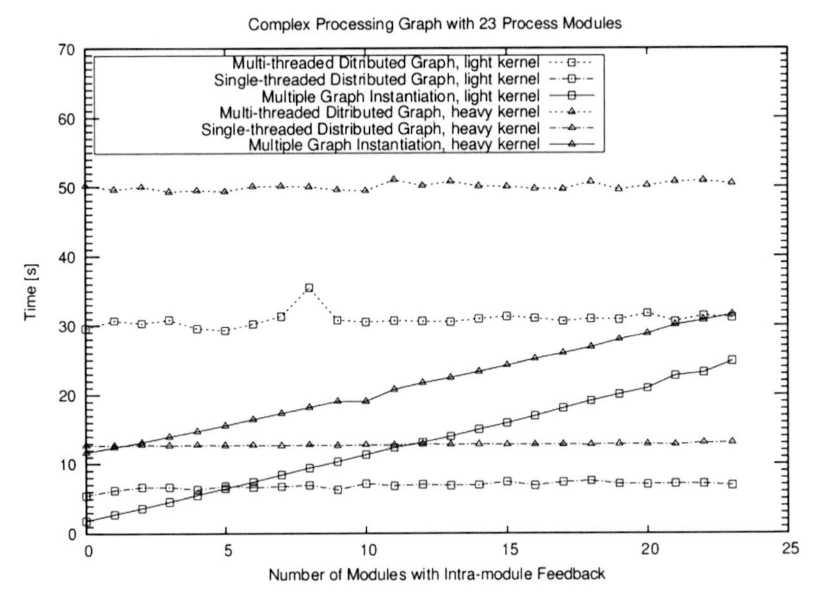

Fig. 7.8 Complex processing graph experiment performed with 0–23 intra-module feedback (s) using either light or heavy kernels for all three concepts

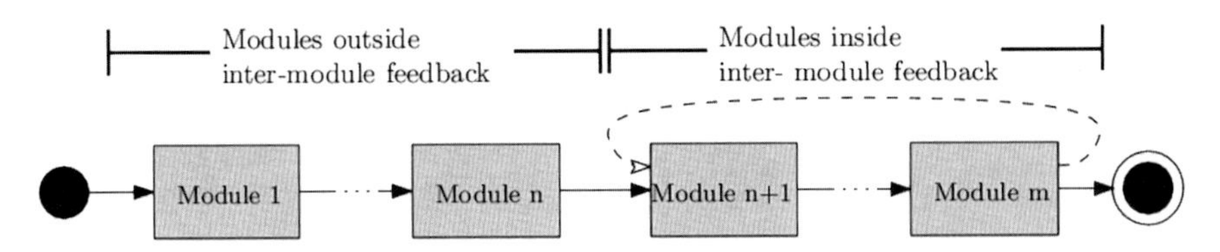

Fig. 7.9 Processing graph used for examining the effect of feedback on scalability (see Sect. 7.4.2)

performs better than multi-threaded Distributed Graph. However, for a large number of intra-module feedback, the single-threaded Distributed Graph outperforms the Multiple Graph Instantiation. This effect is a direct result from the data transfer required for feedback, i.e., the intra-module feedback implementation in the Multiple Graph Instantiation is more expensive than its single-threaded Distributed Graph counterpart (see Sect. 7.3.3).

7.4.2 *Scalability and Feedback*

The Multiple Graph Instantiation approach by default supports the optional functionality of inter-module feedback. To be precise, the implementation does not make any difference between intra- and inter-module feedbacks. We conducted some experiments regarding this feature in order to evaluate the effect of feedbacks on the scalability in terms of the number of GPUs. Therefore, we generated a processing graph, consisting of a linear sequence of modules with an additional inter-module feedback (see Fig. 7.9). For the evaluation, we vary the computational load of modules bridged by the feedback and the ones outside the bridged subgraph.

The results regarding the scalability are shown in Fig. 7.10. As expected, inter-module feedback reduces the performance of our framework. Naturally, bridging the whole graph completely, i.e., having no computational load outside the bridged subgraph, completely destroys the GPU parallelism, since the first processing

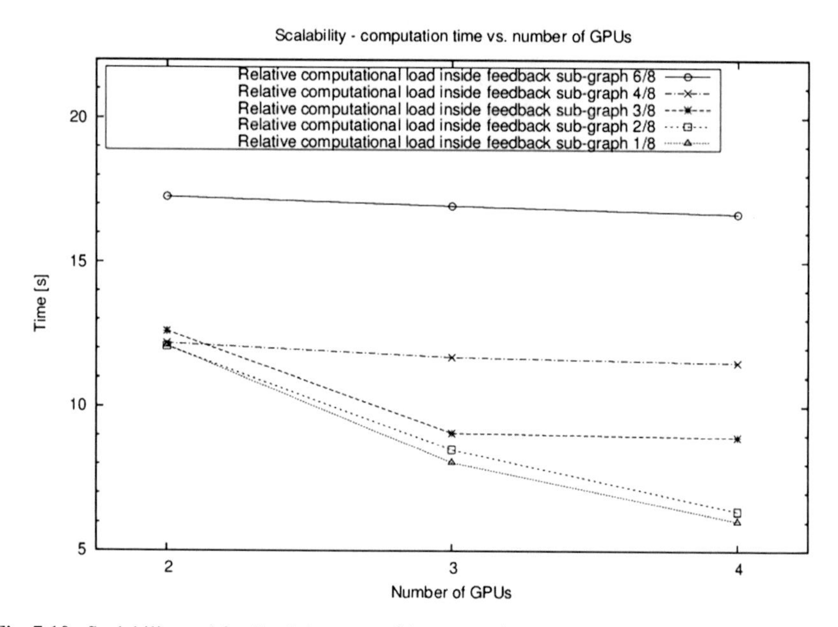

Fig. 7.10 Scalability and feedback in terms of the proportion of the computational load inside and outside the feedback subgraph

module can process $D^0(t_{j+1})$ only after the last module N has generated its result $D^N(t_j)$. The rate of performance degradation is related to the proportion of the time spent within the feedback subgraph and that spent outside the bridged subgraph.

7.5 Applications

In this section, we present two different applications which have been successfully addressed by our framework. The first application deals with information security, whereas the second one is in the field of crystallography.

7.5.1 *Information Security Using Crypto- and Steganography*

Cryptography and steganography form two major groups of methods within the scope of information security. While cryptography is more concerned with hiding the content of a message, steganographic methods try to hide the message itself. To better clarify the difference between the two, one can consider the case of a simple piece of meaningful text communicated between sender and receiver. In case of cryptography, one would encrypt the meaningful text such that each letter is replaced by another, thus leading to an unmeaningful text. In steganography, however, the meaningful text (referred to as cover or cover text) could be written in such a way that the secret message is formed from the first letter of each word. As can be seen in this simple case, the advantage of steganography over cryptography is that it does not attract the attention of those who accidentally access the text, whereas the encrypted text would raise suspicion that there is a secret message hidden in the unmeaningful text. Therefore, cryptographic methods only protect the content of a secret message, while steganography deals with protection of both secret message and communicating parties.

In this section, our framework is exploited to deal with an application where both cryptographic and steganographic methods are involved. The goal is to extract a sequence of secret hidden images from an encrypted cover video. The video is encrypted based on method of [14]. In our implementation, it is assumed that each video frame in the memory is divided into chunks of 8 bytes and corresponding chunks in consecutive frames form a separate sequence of plaintext blocks. Furthermore, in each video frame, a secret image is hidden using least significant bit which is a steganographic transform whereby secret information is written into least significant bits of image pixels, thus causing hard-to-perceive degradations in visual quality of cover image (the interested reader is referred to [15] for a survey of this and other image steganographic methods). Based on these assumptions, our framework first decrypts a video frame by applying method of [14] to obtain the cover

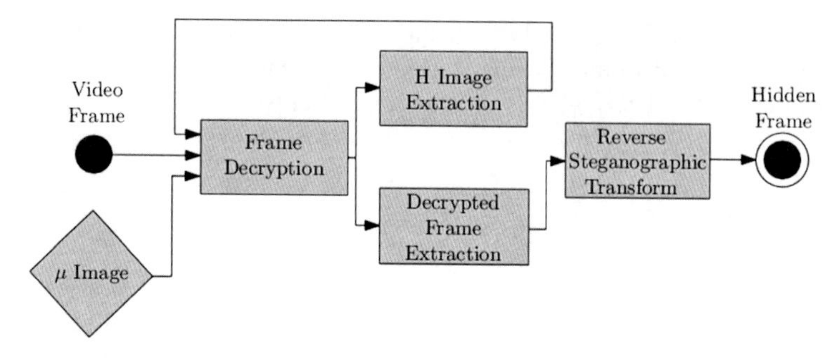

Fig. 7.11 Processing graph used to extract hidden image sequence (video) from encrypted cover video

image and then extracts the hidden image by applying the reverse steganographic transform to the cover image (all processing for this experiment is done on GPUs). Note that in this implementation, the decryption result of each video frame is affected by that of the previous one, thus requiring feedback as shown in Fig. 7.11. Note that H and μ are decryption parameters as defined in [14] and the task of two modules H image and decrypted frame extraction is to separate these two pieces of data which are combined at the output of frame decryption module and provide them on their outputs.

The experiment is done using HD videos of size 1920 * 1080, 24 bpp as cover video. The result would be a video (image sequence) of size 1920 * 135, 24 bpp. This is because from each byte in the input video, only the least significant bit is preserved, thus reducing the size to one eighth. The timing results for 2, 3, and 4 GPUs are shown in Fig. 7.12. Also shown in the figure are timing results for CPU implementation of the same algorithm using 1, 2, 3, and 4 CPU threads to provide the reader with a ground to compare with. Considering the typical frame rate of 1080p HD videos which is between 24 and 60 frames per second, one can easily see that the 4-threaded CPU implementation can only handle frame rates near the lower bound of this range whereas the two-GPU implementation supports frame rates well beyond its upper bound.

7.5.2 Crystallography Using a pnCCD Camera

Considerable amounts of information about crystals are collected through examination by x-ray. There are different types of x-ray sensors which record the result of these examinations. One such sensor is an energy-dispersive CCD with fast readout called pnCCD camera (see Sect. 7.1). The specifications of this camera were mentioned in the introduction. Getting familiar with the operation of the camera, however, needs some basic knowledge of the domain. When x-ray beam is scattered by crystal sample, scattered x-ray photons hit the camera image plane. Depending on the position of incident photons onto the image plane, a number of pixels are

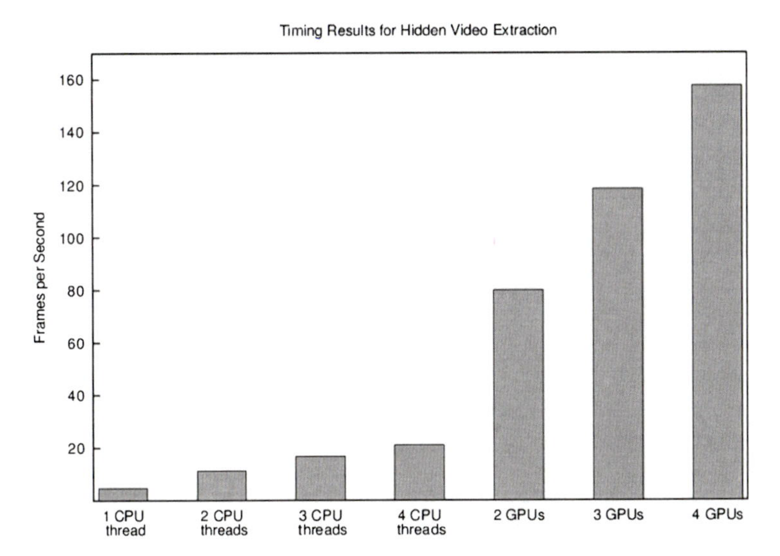

Fig. 7.12 Timing results of hidden video extraction from encrypted HD video using different numbers of CPU threads and GPUs

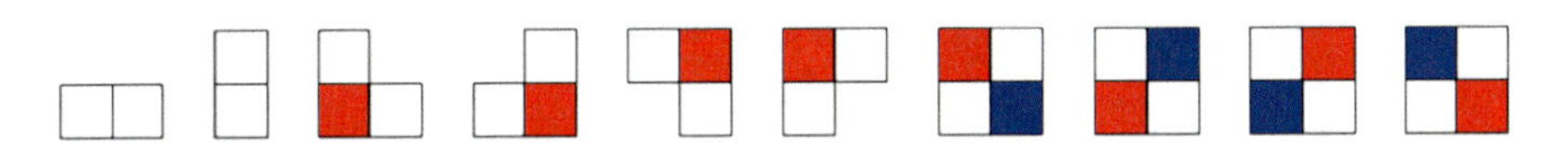

Fig. 7.13 Valid double, triple, and quadruple events: *red and blue pixels* show the highest and lowest pixel values in an event, respectively

illuminated, thus producing nonzero pixel values. Pixels illuminated by a single photon are collectively called an event. Events can consist of 1, 2, 3, or 4 nonzero pixels (the so-called single, double, triple, and quadruple events, respectively). Figure 7.13 shows valid patterns for double, triple, and quadruple events. However, it may happen that in an image we have invalid patterns. These patterns are caused by two or more photons whose event patterns interfere and make a cluster of events. A solution to this problem is to increase the frame rate such that the probability of occurrence of interfering patterns decreases. That is why, pnCCDs support such high frame rates as 400 frames per second. Determining valid events in each frame forms the basis for many other crystallographic experiments which rely on analysis of events.

We have developed kernels for extraction of valid events from pnCCD frames [16] which is based on [17]. The whole processing can be split into two major steps of frame correction and valid event extraction. As Fig. 7.14 shows, first an offset map is subtracted pixelwise from the raw pnCCD frame. During common mode correction, the median value for each row of the image is computed and then subtracted from all pixel values of the corresponding row. The processing continues by "zero" pixel elimination whereby all pixels whose values are less than

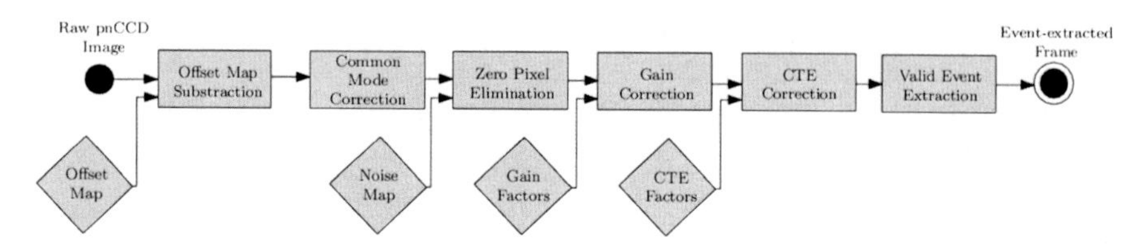

Fig. 7.14 Processing done on each raw pnCCD image to extract valid single, double, triple, and quadruple events

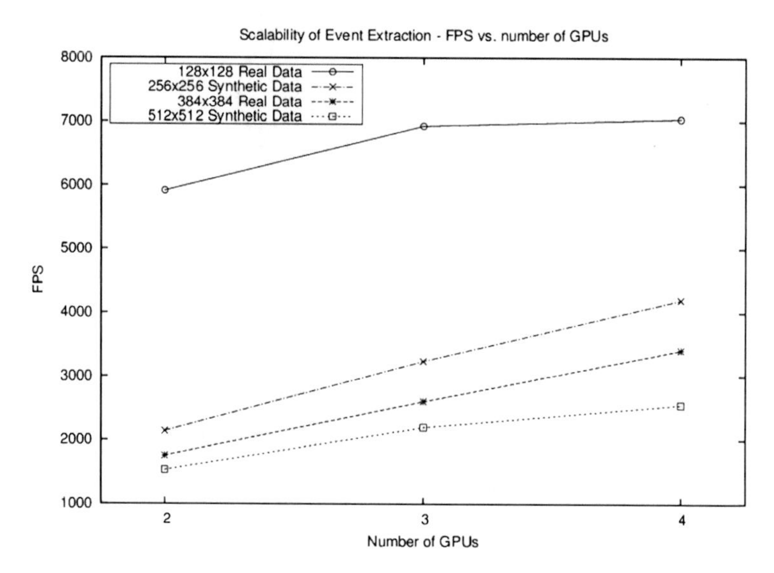

Fig. 7.15 Timing results of valid event extraction for various numbers of GPUs and frame sizes

corresponding pixel values in a noise map image multiplied by a constant factor are discarded. In gain correction, the pixel values in each column are multiplied by a gain factor. In CTE correction for each column, the pixel values are multiplied by a CTE factor raised to the power of the pixel's row index. Now, we have corrected frames which are then used to extract valid single, double, triple, and quadruple events. Figure 7.15 shows the performance and scalability of our framework while working with different numbers of GPUs (2 to 4) and different frame sizes (note that all processing modules in the processing graph run on GPU). To better show the usefulness of GPUs for event extraction, we have implemented a single-threaded CPU version of the mentioned algorithm. The CPU version processes 92 frames of size 384 * 384 per second, whereas this number is 1756 when 2 GPUs are used, thus leaving a lot of computational power for further processing of events (note that event extraction is only a first processing step in many crystallographic applications).

Conclusion
In this chapter, we presented a scalable CUDA-based framework for stateful stream data processing on multiple GPUs in a single node. As described, the framework is designed to be both easy to use and flexible from the user part. The ease of use is achieved by transparent implementation of the framework with regard to synchronization and memory management. This, however, does not limit the flexibility of the framework in the sense that the user still has unlimited freedom to define the CUDA kernels for processing modules as desired.

Still the most important feature of the framework is scalability. For that, the chapter also presents a number of experiments for stateful processing of stream data and examines the effect of feedback in processing graphs on the scalability of the framework with regard to GPUs. Furthermore, the practicality and usefulness of the framework for real-world tasks is demonstrated by two different application scenarios.

Acknowledgments This research was partially funded by the German Ministry for Research and Education (BMBF) under grant No. 05k10PSB.

References

1. Macedonia, M.: The GPU enters computing's mainstream. IEEE Comput. **36**(10), 106–108 (2003)
2. Enmyren, J., Kessler, C.: Skepu: A multi-backend skeleton programming library for multi-GPU systems. In: Proceedings on International ACM Workshop High-level parallel programming and applications, pp. 5–14 (2010)
3. Meyer, B., Plessl, C., Forstner, J.: Transformation of scientific algorithms to parallel computing code: Single GPU and mpi multi GPU backends with subdomain support. In: Proceeding of 2011 Symposium on Application Accelerators in High-Performance Computing (SAAHPC), pp. 60–63 (2011)
4. Chen, L., Villa, O., Krishnamoorthy, S., Gao, G.: Dynamic load balancing on single- and multi-GPU systems. In: Proc. Parallel & Distributed Processing (IPDPS) (2010). doi:10.1109/IPDPS.2010.5470413
5. Chen, L., Villa, O., Gao, G.: Exploring fine-grained task-based execution on multi-GPU systems. In: Proceedings of IEEE International Conference on Cluster Computing, pp. 386–394 (2011)
6. Stuart, J.A., Chen, C.K., Ma, K.L., Owens, J.D.: Multi-GPU volume rendering using MapReduce. In: Proceedings of International ACM Symposium on High Performance Distributed Computing, pp. 841–848 (2010)
7. Schaa, D., Kaeli, D.: Exploring the multiple-GPU design space. In: Proceedings of International IEEE Symposium on Parallel and Distributed Processing (2009)
8. Verner, U., Schuster, A., Silberstein, M.: Processing data streams with hard real-time constraints on heterogeneous systems. In: Proceedings on International Conference on Supercomputing, pp. 120–129 (2011)

9. Yamagiwa, S., Arai, M., Wada, K.: Efficient handling of stream buffers in GPU stream-based computing platform. In: Proceedings on IEEE Pacific Rim Conference on Communications, Computers and Signal Processing, pp. 286–291 (2011)
10. Teodoro, G., Sachetto, R., Sertel, O., Gurcan, M., Meira, W., Catalyurek, U., Ferreira, R.: Coordinating the use of GPU and CPU for improving performance of compute intensive applications. In: Proceedings on Internatyional IEEE Conference on Cluster (2009)
11. Houzet, D., Huet, S., Rahman, A.: Syscellc: A data-flow programming model on multi-GPU. In: Proceedings of International Conference on Computational Science, pp. 1035–1044 (2010)
12. Zhang, Y., Mueller, F.: Gstream: A general-purpose data streaming framework on GPU clusters. In: Proceedings of International Conference on Parallel Processing, pp. 245–254, (2011)
13. Vogelgesang, M., Chilingaryan, S., dos Santos Rolo, T., Kopmann, A.: Ufo: A scalable GPU-based image processing framework for on-line monitoring. In: Proceedings on IEEE 14th International Conference on High Performance Computing and Communications, pp. 824–829 (2012)
14. Wang, X., Bao, X.: A novel block cryptosystem based on the coupled chaotic map lattice. Nonlinear Dyn **72**, 707–715 (2013)
15. Cheddad, A., Condell, J., Curran, K., Kevitt, P.M.: Digital image steganography: survey and analysis of current methods. Signal Process **90**, 727–752 (2010)
16. Alghabi, F., Schipper, U., Kolb, A.: Real-time processing of pnCCD images using GPUs. In: 14th International Workshop on Radiation Imaging Detectors (2012)
17. Andritschke, R., Hartner, G., Hartmann, R., Meidinger, N., Strüder, L.: Data analysis for characterizing pnCCDs. In Proceedings of Nuclear Science Symposium, pp. 2166–2172 (2008)

Chapter 8
The Design of SkyPACS: A High-Performance Mobile Medical Imaging Solution

Tananan Pattanangkur, Sikana Tanupabrungson, Katchaguy Areekijseree, Sarunya Pumma, and Tiranee Achalakul

Abstract Lack of radiologists is a problem that arises in many parts of the world. Radiologists need to work long hours for multiple hospitals. In order to improve the quality of healthcare, SkyPACS is designed. It is a mobile solution that allows radiologists to work more conveniently. SkyPACS is a low-cost and customizable medical image viewer that can be used for prognosis. The solution is designed to be an assistive technology with the focus on simplicity, flexibility, and user experiences. The architecture of SkyPACS is designed based on service-oriented Model-View-Controller. The customers can freely choose the back-end services: cloud computing and storage on public cloud, private server, or hybrid system. The compute-intensive modules are deployed on a GPU server taking advantage of data parallel with CUDA library. The main features include all standard tools for viewing and diagnosis in 2D and 3D, convenient tools for collaborations, and case management. In addition, advanced functions such as automatic tumor detection and reconstruction and bone/skin/muscle segmentation are provided. This paper describes the details of SkyPACS's design, as well as its implementation and initial deployment. We believe that SkyPACS will soon be available to a broad range of users in Thailand and AEC's countries and will be able to reduce the cost of the healthcare platform in the near future.

Keywords Medical imaging • Healthcare solution • Medical image mobile solution • Cloud computing • Cloud storage • GPU server

---

T. Pattanangkur (✉) • S. Tanupabrungson (✉) • K. Areekijseree • S. Pumma • T. Achalakul
Computer Engineering, King Mongkut's University of Technology Thonburi, Bangkok, Thailand
e-mail: tananangpatt@gmail.com; sikana.tanu@gmail.com; katchaguy.are@gmail.com; sarunya.pumma@gmail.com; tiranee@cpe.kmutt.ac.th

Y. Cai, S. See (eds.), *GPU Computing and Applications*,
DOI 10.1007/978-981-287-134-3_8

8.1 Introduction

In the year 2012, the statistics published by TEH & Associates [1] showed that in every million death, over 120,000 cases are caused by medical error, which was almost four times higher than the death caused by road accidents. Prognosis based on medical imaging is likely to reduce the rate as internal physical anomaly can be visually studied prior to treatments. Medical imagery, such as ultrasound, computerized tomography (CT), and magnetic resonance imaging (MRI), then becomes important tool in diagnosis and is embraced across the global healthcare enterprises.

A hospital around the world nowadays scans a large number of patients each day. For example, a Thailand's university hospital may produce over 1,000 image series from more than 40 patients in a single day [2]. These images have to be diagnosed by the radiologists. However, radiology has not been widely studied in many parts of the world. There are only 4.2 % of medical doctors majoring in radiology in Thailand. The percentage is even much lower in Central and South America where there are less than 1 % in countries like Honduras, El Salvador, Cuba, and Argentina [3]. This insufficiency in human resources may cause the delay in patient treatment. Moreover, the backlog can only get worse as the number of medical cases is growing much faster than the number of radiologists. As a result, it is necessary to facilitate the radiologists so that they can work for multiple healthcare institutes more conveniently with more appropriate number of working hours.

In this chapter, we propose a software solution that can alleviate the mentioned problems. The software is called SkyPACS. SkyPACS is a low-cost and customizable mobile solution for radiologists and medical doctors to view and manipulate DICOM[1] images of any types in both 2D and 3D planes. The solution is an assistive technology with the focus on anytime-anywhere working concept. SkyPACS can also be integrated to any existing Picture Archiving and Communication System or PACS[2] [4].

During the design of SkyPACS, some challenges arise. First, in order to produce a true and natural perception of human anatomy, 3D visualization is needed. 3D visualization in real time, however, is compute intensive and the use of high-performance computing machines is not of low cost. Second, seamless integration to existing systems is difficult since multiple platforms are deployed across hospitals in Thailand. Such integration constrains architecture design choices to client–server with web-based interface. Third, the amount of image data grow so quickly that a cost-effective storage space that can grow on demand may become a necessity. Fourth, different hospitals may have different workflows; creating a one-size-fits-all product is unlikely. Lastly, security is a big issue in patients' data; there is the need for the software to leave zero footprint on mobile devices.

---

[1] Digital Imaging and Communications in Medicine or DICOM is a universal medical image used in the standard PAC system.

[2] Picture Archiving and Communication System or PACS is a storage and management system for medical image in the standard format, namely, DICOM.

In this work, we have surveyed and selected technologies that are appropriate to overcome these challenges. For the better understanding of the readers, Sect. 8.2 presents an imagery procedure example of radiology departments in Thailand. Then, the software features and design framework are discussed in Sects. 8.3 and 8.4, respectively. Section 8.5 briefly describes the software implementation and deployment. Section 8.6 offers comparisons between our mobile solution and some existing packages. Concluding remarks are then presented in final section.

8.2 Imagery Procedure

In order to allow the solution to be practical, the flow of the imagery procedure from the scanner all the way to the doctors' desktops has been studied extensively with the collaboration of radiology departments in Thailand.

The information flow of the radiology departments is managed by the Radiology Information System or RIS [5]. RIS is responsible for all information involving medical image prognosis, i.e., patient tracking, image case assignments, diagnosis reports, and case transfers. It directly connects to a central system, called Hospital Information System or HIS [6]. Master data, such as patient data, registration, and scheduling, are queried from HIS and stored in RIS using HL7[3] [7] standard. In addition to RIS, radiology process includes another important system called Picture Archiving and Communication System or PACS. PACS can be divided into PACS server and client. PACS client is basically a medical image viewer that communicates with PACS server.

PACS server, on the other hand, serves as the image scan repository for the hospital. On the server, all images are stored in the DICOM format with metadata (image properties, patient and study information, and acquisition information) and image pixels (in bits). Client and server communicate through Query/Retrieve image communication protocol in DICOM standard. PACS client has to select an appropriate image query level which can be arranged in a hierarchical order as follows: patient, study, series, and images. The relationship between levels is one to many; for example, one patient can have multiple studies and one study may contain multiple series. These level definitions are compatible with most PACS.

The workflow of imagery process is shown in Fig. 8.1. Once the patient is scanned through MR/CT scanners, a set of images in DICOM format will be stored in PACS. At the same time, the information of irradiation will be automatically saved in RIS. After the scanning process, the technician will assign the study to a radiologist via RIS management portal. The radiologist will be notified about the assigned study when he/she opens the PACS client. Radiologists can choose the study from the assigned study list for diagnosis. The PACS viewer then fetches

[3] Health Level Seven or HL7 is the global standard for exchanging information between medical applications.

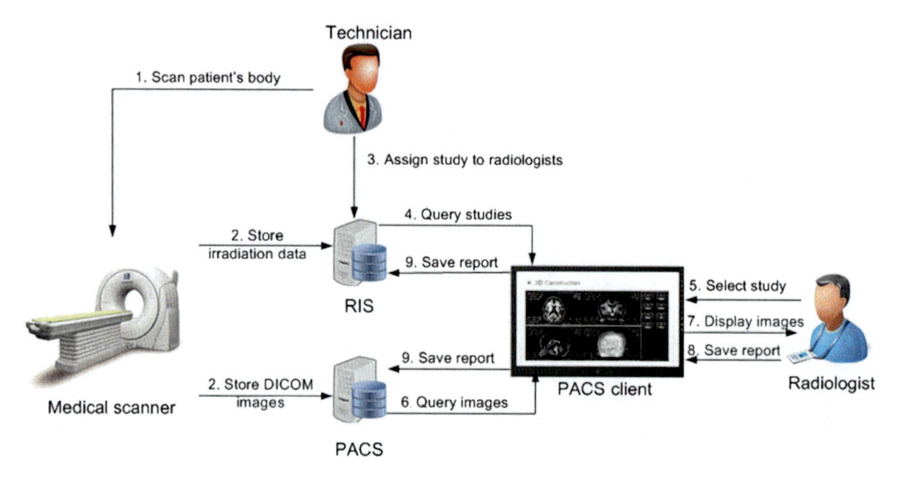

Fig. 8.1 Imagery workflow

DICOM images from PACS, extracts images and metadata, and displays images with information on the display device. A set of images from different studies or series of the same patient can be fetched simultaneously for display. After the prognosis is completed, the diagnosis report is written and kept in both RIS and PACS. Every radiologist and medical doctor who has the authority to diagnose or medicate the same study can view the existing prognosis report. Furthermore, the image studies are often transferred among radiologists for second opinions.

SkyPACS is designed based on the mentioned workflow. The mobile solution is introduced as an alternative to the current desktop-based PACS client used at most hospitals in Thailand. The following sections described SkyPACS features and its design.

8.3 Features of SkyPACS

SkyPACS can be viewed as a mobile extension to the PACS. The software is service oriented and can work with any PACS server. The main advantage of the software is that the users have the flexibility to choose back-end services: cloud computing and storage on public cloud, private server, or hybrid system. A full Software-as-a-Service or SaaS option is also possible. On the front-end side, devices on any type including iPad, Android tab, Windows 8 tablet, and desktop machines can access SkyPACS through the Internet. Main features of SkyPACS include:

- Dashboard: The case management module which provides the patient information in relations to PACS and RIS. The list of image studies is provided for a specific user based on RIS access right setting. The module can notify doctors when new cases enter the workflow. The doctor can produce text-based reports and email them through the provided UI. Moreover, the module also facilitates doctors in referring cases when a second opinion is needed.

- 2D Viewer: This module is designed to display medical images retrieved from the scanners. Several diagnostic tools are provided including distance measurement, area calculation, standard image enhancement, album viewers, comment authoring, zoom, slice selection, and screen splitting. The screen splitting can be used for comparing images from different studies or series of the same patient.
- 3D Viewer: In this viewer, screens are split into four parts to display anatomy images in axial (top to bottom), coronal (front to back), and sagittal (left to right) planes. Coronal and sagittal images are automatically generated by using the MPR[4] technique when the viewer is loaded and the bottom right window displays the corresponding 3D object. The module interface also allows users to segment the anatomy into muscle, skin, and bone before 3D reconstruction for better visualization. Moreover, the 3D model can be printed directly from the application.
- SkyLink: This is a simple collaboration tool for the users in near proximity to share cases. Cases can be passed along with a simple swipe on the tablet screen, if the receiver has access right to the case file.
- SkySync: This is another tool for collaborative diagnosis. Once the tablets are synced, the users will see the same screen and can work on the images together in a similar fashion as the Google Doc service.
- Brain Tumor Detection: With this feature, SkyPACS can automatically investigate image slices in 2D and make suggestions on where the tumors might be located. Techniques used are a combination of image processing and a rule-based system. Rules given by doctors and templates of organs are used as parts of the decision-making process. The inference engine does reason from the knowledge base like a human would. Once suggestions are made, the doctors can confirm the tumor location and the tumor can then be reconstructed and shown in 3D with the calculated volume.

Sample of screenshots from SkyPACS' features listed above are given in Appendix A.

8.4 Software Design

Our design emphasizes the flexibility as SkyPACS must integrate with multiple PACS servers that run on different operating systems and platforms. The Model-View-Controller (MVC) [8] software architecture is adopted in order to separate the

[4] Multiplanar Reconstruction or MPR constructs the volume by stacking images that retrieved from medical scanner, which is axial slices, together and cuts the volume orthogonally in a different plane to obtain the coronal and sagittal slices.

data and logic from the user interface. The change in one must not affect the others. "Model" represents the medical image data/patient information in the repository. A layer of model services (SkyPACS's main business logic) is provided including functions, such as information retrieval and storing, image extraction, image manipulations, and 3D reconstructions. These services update states of the model. "View" is the output representation in the form of image strings, information in text form, and HTML rendering code. Basically, "View" shows the model states to the user through the interfaces. "Controller" sends commands to "View" to change the presentation and also update the model states according to users' commands. In other words, "Controller" receives user commands from the interface and initiates responses by interacting with "Model," changing its state, and presents the new "View" to the users. Model services are a collection of programs, while the controller services are implemented in the form of web services. In our design, the three components are encapsulated in different layers. Adopting MVC in this service-oriented manner allows SkyPACS to utilize private server with GPU and public cloud storage at the same time. Front-end and back-end services can be selected according to the legacy system already in place at each hospital.

In addition, SkyPACS utilizes the thin client approach, meaning that almost the entire model, view, and controller logics are placed on the server side. The client sends HTTP requests to the controller and then receives an updated webpage in return. Figure 8.2 shows the service layers of SkyPACS along with the service invocation steps. Notice that some controller services are executed on the client through HTML5 technology (along with JavaScript and CSS). These services are related directly to users' commands given through the UI and are left on the client to

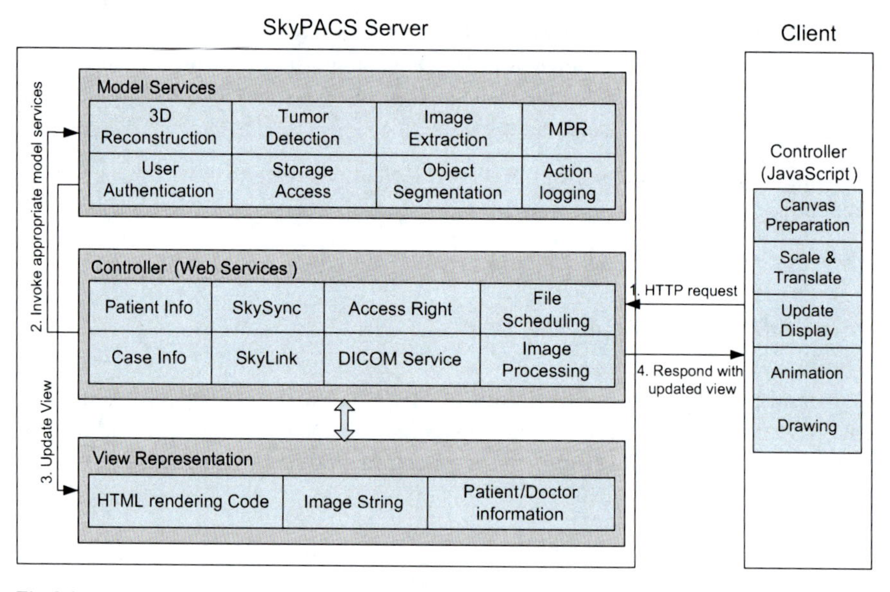

Fig 8.2 Software architecture

reduce communications between the client and server. With HTML5, SkyPACS can be executed on the standard web browser with no plug-in component required.

With the software architecture shown in the figure above, SkyPACS leaves a zero footprint on the mobile devices. Everything goes through the server, which is located behind the hospital's firewall. The client device only caches images when SkyPACS is in operation. Once the software is closed or left inactive for a period of time, everything will be wiped clean. If a doctor loses his or her tablet, patient information cannot be released. The following subsections describe the design of the two key modules in SkyPACS services, which are 3D reconstruction and PACS storage management.

(a) 3D Reconstruction with GPU Computing

There are two methods typically used in reconstructing objects, which are direct and indirect volume rendering. Using indirect technique, such as Marching Cube [9], an actual 3D model will be created, but the computation is so expensive that an interactive, real-time display becomes a challenge. In order to reduce the time, direct volume rendering, i.e., Ray Casting [10], Shear Warp [11], and Splatting [12], can be used. These techniques create an illusion of a 3D object from a series of 2D images for visualization purpose only. No model is generated. However, with these direct techniques, the processing time required on a typical quad-core server is still in the order of several minutes, which is not sufficiently fast for a near real-time experience. To overcome such a problem, SkyPACS provides data-parallel Ray Casting that can be executed on the graphic processing unit (GPU). NVIDIA's GPU is an inexpensive platform that is highly parallel and is built based on the "many-core" technology. By exploiting the relatively inexpensive GTX780 GPU card and CUDA library, SkyPACS is able to deliver the 3D perception of a large image set in under 5 s. The GPU computing module in SkyPACS can be illustrated in Fig. 8.3.

From the figure, notice that the GPU is installed on the server side and the reconstruction service can be called by a web-based client application through our designed application interfaces (APIs). The host (CPUs) is responsible for DICOM file fetching and extraction. Once the DICOM file is fetched, it is extracted into a set of 8-bit grayscale image files. The header information including image dimension, thickness of 2D slices, pixel spacing[5], slice order, and patient's orientation[6] is extracted into SkyPACS's database. Slice order and patient's information are then used to register images by sequentially stacking the slices. Distances between slices are determined using the

[5] Pixel spacing is an attribute which indicates the physical distance between two pixels. It consists of two values, row and column spacing in millimeter.

[6] Patient's orientation specifies the position of the patient. When facing the front of the imaging equipment, Head First is defined as the patient's head being positioned toward the front of the imaging equipment, while Feet First is defined as the patient's feet being positioned toward the front of the imaging equipment.

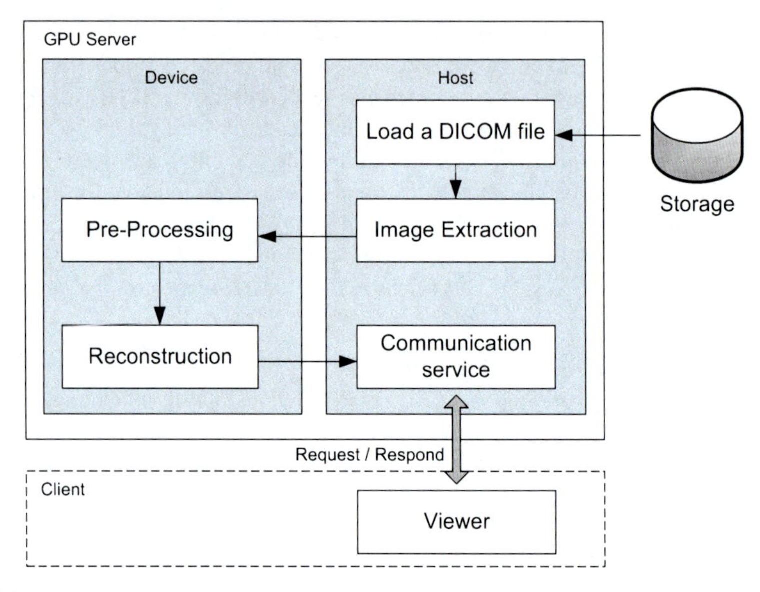

Fig 8.3 3D Reconstruction workflow

extracted thickness and pixel spacing of each slice. The image slices are then sent to the device (GPUs) on the same server. The device is responsible for preprocessing and reconstruction using Ray Casting. The preprocessing step includes normalization and level-contrast adjustment. Depending on users' actions, template matching, and more advanced AI-based algorithms may also be executed for bone/muscle/skin segmentation and brain tumor detection. Once images of the 3D perception are generated, they are sent back to the host, which in turn forward these images to the viewer module on the client device. With this workflow, all the heavy computations are off-loaded from the client device, allowing inexpensive tablets to smoothly run our software as long as there is a good broadband connection. In addition, an actual 3D model will never be generated unless a user chooses to print an object with a 3D printer.

(b) PACS Storage Abstraction

Medical image files are large and patient's data are needed to be kept for at least 5 years after a case becomes inactive. The file storage that serves PACS then needed to be extended frequently causing tremendous overhead to the hospitals. On average, a hospital in Thailand adds around 8 terabytes of storage per year. To remedy the problem, SkyPACS adopts storage abstraction concept where repository layer is abstracted from the software and files can be transferred back and forth automatically between local storage and the cloud. The local storage can be any legacy storage of a hospital, and the cloud can be any public cloud, such as Microsoft Azure or Amazon EC2. These cloud storages can flexibly be extended or shrunk on demand.

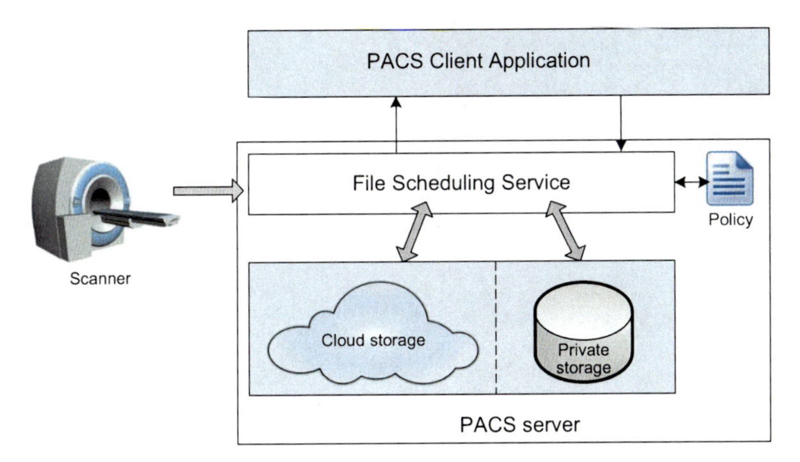

Fig 8.4 A file scheduler

SkyPACS implements a file scheduler as a service to be called by any PACS client application or viewer. Figure 8.4 illustrates the scheduling service. When a scanner or a client system needs to write image files, a file scheduling service will automatically select appropriate disks to store data according to the policy defined by a system administrator through a provided GUI.

One example policy is for an administrator to set a threshold value that specifies when image files should be transferred from a local repository to the cloud. The file selection is performed based on the Least Recently Used or LRU algorithm. In other words, the least recently accessed files will be transferred first, while the most recently accessed files will always be stored locally. When the read access is required, a file scheduler will locate, retrieve, and forward automatically the requested files. Thus, PACS server will be able to use the local storage in combination with the cloud without the knowledge of the physical location of each file. The abstraction layer allows the repository management to be flexible. Moreover, the policy can be changed without affecting file-accessing workflow.

8.5 Implementation and Deployment

SkyPACS is implemented as a 3-tier service-oriented application. The interface responsible for interacting with end users is web based with no installation required on the client side. Touch screen input and gestures are carefully developed for the simplicity and ease of use. We emphasize the use of an open platform with HTML5, JavaScript, and CSS for the front-end modules. These technologies are compatible on most browsers and tablets. The core business logic of SkyPACS is implemented on .NET framework. The web services and service protocol are built based on Windows Communication Foundation or WCF. On the back-end computing,

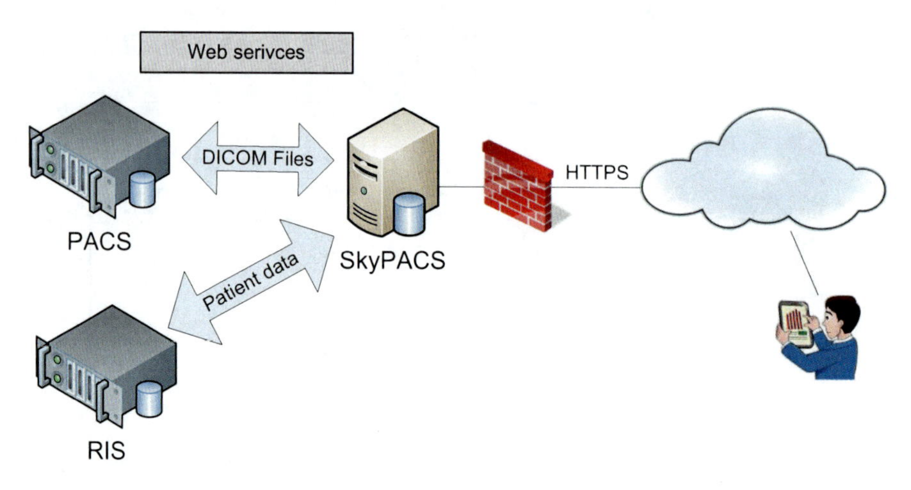

Fig 8.5 Network connection

CUDA-C and C++ are used for 3D reconstruction and other compute-intensive services. SkyPACS server runs Windows operating system with Internet Information Service or IIS web server as this is the standard platform used in Thai hospitals. The SkyPACS storage server is implemented with MySQL and open source DICOM toolkit.

SkyPACS has currently been deployed at one of the MRI centers and is scheduled to be deployed at another university hospital in Thailand in the last quarter of 2013. With data security as the main concern, the implementation is done in such a way that all patient data and case files are streamed through the encrypted channels on demand. We deployed a SkyPACS server at the customer site and open a series of connection channels between SkyPACS and PACS server. The number of channels created depends on the number of concurrent users specified by the customer. When an end user requests data, SkyPACS queries RIS for patients' information and PACS for DICOM files. The information is then stored in SkyPACS data server, which sits behind a firewall. A dedicated communication channel between PACS and SkyPACS server is then assigned to each user session. Requests/responses are then carried out using the channel until the user terminates the application. If the session time is over, the communication channel will also be reassigned. Figure 8.5 illustrates the network connection.

8.6 Product Comparisons

This section compares SkyPACS with some commercial medical imaging software packages available in Thailand, namely, RadiAnt [13], Synapse Mobility [14], and OsiriX HD [15]. Similar to SkyPACS, these mobile solutions were designed to be a viewer of DICOM files and offer standard tools such as zooming, panning, marking, and image manipulation tools.

RadiAnt is a Windows-based solution designed to be a stand-alone viewer. Connection to any PAC systems will be a challenge. The software requires the user to manually provide the data through CD/DVD media. Image data are stored in the device's storage. Without a predefined method to pull data from PACS, RadiAnt cannot be seamlessly integrated to the hospital IT platforms.

Synapse Mobility is a web-based solution developed to be an extension of *Synapse product suite* which is a clinical workstation solution. Once the data are requested through Hypertext Transfer Protocol or HTTP, they will be sent over the Internet and cached in the device in a similar fashion as any web application does. Synapse Mobility requires that a hospital uses Synapse product suite, which is one of the solutions with a very high cost.

OsiriX HD is an iOS application developed to be both stand-alone and extension solutions. User can either manually provide the data or connect the application to any standard PAC system. Once data are presented, they will be stored in the device's storage. Moreover, OsiriX HD is restricted to iOS platform.

In our study, we compare the products in four dimensions: data security, supporting platform, PACS compatibility, and cloud integration. Details are below.

The handheld device presents more risk of data being stolen than the desktop machine located in the hospital. This is an important issue since the sensitivity of medical data and patient's record is very high. Leaving a zero footprint with no plug-ins or image data on the client device is necessary in many usage scenarios. From the four packages, only Synapse Mobility and SkyPACS were implemented based on this concept.

As there are several popular platforms for mobile devices nowadays, portability across platform is important. RadiAnt and OsiriX HD are restricted to a specific platform making them less flexible. Synapse Mobility and SkyPACS then have an advantage.

Most radiology departments have already installed a PAC system; the integration with the existing PACS is expected for a mobile extension. All packages but RadiAnt offer an option to connect to PACS through the standard DICOM protocol. Among the 3 packages, Synapse Mobility restricts the integration to Synapse PACS only. Unless the hospital deploys the Synapse workstation, this mobile extension is not available.

In order to effectively manage PACS storage and 3D image computation, cloud integration has been studied. From the survey, RadiAnt and OsiriX HD are native applications and are required to operate on the device's processor; cloud integration is unlikely. Synapse Mobility also requires the specific PACS and cloud option is not currently available. SkyPACS is differentiated from the others due to the fact that SkyPACS's back-end services can be customized and integrated to any server platform.

To summarize, SkyPACS was designed by compiling benefits from the product survey and extending some features to maximize the capability of the application.

Conclusion

SkyPACS is a mobile solution that is designed to be fully service oriented. Front-end and back-end services are encapsulated and thus independent of one another. The software emphasizes the ease of use as well as the ease of integration. The main advantage of SkyPACS is that it can be integrated with any PAC system at any healthcare institute. Product customization is possible at a low cost. Virtualization on the cloud and computing on the GPU are also fully utilized in the design. In summary, SkyPACS has been developed based on the cutting-edge technology in the field of mobile and cloud computing. The road map of the development efforts will include the performance improvement in the 3D domain. More advanced features will also be developed including blood vessel reconstruction, computation staining, as well as automatic mobile offloading. With our on-going research works, we believe that we will be able to continuously fine-tune and improve the user experiences in the future.

Acknowledgments The authors would like to thank many of the people who help turning our research works into a commercial product: J.F. Advance Med Co., Ltd. for providing the facilities to complete the installation at the first test site, the National Innovation Agency (NIA) for supporting during the initial stage of the development, NVIDIA and Smart Technology Co., Ltd., for the equipment loans, Microsoft for providing the opportunity for students to showcase the product through the Imagine Cup competition. Last but not least, we would like to thank King Mongkut's University of Technology Thonburi for funding and supports throughout the year.

Appendix A: Screenshots

(a) Windows 8 Version

a

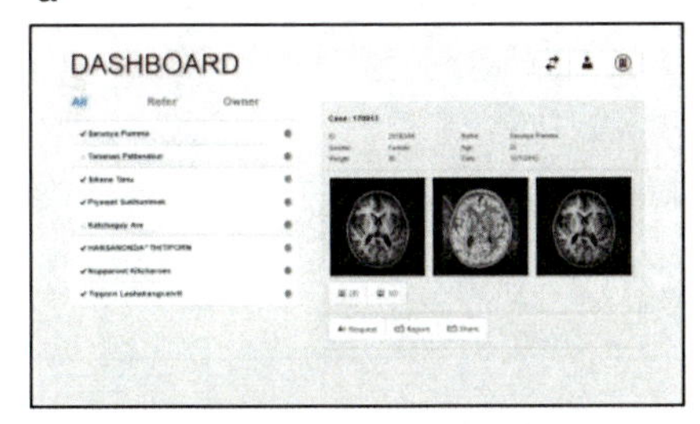

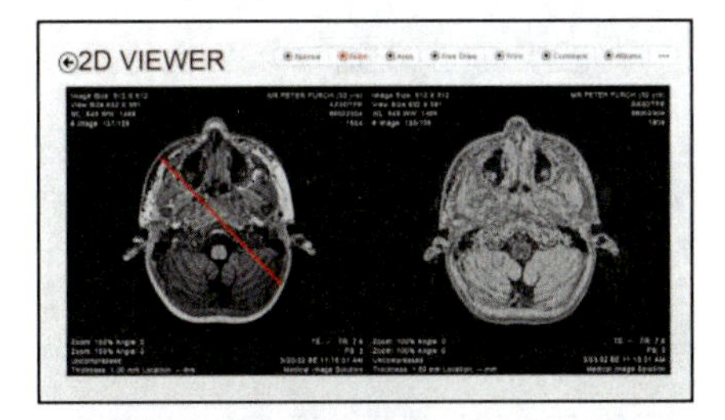

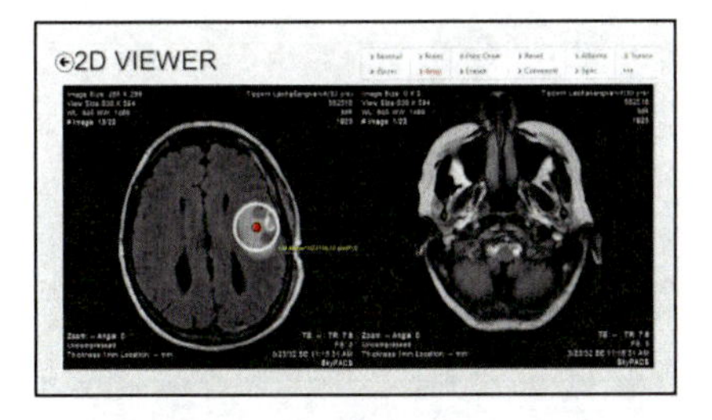

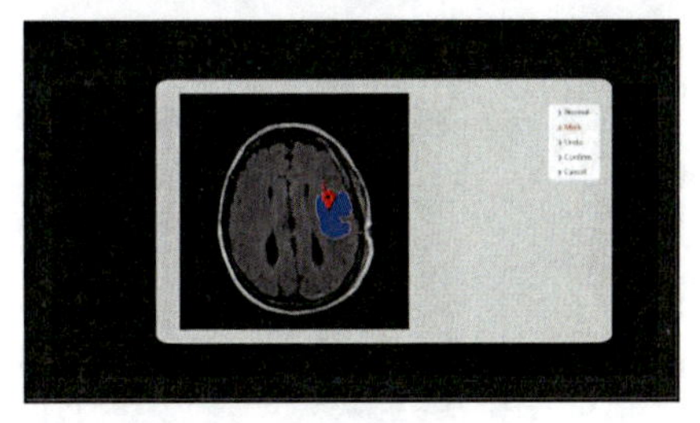

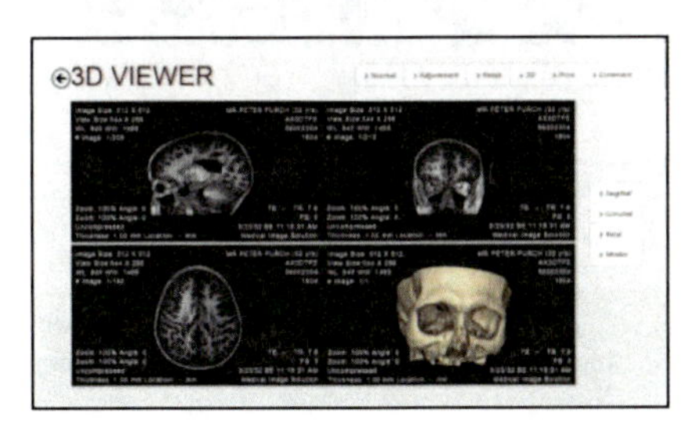

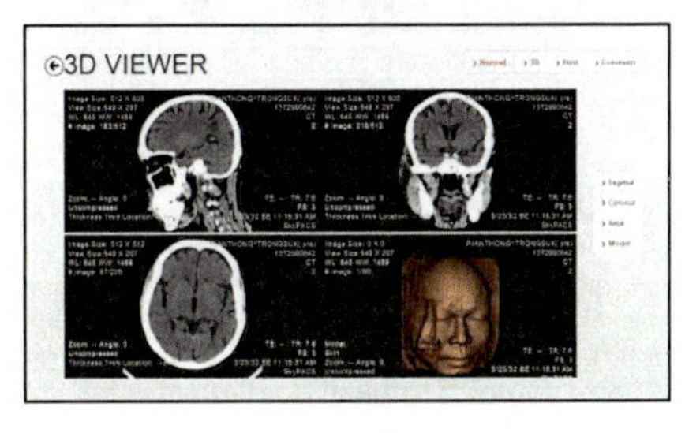

b

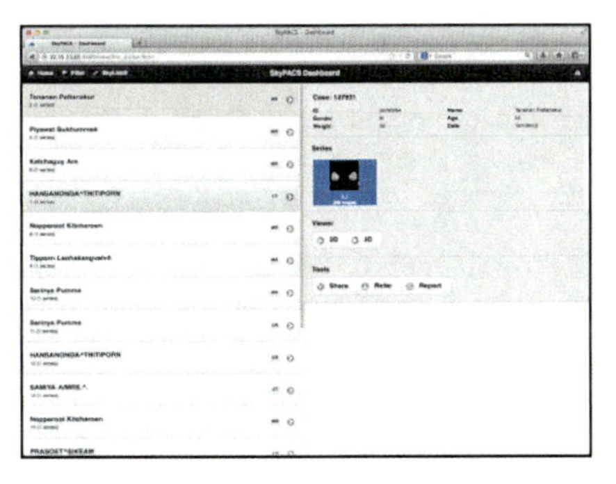

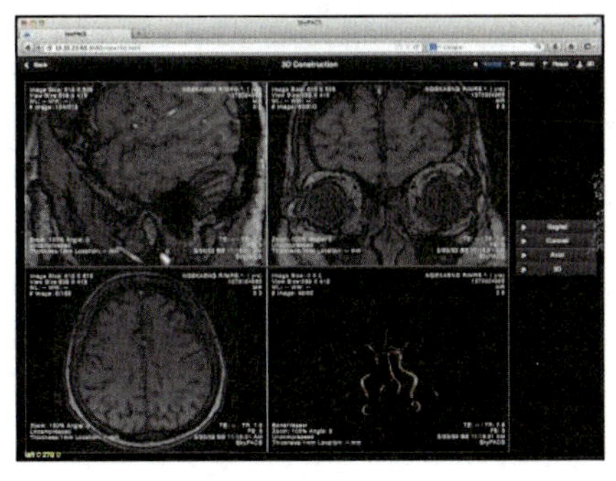

(b) Web-Based Version

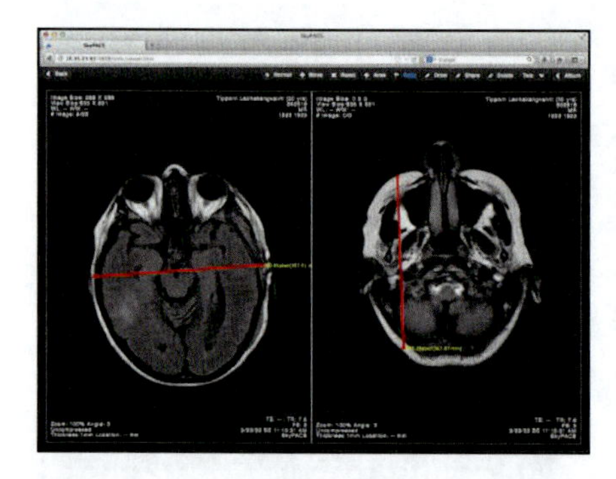
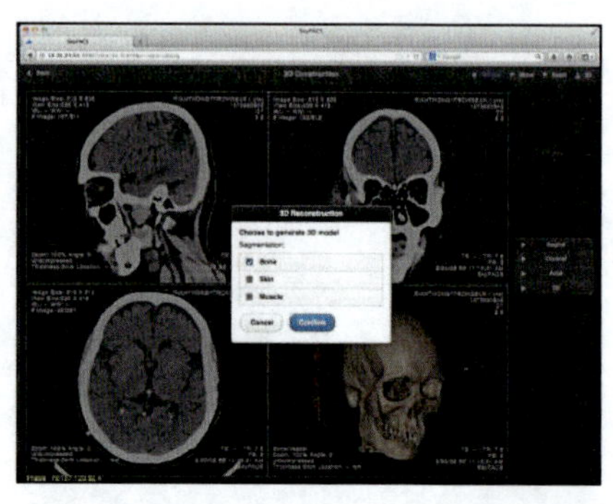
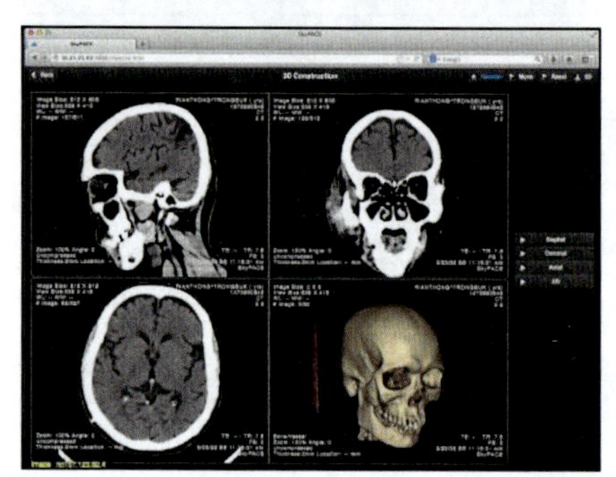
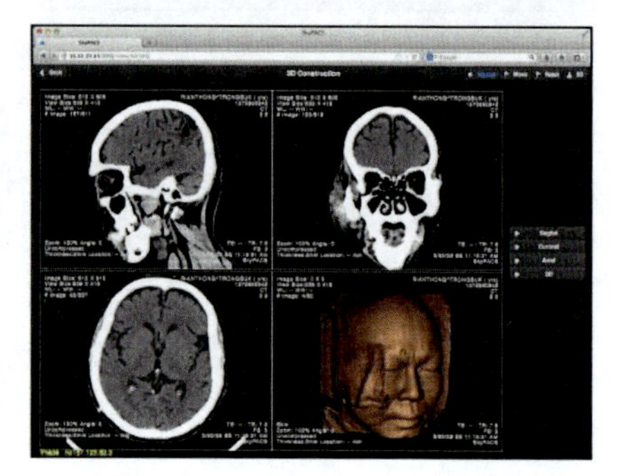

References

1. TEH and Associates: 10% of Hospital Admissions are associated with an ad-verse event. http://www.tehandassociates.com (2013). Accessed 26 July 2013
2. Nakornping Hospital: Radiology patient statistic. http://www.nkp-hospital.go.th/institute/x-ray/data3.php (2013). Accessed 26 July 2013
3. Radiological Society of North America: Radiologists Worldwide Face Similar Training and Staffing Issues. https://rsna.org/NewsDetail.aspx?id=8001 (2013). Accessed 29 July 2013
4. Robert, H.C., Choplin, Jobannes, M.B., Douglas Maynard II, C.: Picture archiving and communication systems: an over-view. Radiographics **12**, 127–129 (1992)
5. Bernard, C., Lawrence, S.: Implementation of a radiology information system/picture archiving and communication system and an image transfer system at a large public teaching hospital? Assessment of success of adoption by clinicians. J. Telemed. Telecare **10**, 27–29 (2004)
6. Kiyonari, I., Tokuo, U., Hajime, H., Hiroshi, K., Toshitsune, H., Takahiro, K., Hiroshi, T., Michitoshi, I.: Time and flow study results before and after installation of a hospital information system and radiology information system and before clinical use of a picture archiving and communication system. J. Digit. Imaging **10**(1), 1–9 (1997)
7. Health Level Seven International: About HL7. http://www.hl7.org/about/index .cfm (2013). Accessed 29 July 2013
8. Buschmann, F.: Pattern-Oriented Software Architecture. Wiley, New York, NY (1996)

9. William, E., Lorensen, Harvey, E.: Cline: marching cubes: A high resolution 3D surface construction algorithm. Comput. Graph. **21**(4): (1987)
10. Marc, L.: Volume rendering. IEEE Comput. Graph. Appl. **8**(5), 29–37 (1988)
11. Philippe, L., Marc, L., 1994. Fast Volume Rendering Using a Shear-Warp Factorization of the Viewing Transformation. In: Proc. SIGGRAPH '94, pp. 451–458
12. Lee, AW., 199. SPLATTING: A Parallel, Feed-Forward Volume Rendering Algorithm. University of North Carolina, Chapel Hill Chapel Hill, NC
13. Medixant: RadiAnt DICOM Viewer. http://www.radiantviewer.com (2013). Accessed 29 July 2013
14. Fujiflim: Fujuflim's Synapse? Mobility Extends Image Reach. http://www.fujifilmusa.com/press/news/display_news?newsID=880037 (2013). Accessed 29 July 2013
15. OsiriX: OsiriX HD User Manual. http://www.osirix-viewer.com/Manual/index.html (2013). Accessed 29 July 2013

Chapter 9
Collision Detection Based on Fuzzy Scene Subdivision

David Mainzer and Gabriel Zachmann

Abstract We present a novel approach to perform collision detection queries between rigid and/or deformable models. Our method can handle arbitrary deformations and even discontinuous ones. For this, we subdivide the whole scene with all objects into connected but totally independent parts by a fuzzy clustering algorithm. Following, for every part, our algorithm performs a Principal Component Analyses to achieve the best sweep direction for the sweep-plane step, which reduces the number of false positives greatly. Our collision detection algorithm processes all computations without the need of a bounding volume hierarchy or any other acceleration data structure. One great advantage of this is that our method can handle the broad phase as well as the narrow phase within one single framework. Our collision detection algorithm works directly on all primitives of the whole scene, which results in a simpler implementation and can be integrated much more easily by other applications. We can compute inter-object and intra-object collisions of rigid and deformable objects consisting of many tens of thousands of triangles in a few milliseconds on a modern computer. We have evaluated its performance by common benchmarks.

Keywords Collision detection • Fuzzy clustering • Physics-based animation • Computer animation • Cloth simulation

9.1 Introduction

Collision detection between rigid and soft bodies is important for many fields of computer science, e.g., for physically based simulations, medical applications like virtual surgery, and cloth simulation. The underlying collision detection needs to check if collisions occur between a pair of objects as well as self-collisions among deformable objects. In many applications, an additional requirement is that the

D. Mainzer (✉)
Clausthal University, Clausthal-Zellerfeld, Germany
e-mail: dm@tu-clausthal.de

G. Zachmann
University of Bremen, Bremen, Germany
e-mail: zach@cs.uni-bremen.de

Y. Cai, S. See (eds.), *GPU Computing and Applications*,
DOI 10.1007/978-981-287-134-3_9

collision detection has to be calculated within milliseconds. Penalty-based physical simulations, for example, typically perform a number of iterations for a single rendering frame, requiring collision detection at $n \times 30$ Hz, if the scene is rendered at 30Hz.

There exist various approaches that propose spatial subdivision for collision detection or approximate the surface of rigid and soft bodies. These algorithms employ axis-aligned bounding boxes (AABB) [1], oriented bounding boxes (OBB) [2], or inner sphere trees (IST) [3] to reduce the computation time.

Most of the earlier efficient collision detection algorithms were sequential ones, which are perfect for devices that can execute only one instruction at a time. The current trend in computer architecture focuses on multi-core CPUs and many-core GPUs, and so many parallel collision detection algorithms have been proposed in the last years. The collision detection algorithm we present in this chapter is a fast, fully GPU-based algorithm that can exploit data and thread-level parallelism.

Modern GPUs can be thought of as many-core stream processors, and such streaming architectures have significant implications on algorithm design, especially when applied to general-purpose tasks because they were initially designed for graphics manipulations. Because of this, many prior GPU-based collision detection algorithms [4–6] or hybrid combinations of CPU and GPU [7–9] have been developed. A lot of well-known culling methods for collision detection algorithms exist, which include *Sort and Sweep* [10], also known as *Sweep and Prune* [11], to limit the number of pairs of primitives that need to be checked for collision. Without using these culling methods, a huge amount of computation time is wasted and additional memory access is needed, which takes a lot of time especially when accessing global memory on GPUs.

9.1.1 Our Contributions

Our novel *Collision Detection Based on Fuzzy Scene Subdivision* algorithm is designed for interactive and exact collision detection in complex environments and can handle object movement and deformation at the same time. To achieve these features, our algorithm subdivides the whole scene, with all objects, into independent, overlapping parts in the first step. For the segmentation process, we implemented a GPU-based clustering algorithm called *fuzzy C-means* (see Sect. 9.4). For all clusters, we can execute the collision detection steps independently, and this offers the possibility to distribute the collision detection computation for the clusters to different GPUs. To reduce the number of false positives, we use an adapted version of the *Sweep and Prune* approach in combination with *Principal Component Analysis* (see Sect. 9.3). This has the advantage that our algorithm does not need to distinguish between a broad and narrow phase.

Our novel approach is as fast as state-of-the-art collision detection algorithms but with the additional advantage that our collision detection can be distributed easily to more than only one GPU, because we subdivide the whole scene into independent but connected parts; thus, it scales very well with the number of GPUs. Also, our collision detection algorithm works directly on all primitives (e.g., triangles) of the whole scene, which results in a simpler implementation and can be implemented much more easily by other applications. In addition to that, working on all primitives directly avoids approximate errors.

9.2 Previous Work

Since collision detection is a fundamental technique in many simulations, it has been extensively investigated by researchers over the last decades. As a result, a large number of different techniques for collision detection queries and handling exist [12]. In this section, we focus on those approaches only, which can handle collisions between deformable objects.

9.2.1 Approaches Using Bounding Volume Hierarchies

Using Bounding Volume Hierarchies (BVH) is the most common approach to speed up collision detection of rigid and deformable objects [13]. Govindaraju et al. [7] used precomputed chromatic decomposition of a mesh to check for collisions between nonadjacent primitives. A limitation of this approach is that the connectivity of the mesh has to be fixed. Consequently, this approach is not applicable when you want to simulate ripping or cutting a virtual object, which has main importance in simulations like virtual surgery and advanced cloth animation. Greß et al. [5] used stenciled geometry images to generate GPU-optimized BVH in real time. This approach is optimized for collision and self-collision detection for NURBS models or other types of rigid or deformable parameterized surfaces. This approach is limited to a few thousand NURBS patches. Kim et al. [8] presented a hybrid CPU-GPU parallel continuous collision detection (HPCCD) method. HPCCD is based on a BVH and performs efficient reconstructions for selective parts of the BVH. Because they do the BVH reconstruction on the CPU, there is a significant communication between GPU and CPU. A GPU-based linear BVH approach was presented by Lauterbach et al. [14]. Their approach used thread and data parallelism to perform fast hierarchy operations. The linear BVH (LBVH) is used to check for collisions between two disjoint objects as well as self-collisions for deformable objects. Updating these LBVH over more than one GPU is difficult and leads to a huge communication overhead. Tang et al. [15] presented a GPU-based streaming algorithm for collision detection between deformable models. Their approach used BVH as culling technique and reduces the computation to generating different streams. This technique cannot be easily extended to use more than one GPU.

9.2.2 GPU-Based Collision Detection

Most modern collision detection algorithms using BVH are GPU based. However, there are some approaches which use distance fields, space subdivision, or image-space techniques to improve their performance. Teschner et al. [16] presented a new approach to collision and self-collision detection of dynamically deforming objects that consist of tetrahedrons. This proposed algorithm employs a hash function for compressing a potentially infinite regular spatial grid. This hash function maps 3D cells to a hash table, thus realizing a very efficient spatial subdivision. This approach is limited to objects that consist of tetrahedrons only. Heidelberger et al. [17] proposed a simple and efficient algorithm based on Layered Depth Images (LDI). They use a discrete representation of the intersection volume which allows for volume-based collision queries. The accuracy of this method corresponds with the LDI resolution and the depth-buffer resolution. Because the LDI provides only a discrete representation of the underlying objects, in some cases, collision may be missed. Morvan et al. [6] presented an algorithm for proximity queries between a closed rigid object and an arbitrary mesh, for example, deformable, polygonal mesh. They sampled the distance field of the rigid object over the arbitrary mesh. One downside of this approach is that one object has to be a rigid body and so they cannot simulate collisions between two soft bodies, for example. A hybrid CPU-GPU collision detection technique based on spatial subdivision was presented by Pabst et al. [9]. They prune away non-colliding parts of the scene by using an adapted highly parallel spatial subdivision method. Mainzer and Zachmann [18] presented a new approach to collision and self-collision detection which is completely GPU based. Therefore, they subdivide the scene into independent parts by fuzzy clustering. However, the thread and memory management can be improved which results in a less memory-consuming implementation.

9.3 Sweep-Plane Technique Using PCA for Collision Detection

Due to the fact that our collision detection approach treats all objects in a scene at the same time, we do not differentiate between individual objects in the rest of this chapter. Furthermore, we tread all primitives, whether from the same or from a different object, as equals which ensures that our approach detects inter-object and intra-object collisions. A majority of computer animation and simulation use triangles as their fundamental modeling primitive, and therefore, we choose triangles as primitive for our collision detection approach too. However, our approach can be extended to use other primitives easily.

During the collision detection process, we use an adapted version of the standard Sweep and Prune approach, a 1D version, hereafter referred to as *sweep-plane*

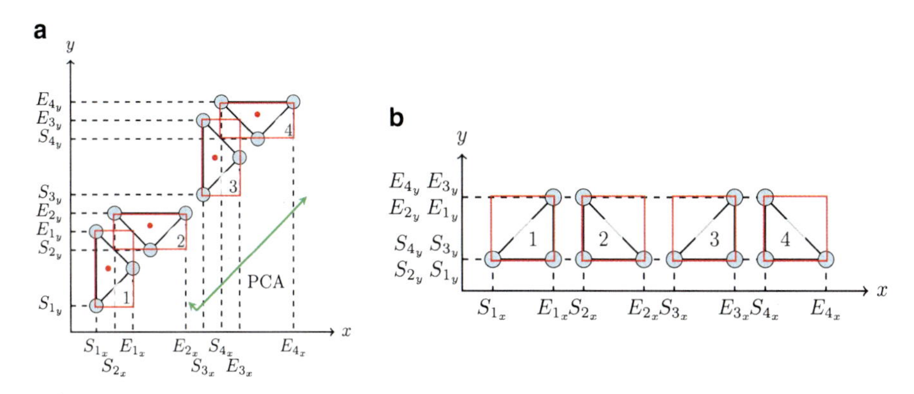

Fig. 9.1 Improvement of sweep-plane approach via Principal Component Analysis. (**a**) The initial scene consisting of a number of triangles with corresponding bounding boxes and the result of the Principal Component Analysis. As can clearly be seen, the bounding boxes of triangles 1 and 2 and triangles 3 and 4 intersect. (**b**) Initial scene from Fig. 9.1a, rotated so that the direction of the first component of the Principal Component Analysis points along the x-axis. As can clearly be seen, in this example, the number

technique. We compute the bounding box for every primitive. Each bounding box spans an interval $[S_i, E_i]$ for each primitive T_i on the x-axis. Sorting all intervals along the x-axis provides information about possible colliding bounding boxes because two bounding boxes collide if one of the four cases $[S_a, S_b, E_b E_a]$, $[S_b, S_a, E_a E_b]$, $[S_a, S_b, E_a E_b]$, or $[S_b, S_a, E_b E_a]$ occurs (see Fig. 9.1).

Figure 9.1a depicts an example of a downside of using bounding volumes, like AABBs or OBBs. If, for example, primitives are moving, then in a significant amount of cases, a huge number of false positives may occur, when we choose any of the *fixed* world coordinate axes as sweep direction. In our case, the best sweep direction is the one which allows projection to separate the primitives as much as possible. In order to achieve the best sweep direction, even if the primitives move through 3D spaces, we compute the *Principal Component Analysis* (PCA) [19, 20] in every frame, because the direction of the first principal component maximizes the variance of primitives, after projection [20].

The type of covariance analysis we perform is commonly used for dimension reduction and statistical analysis of data [13]. As data points we use the centroid C_i of every primitive in the scene. The covariance matrix $\boldsymbol{Cov} = [h_{ij}]$ for all centroid points $C_1, C_2, \ldots, C_n$ is given by

$$h_{ij} = \frac{1}{n} \sum_{k=1}^{n} (C_{k,i} - \text{mean}_i) \cdot (C_{k,j} - \text{mean}_j), \tag{9.1}$$

with mean_i and mean_j the mean of the ith and the jth coordinate value of all the centroid points, respectively.

In Fig. 9.1b, we move the direction of the first principal component on the x-axis. Now, we compute the bounding box intervals $[S_i, E_i]$ and use the x-axis, more specifically the direction of the first component of the PCA, respectively, as sweep direction. Comparing Fig. 9.1a with Fig. 9.1b depicts the advantages of using the first principal component as sweep direction. The number of false positives greatly reduces.

As a consequence, combining sweep-plane and PCA reduces the number of primitive pairs tested for intersection and thus significantly reduces the calculation time.

9.3.1 Thread Management

In this section, we depict how we determine the minimal number of working (CUDA) threads, which are needed for identifying all possible colliding pairs. Additionally, we compute the worst-case memory usage, i.e., the space needed to store all possible colliding primitives, at the same time.

In the first step, we sort all start (S_i) and end (E_i) points of the bounding box intervals along the longest principal axis. Additionally, an array "Type" with the information if at position j is a start ($S_j \rightarrow \text{Type} == 1$) or an end ($E_j \rightarrow \text{Type} == 0$) point is created at the same time (see Fig. 9.2 upper part).

On account of the fact that we want to avoid counting overlapping bounding boxes twice, we only consider the start point (S_i) of the bounding box intervals i. If this is not taken into account, and we consider both the start (S_i) and end point (E_i) of the bounding box interval, for example, in the case of $[S_a, S_b, E_a, E_b]$, we will receive two intersections. Primitive a intersects with primitive b, and vice versa. So, when we consider the start point (S_i) solely, we will get an intersection between primitive a and b only, because S_b is in the interval $[S_a, E_a]$, while S_a is not in the interval $[S_b, E_b]$.

To identify the number of working threads needed to do all intersection tests for a primitive, we need the amount of bounding box intersections between the bounding box of a primitive and all other bounding boxes for all primitives. Therefore, a very suitable solution is the prefix sum algorithm from the Thrust[1] library using the "Type" array as input (see Fig. 9.2 upper part).

The resulting array pT can be used to compute the working threads needed for a primitive to do all possible intersection tests. Therefore, we calculate $pT[E_i] - pT[S_i] - 1$ for a primitive i which generates the number of threads needed for each primitive. The total amount of threads is equal to the number of the worst-case memory usage required to store all possible colliding primitive pairs.

[1] http://thrust.github.com

Position	*0*	*1*	*2*	*3*	*4*	*5*	*6*	*7*	
Bounding Box ID (Start/End)	S_A	S_C	S_B	E_C	E_A	E_B	S_D	E_D	...
Type (Start/End)	1	1	1	0	0	0	1	0	...
prefix sum of Type (pT)	0	1	2	3	3	3	3	4	...

Triangle ID	*A*	*B*	*C*	*D*
Start position (S)	0	2	1	6
End position (E)	4	5	3	7

Triangle ID	*A*	*B*	*C*	*D*	
$pT[E] - pT[S] - 1$	$3 - 0 - 1$	$3 - 2 - 1$	$3 - 1 - 1$	$4 - 3 - 1$	
number Threads	2	0	1	0	$= 3$

Fig. 9.2 Determination of the minimal number of threads needed to identify all possible colliding primitive pairs and the worst case memory usage to store all these pairs

9.4 Object Subdivision Using Fuzzy C-Means

Using the first principal component as sweep direction only will nevertheless produce false positives, because of the dimensional reduction in the sweep-plane step. The sweep-plane technique, used to separate the primitives, projects all 3D bounding volumes to 1D points. This means, for example, that in some cases, primitives of the *front side* and primitives of the *backside* of an object will be recognized as potentially colliding pairs, even if there is a large distance between them. This recognition will result in an amount of unwanted false positives.

To eliminate this kind of false positives, we subdivide the scene (see Fig. 9.3 for some examples) into connected components using *fuzzy C-means* (FCM) algorithm [21, 22]. We use a fuzzy clustering algorithm because the primitives, which are located on the border between two clusters, have to be in both clusters. If adjoining clusters are not connected, then in some cases, collisions across the border of the clusters would not be taken into account (see Fig. 9.4).

The FCM algorithm is a soft, or fuzzy, version of the well-known k-means clustering algorithm. In the classic k-means clustering algorithm, every data point is associated with only the nearest cluster center point. In the fuzzy version of the k-means algorithm, fuzzy C-means, every data point has a membership value in the range of 0 and 1 for every cluster. The algorithm tries to minimize the total error, which is the sum of the squared distances of each data point to each cluster center, if we use the Euclidean distance, weighted by the membership of the data point to each cluster, for all data points.

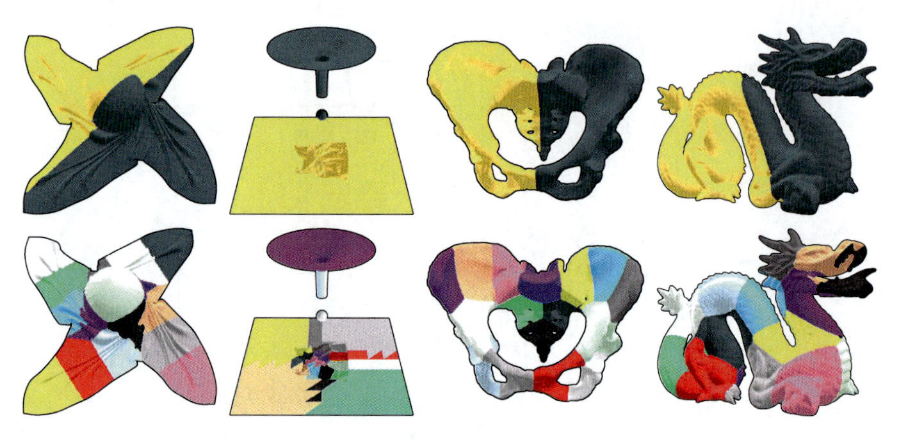

Fig. 9.3 Examples of some high-detail objects, partitioned by fuzzy C-means into two (*top row*) and 16 clusters, respectively. From *left to right*: cloth on ball (92k triangles), funnel (18k triangles), model of the female pelvis (200k triangles), and dragon (202k triangles)

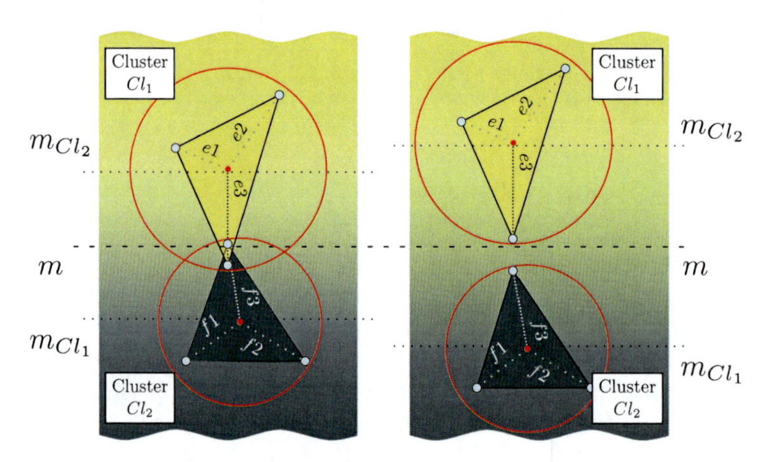

Fig. 9.4 The figure shows two adjoining clusters with two triangles, one colored in *yellow* and one in *gray*. The *yellow triangle* is completely assigned to the yellow cluster Cl_1, and the *gray triangle* is completely assigned to the gray cluster Cl_2. On the *left side* of the figure, we choose the overlap $d(m, m_{Cl_1}) == d(m, m_{Cl_2}) < \|f_3\|_2 < \|e_3\|_2$. Accordingly, like you can see in the figure, it is possible that the *yellow triangle* intersect with the *gray one*. In this case this collision will not be recognized by our collision detection. On the *left side* of the figure, we increase the overlap such that $d(m, m_{Cl_1}) == d(m, m_{Cl_2}) > ||f_i||_2,\ i = 1,2,3$ and $d(m, m_{Cl_1}) == d(m, m_{Cl_2}) > ||e_i||_2,\ i = 1,2,3$. As a result it is impossible that triangles, which are completely assigned to a different cluster, can intersect

Another advantage is that the fuzzy C-means algorithm can be run incrementally thus exploiting temporal coherence that is inherent in most real scenes. For the next iteration, the algorithm uses the last computation result as starting point and iteratively minimizes the total error with the new data points. This approach takes advantage of the fact that the scene changes not very much from one frame to the next one.

Assuming we want to subdivide the scene into c clusters, we compute a sum of dispersion between the data points x_i and a set of prototypes (cluster center points) $v_1, v_2, \ldots, v_c$

$$Q = \sum_{i=1}^{c} \sum_{k=1}^{n} u_{ik}^{t} d(x_k, v_i) \tag{9.2}$$

with $d(x_k, v_i)$ being a given fixed distance function (e.g., Euclidean distance or any l_p -Norm in general) between the data points x_k and v_i, the center point of cluster i.

Furthermore, Eq. 9.2 contains the fuzziness factor t, $t > 1$, and a partition matrix $\boldsymbol{U} = [u_{ik}]$, $i = 1, 2, \ldots, c$, $k = 1, 2, \ldots, n$, which allocate the data points to the clusters. A fuzziness factor $t = 1$ means that the algorithm is doing a hard clustering, like fuzzy k-means, and if $t \to \infty$, the membership will be equal in all clusters. The fuzzy clustering algorithm will iteratively optimize Eq. 9.2. In each iteration, all elements u_{ik} of the partition matrix $\boldsymbol{U}$ are updated using Eq. 9.3:

$$u_{ik} = \frac{1}{\sum_{j=1}^{c} \left(\frac{d_{ki}}{d_{ji}} \right)^{\frac{2}{t-1}}} \tag{9.3}$$

In the next step, the algorithm updates the cluster centers v_k:

$$v_k = \frac{\sum_{i=1}^{n} u_{ik}^{t} \cdot x_i}{\sum_{i=1}^{n} u_{ik}^{t}} \tag{9.4}$$

The algorithm repeats these steps until the center points converge. In the initialization phase, we choose the stop criterion much smaller than during runtime. We also limit the number of iterations for the clustering process to a fixed number at runtime because it is not necessary to get a perfect clustering. These properties ensure that the time, needed for clustering, will not rise dramatically when the scene changes drastically.

9.5 GPU-Based Collision Detection

In this section, we show how our method combines all previously introduced techniques. Algorithm 9.1 provides a short overview of the pipeline of our collision detection approach with the main procedures, which are mapped to a set of computation kernels.

First of all, we subdivide the whole scene into independent, overlapping parts by fuzzy clustering. Thus, we use the centroid of all primitives to decide to what cluster a primitive belongs to. Using a well-chosen stop criterion and a maximum number of iterations for the clustering process limits the time needed for clustering, even when the scene changes significantly. The stop criterion determines when the clustering process has reached an almost steady state, which means the movement of the cluster center point of all clusters is smaller than the predefined criterion.

Algorithm 9.1 GPU-based Collision Detection
Each line is mapped to a massively parallel computation kernel

Input: primitives of all objects
Output: intersecting pairs of primitives
1: subdivide scene into c clusters using fuzzy C-means
2: **for all** clusters **do in parallel**
3: compute and apply PCA
4: sort AABBs along longest principle axis
5: collect all overlapping intervals
6: **for all** overlapping intervals **do in parallel**
7: **if** AABB intersect along y-axis **then**
8: do primitive-primitive intersection test
9: **if intersection**
10: **return primitive pair**
11: **end if**

Now, we can do the following steps for every cluster independently. As described in Sect. 9.3, we do a PCA using the centroid of the primitives of the cluster. The result of the PCA is applied to the primitives of the cluster, which means that the direction of the first component of the PCA points along the x-axis (step "clustering and PCA" in Figs. 9.6 and 9.7).

We now use the x-axis as sweep-plane direction because this direction maximizes the variance of primitives after projection. Therefore, we compute the bounding box of all primitives of this cluster. We calculate the bounding box for the x-dimension and y-dimension in the same step. In this way, we can exploit the fact that we can get completely coalesced memory access, which results in a lower computation time (step "compute AABBs" in Figs. 9.6 and 9.7). We have coalesced memory access because, for example, primitive k will be adapted by the thread with tid $= k$, which can read all vertices from position k and write the result to memory at position k, and consequently, there is no discontinuous read or write access to the memory. We do not compute the bounding box for the z-dimension because our approach only uses the x- and y-dimension for the bounding box intersection test. We explain the fact why we omit the z-dimension bounding box intersection test in the following section.

After computing the bounding boxes for all primitives of this cluster, we sort them along the x-axis using a highly tuned radix sort algorithm from the Thrust library.

The next challenge is to collect all bounding box intervals which intersect in the x-dimension. In order to avoid counting overlapping bounding boxes twice, which would increase computation time and memory needed for the collision detection, we only consider the start point (S_i) of a bounding box interval. In order to decide the required memory and the position of all possible colliding pairs, we use the prefix sum (or so-called scan) algorithm from the Thrust library. This step (see "collect overlapping intervals" in Figs. 9.6 and 9.7) takes up the most computation time in our collision detection algorithm. The problem is that it is not possible to access the memory completely coalesced, which slows down the computation process.

After collecting all possible colliding pairs, whose intervals overlap in x-dimension, we verify whether the bounding boxes of both primitives overlap in the y-dimension or not. We omit a bounding box overlap test for the z-dimension, because it takes more time to read the bounding box information from memory and to compare the values, than using the primitive vertices, which may potentially be needed further in case both primitives intersect, to test if the primitives overlap in the z-dimension. In the case of using a complex polygon as primitive, the algorithm will not omit the z-dimension bounding box test. If both primitives overlap in all three dimensions, the algorithm performs a primitive-primitive intersection test.

Our collision detection algorithm computes all colliding primitive pairs and, if needed, the intersection point or line, respectively.

9.5.1 Accuracy and Limitations

Our collision detection algorithm will recognize all intersections between all primitives. Therefore, our approach performs bounding box intersection tests with all primitives of a cluster, to detect all colliding primitive pairs. However, in the case of significant differences in the size of the primitives, it could happen that a primitive is completely assigned to one cluster, but collides with a primitive which is completely assigned to an adjoining cluster. The reason for this is that our approach uses the centroid, which represents a primitive for the clustering process. To prevent this, we have to decrease the membership value in the clustering step. This results in a higher degree of overlap between adjoining clusters (see Fig. 9.4). The size of the overlap has to be at least as large as the overall maximum distance from primitive's centroid to one of its vertices:

$$\max_{i=1,2,\ldots,n}\left(\max_{k=0,1,\ 2}\left(\|C_i - \mathrm{vertex}_{i,k}\|_2\right)\right) \tag{9.5}$$

From this follows one small restriction for our approach. The large overlap between clusters can affect the performance in some scenarios, because of a higher number of collision computations. This limitation can be avoided by virtually subdividing huge primitives. The virtual primitives are used for clustering and sorting instead of the initial primitive.

If the size of all primitives is more or less equal, then our algorithm chooses a membership value so that the overlap between adjoining clusters consists of exactly two primitives.

9.6 Results

We have implemented our collision detection algorithm on an NVIDIA GeForce GTX 480 using the CUDA toolkit 5.0 as development environment. Because our collision detection algorithm is purely GPU based, components like CPU and RAM do not have effect on the running time. However, for the sake of completeness, we will provide the key data of our system. Our collision detection algorithm is implemented in C++/CUDA. The platform for benchmarking consists of a PC running Gentoo Linux with an Intel Core i5-2500K 3.30 GHz CPU and 8 GB of memory. For sorting and prefix computation steps, we used Thrust, a parallel algorithm library.

9.6.1 Benchmarking

To evaluate the performance of our collision detection algorithm in different situations, we choose some often used collision detection benchmarks to compare our results against other approaches. Experiments have shown that subdividing the scene into two respectively four clusters, when the objects are far apart from each other, for a single GPU provides the best performance. Therefore, in the following benchmarks, we subdivided the scene into two clusters.

In Table 9.1, we show the average collision detection time needed for all benchmarks compared with state-of-the-art collision detection algorithms. Our approach is slightly slower than the CStreams [15] technique, but this approach cannot be easily extended to more than one GPU. Comparing our approach to the hybrid CPU-GPU collision detection techniques [8, 9] and the multi-core collision detection approach [23] shows that our technique performs better.

9.6.1.1 Cloth on Ball

In this benchmark, a cloth (92k triangles) drops down on a rotating ball (760 triangles) (see Fig. 9.5 upper row). Thereby, the cloth has a huge number of self-collisions. This benchmark is subdivided into 93 frames. Our collision detection algorithm needs for this benchmark 20.24 ms in average (see Table 9.1).

Figure 9.6 shows that the collision detection time needed to compute all collisions from frame 60 onward increases because the number of self-collisions increases heavily like you can see on Fig. 9.5 (upper row). Our collision detection algorithm needs more time to collect all possible colliding triangles and has to do more intersection tests between them. The benchmark, provided by the UNC Dynamic Scene Benchmarks collection, contains self-intersecting triangles, which means that real collisions occur, like you can see at frame 93.

Table 9.1 Collision detection computation times in milliseconds. The timings include both external and self-collision detection. CStreams (CSt.), GPU-based streaming algorithm for collision detection [15]; Pab., a hybrid CPU-GPU collision detection technique based on spatial subdivision [9]; HP, a hybrid CPU-GPU parallel continuous collision detection [8]; MC, a multi-core collision detection algorithm running on a 16 core PC [23]

Bench.	Our	CSt.	Pab	HP	MC
Cl. on ball	20.24	18.6	36.6	23.2	32.5
Funnel	6.53	4.4	6.7	–	–

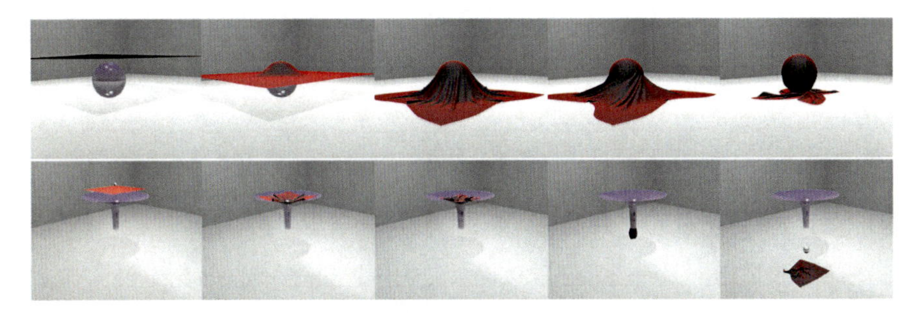

Fig. 9.5 The upper row shows the frames 0, 10, 40, 60, and 93 of the cloth on ball benchmark. The lower row shows the frames 0, 125, 200, 375, and 500 of the funnel benchmark

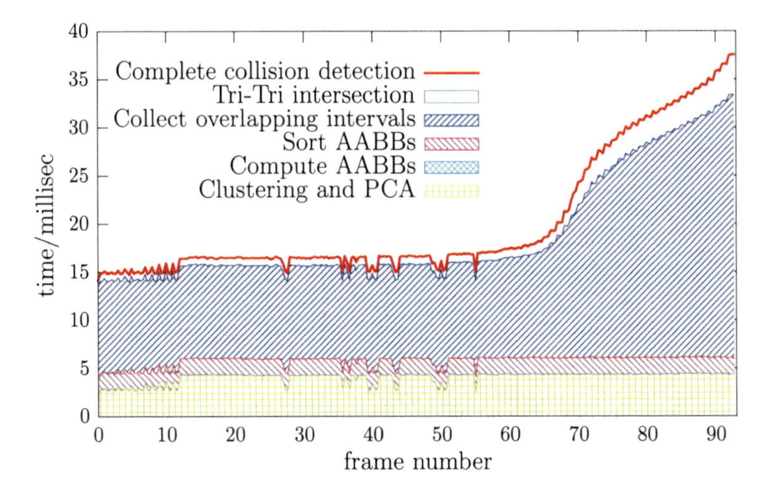

Fig. 9.6 Collision detection time needed for cloth on ball (92k triangles) benchmark

9.6.1.2 Funnel

A cloth (14.4k triangles) falls into a funnel (2k triangles) and passes through it, due to the force applied by a ball (1.7k triangles). The ball slowly increased in volume over the time (see Fig. 9.5, lower row). Our collision detection algorithm needs for this benchmark 6.53 ms in average (see Table 9.1).

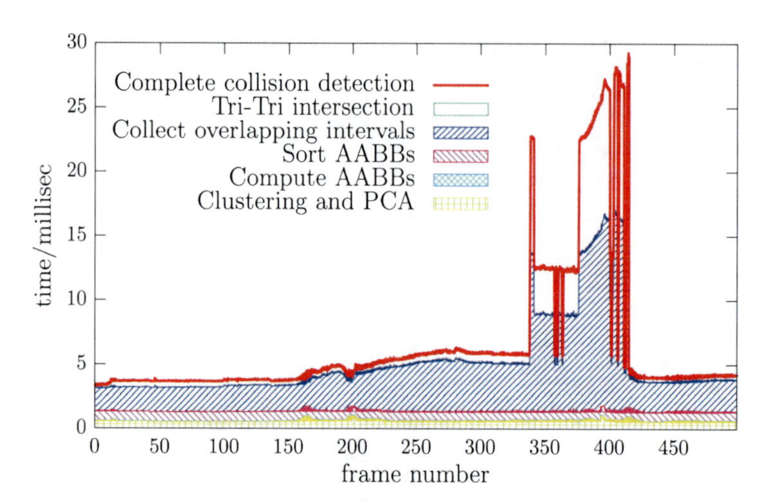

Fig. 9.7 Collision detection time needed for funnel (18.5k triangles) benchmark

Figure 9.7 depicts that the collision detection time needed to compute all collisions increases slightly between frame 150 and frame 345. In these frames, the cloth hit the funnel and slides a little bit into the funnel. From frame 345 onward, the ball pushes the cloth through the funnel and produces a huge number of self-collisions which results in a higher computation time needed for collision detection.

Conclusions and Future Work

We presented a novel, accurate, and fast collision detection algorithm which is completely GPU based and does not require additional communication between host (CPU) and device (GPU). Our *Collision Detection Based on Fuzzy Scene Subdivision* technique can perform collision queries between rigid and/or deformable models consisting of many tens of thousands of triangles in a few milliseconds. One great advantage of this is that our method can handle the broad phase as well as the narrow phase within one single framework. Arguably, our method is much easier to implement than many other GPU-based deformable collision detection approaches, because we do not need any BV hierarchy or other acceleration data structure. Our results show that our collision detection algorithm is as fast as state-of-the-art approaches. However, because of the subdivision process, our collision detection approach can be distributed easily to more GPUs.

A multi-GPU version of our algorithm is currently being implemented to evaluate the speed improvement. We believe that we can further improve the performance of our algorithm by improving the PCA process, to reduce the number of false positives, even when the objects are deform intensive or closely intertwined. An interesting extension would certainly be to handle

(continued)

triangles which sizes significantly differ. To realize this, we can use virtual subdivision for the degenerated triangles. Finally, we will extend the approach to perform other proximity queries, including distance and penetration depth or volume queries.

Acknowledgments The cloth on ball and funnel simulation benchmarks are courtesy of the UNC Dynamic Scene Benchmarks collection and were provided by Naga Govindaraju, Ilknur Kabul, Stephane Redon, and Simon Pabst.

References

1. Van den Bergen, G.: Efficient collision detection of complex deformable models using AABB trees. J. Graph. Tools **2**(4), 1–13 (1997)
2. Gottschalk S., Lin M. C., Manocha D.: OBBTree: A hierarchical structure for rapid interference detection. Comput. Graph. **30** (Annual Conference Series): 171–180 (1996)
3. Weller R., Zachmann G.: Inner sphere trees for proximity and penetration queries. In: Robotics: Science and Systems Conference (RSS), Seattle, WA, USA, June/July 2009
4. Gress A., Zachmann G.: Object-space interference detection on programmable graphics hardware. In: Lucian M. L., Neamtu M., (eds.) SIAM Conf. on Geometric Design and Computing, Seattle, Washington, 13–17 November 2003, pp. 311–328. Nashboro Press (2003)
5. Gress A., Guthe M., Klein R.: Gpu-based collision detection for deformable parameterized surfaces. Comput. Graph. Forum **25**: 497–506 (2006)
6. Morvan T., Reimers M., Samset E.: High performance GPU-based proximity queries using distance fields. In: Computer graphics Forum, vol. 27, Wiley Online Library, pp. 2040–2052 (2008)
7. Govindaraju N., Knott D., Jain N., Kabul I., Tamstorf R., Gayle R., Lin M., Manocha D.: Interactive collision detection between deformable models using chromatic decomposition. ACM Trans. Graph. **24**: 991–999 (2005)
8. Kim D., Heo J., Huh J., Kim J., Yoon S.: Hpccd: Hybrid parallel continuous collision detection using cpus and gpus. In: Computer Graphics Forum, vol. 28, Wiley Online Library, pp. 1791–1800 (2009)
9. Pabst S., Koch A., Strasser W.: Fast and Scalable CPU/GPU Collision Detection for Rigid and Deformable Surfaces. In: *Computer Graphics Forum*, vol. 29, Wiley Online Library, pp. 1605–1612 (2010)
10. Baraff D.: Dynamic simulation of non-penetrating rigid body simulation. PhD thesis, PhD thesis, Cornell University, (1992)
11. Cohen J., Lin M., Manocha D., Ponamgi M.: I-COLLIDE: An interactive and exact collision detection system for large-scale environments. In: Proceedings of the 1995 symposium on Interactive 3D graphics, ACM (1995)
12. Teschner M., Kimmerle S., Heidelberger B., Zachmann G., Raghupathi L., Fuhrmann A., P. Cani M., Faure F., Magnenat-Thalmann N., Strasser W., Volino P.: Collision detection for deformable objects. Comput. Graph. Forum, 61–81 (2004).
13. Ericson C.: Real-time collision detection. Morgan Kaufmann, San Francisco, CA (2004)
14. Lauterbach C., Mo Q., Manocha D.: gproximity: Hierarchical GPU-based operations for collision and distance queries. In *Computer Graphics Forum* (2010), vol. 29, Wiley Online Library, pp. 419–428

15. Tang M., Manocha D., Lin J., Tong R.: Collision-streams: fast gpu-based collision detection for deformable models. In: Symposium on Interactive 3D Graphics and Games, ACM, pp. 63–70 (2011)
16. Teschner M., Heidelberger B., Müller M., Pomeranets D., Gross M.: Optimized spatial hashing for collision detection of deformable objects. In: Proceedings of vision, modeling, visualization VMV'03, pp. 47–54 (2003)
17. Heidelberger B., Teschner M., GROSS M.: Real-time volumetric intersections of deforming objects. In: Proceedings of Vision, Modeling and Visualization, vol. 3, (2003)
18. Mainzer D., Zachmann G.: CDFC: Collision Detection Based on Fuzzy Clustering for Deformable Objects on GPU's. In: WSCG 2013 - POSTER Proceedings, Plzen, Czech Republic, 7, vol. 21, pp. 5–8, Poster (2013)
19. Jolliffe I.: Principal component analysis. Wiley Online Library, (2005)
20. Liu F., Harada T., Lee Y., Kim Y. J.: Real-time collision culling of a million bodies on graphics processing units. In ACM Trans. Graph. **29**: 154 (2010)
21. Bezdek J.: Pattern recognition with fuzzy objective function algorithms. Kluwer Academic, Norwell, MA (1981)
22. Pedrycz W.: Knowledge-based clustering: from data to information granules. Wiley-Interscience (2005)
23. Tang, M., Manocha, D., Tong, R.: MCCD: multi-core collision detection between deformable models using front-based. Decomposition. Graph. Models **72**(2), 7–23 (2010)

Chapter 10
Smoothed Particle Hydrodynamics Applied to Cartilage Deformation

Philip Boyer, Sean LeBlanc, and Chris Joslin

Abstract Modelling of the cartilage within the acetabulum is necessary for determination of stresses in preoperative simulation of femoral acetabular impingement (FAI), a condition that is considered a primary cause of osteoarthritis. Presented is a previously proven method for elastic solid deformation using smoothed particle hydrodynamics (SPH). Smoothed particle hydrodynamics is a mesh-free method that has advantages in computational speed and accuracy over other graphical methods and as such is attractive for medical simulations that require high degrees of precision and real-time operability. A complete formulation of the method of polar decomposition as devised for smoothed particle hydrodynamics is outlined with the inclusion of a corotational formulation for accurate rotation handling. Modifications to the existing method include boundary and collision handling using an adapted virtual particle method, as well as an algorithm for parallel implementation on the GPU using NVIDIA's CUDA framework. The method is verified through testing with a range of material parameters within the provided elastic solid framework. Employing CUDA for calculations is found to dramatically increase the computational speed of the simulation. The results of an indenter analysis of cartilage modelled as a purely elastic solid are presented and evaluated, with the conclusion that with further refinement the presented method is promising for use in cartilage simulations.

Keywords SPH (smoothed particle hydrodynamics) • FAI (femoral acetabular impingement) • Cartilage • CUDA

10.1 Introduction and Background

Femoroacetabular impingement (FAI) is a condition in which abnormal bony alterations in the form of osseous growths on the femoral head or overcoverage of the proximal femur result in supraphysiological motion or high impact within the acetabulum. Stresses induced in an FAI state occurring during contact between the femur and the anatomical hip have been recognized as a possible progenitor to

P. Boyer (✉) • S. LeBlanc • C. Joslin
Department of Systems and Computer Engineering, Carleton University, Ottawa, ON, Canada
e-mail: philipboyer@cmail.carleton.ca

Y. Cai, S. See (eds.), *GPU Computing and Applications*,
DOI 10.1007/978-981-287-134-3_10

osteoarthritis [1, 2]. Hip impingement is of particular interest in total hip arthroplasty where treatment includes identifying areas of potential conflict between the remaining structure and the implant to reduce the accumulation of wear and restore normal range of motion. For this reason, abnormal physiology resulting in increased stresses should be identified preoperatively.

A commonly proposed methodology to accomplish this is to create a computer simulation based upon the real anatomy of an FAI patient, and these investigations have for the most part been based on computed tomography (CT) scans in a non-interactive simulated environment [3, 4]. One primary concern with these simulations is that the cartilage between the femoral head and the acetabulum is a soft tissue that does not appear in CT images, and despite being an essential component in the accurate determination of stresses within the joint, it has been largely ignored. Magnetic resonance imaging (MRI), on the other hand, can provide detailed imagery of the cartilage and other soft tissues within the acetabulum that can be used to create a more robust computer model. This is important since osteoarthritis, as the greatest cause for concern in impingement cases, is attributed to the degradation and breakdown of cartilage, and it is ultimately the treatment and prevention of this result that should be pursued.

In order for a computer simulation of FAI to be practical in a clinical environment, it should provide a level of interactivity that allows for fast or even real-time response to changes in preoperative planning. Inclusion of soft tissues such as cartilage in a simulation introduces a level of complication that is difficult to implement in an interactive environment due to performance constraints in modern computing. Therefore it is necessary to identify a method for simulating soft tissues that provides high performance and simultaneously gives reasonable response to deformation in the physiological range.

Deformable models have undergone a renaissance since the introduction of what are considered the first physically based models grounded on mathematical principles by Terzopoulos [5]. A variety of methods are now available for this purpose, including but not restricted to point-based mass-spring [6–8], the boundary element method (BEM) [9], the mass tensor model (MTM) [10, 11] and the finite element method (FEM) [12–14]. Unfortunately, none of these procedures come without limitations. Mass-spring models are non-volumetric and must undergo a situation-specific parameterization procedure to arrive at only a rough approximation of solid behaviour. BEM draws the calculations of the interior of a solid model to its surface, so that only a surface discretization is required; however this means that internal behaviour, anisotropy and other non-homogeneities cannot be simulated. MTM is able to incorporate some non-linear behaviours that are essential to soft tissue simulation by combining FEM and mass-spring, but are slower and less stable than mass-spring and less accurate than pure FEM [15]. FEM, considered the gold standard for accuracy in solid simulation for most engineering disciplines, is noted for its slow computational performance, especially with the incorporation of the non-linearities associated with soft tissues into a simulation. However, without non-linearity, large deformations (i.e. $>10\ \%$) cannot be represented in a realistic manner.

Meshless, particle-based simulations such as smoothed particle hydrodynamics (SPH) offer an attractive alternative to the more conventional deformable models used in soft tissue simulation due to their speed and ability to represent complex physical phenomena, including large deformations [16–18]. SPH was originally developed by Gingold and Monaghan in 1977 to simulate stellar formation [19], but has since found most frequent use in fluid dynamics simulations incorporating precise implementations of the Navier–Stokes equations [20–22], as well as solid fracture mechanics and plasticity [23, 24]. A specific advantage of SPH is that because each particle carries its own parameters in a Lagrangian formulation, the functions as applied to each individual particle can be computed completely in parallel using NVIDIA's GPU platform CUDA, resulting in a significant increase in computational speed and bringing us closer to the goal of a real-time environment for soft tissue simulation in FAI. Although using CUDA to perform SPH calculations in parallel is not new, rarely in the available literature has it been applied to elastic solid deformation.

Previous research into soft biological tissue using SPH has been relatively limited. The work by Qin used fluid SPH to represent blood flow through arterial walls composed of mass-springs, and Mesit and Guha created a similar system for an unidentified soft body composed of a mass-spring shell with internal gaseous pressure [25, 26]. It is only with the recent implementations of stable solid models based on elasticity theory that SPH has become feasible for realistic representation of soft tissues without requiring the use of a mass-spring mesh to constrain an SPH fluid. Hieber et al. used the elasticity theory of SPH to represent a virtual liver; however their simulations were limited to small deformations in two dimensions [27]. Solenthaler devised an SPH approach to approximate the Jacobian of the deformation field, but the model was unable to separate rotation of the particle positions from their original configuration and the true strain, resulting in non-physical forces that restrained objects from rotating [28]. The model employed in this chapter to simulate cartilage is based on the work by Becker, which extended the work of Solenthaler by using a corotational approach to enforce accurate rotations, in addition to the handling of large deformations [29].

Cartilage, like most soft tissues, has a long history of the introduction of ever more complex and accurate constitutive models that incorporate such features as poroviscoelasticity, permeability and collagen networks [30–32]. In this chapter, cartilage is treated as a simple elastic solid with the near future goal of incorporating the more complex physiological phenomena known to be active during cartilage deformation.

The first section to follow provides an overview of the calculations required during the elastic solid deformation procedure as well as a brief outline of the method used to enforce boundaries and collision handling. A high-level algorithm details the associated implementation. Results of the procedure are given in the form of tests of the capacity of the model to handle a wide range of deformations and rotations, followed by an investigation of its application to cartilage modelling with a simple indenter test.

10.2 Materials and Methods

The foundation of SPH as defined by Gingold and Monaghan [19] is the smoothing function, which in its generalized form is

$$A_1(r) = \int A\left(r'\right) W\left(r - r', h\right) dr' \tag{10.1}$$

where A is a function of the spatial coordinates, W is the smoothing kernel and dr is a differential volume element. The integral is calculated by a summation over all of the particles in a neighbourhood, which is predefined by h, the kernel smoothing length.

The kernel used in this chapter was developed by Solenthaler [28] for the specific case of elastic deformations

$$W(r,h) = c\frac{2h}{0}\cos\left(\frac{(r+h)\pi}{2h}\right) + c\frac{2h}{\pi} \quad 0 \le r \le h \tag{10.2}$$

$$c = \frac{\pi}{8h^4\left(\frac{\pi}{3} + \frac{8}{\pi} + \frac{16}{\pi^2}\right)}, \tag{10.3}$$

where c is a constant that can be pre-calculated and W is equal to 0 for all other values of r, which is the magnitude of the difference in positions of a particle and its neighbour. The neighbourhood of each particle is pre-calculated and remains constant throughout the simulation. These equations form the basis of the elastic solid deformation procedure to follow.

10.2.1 Elastic Solid Forces

In contrast to the approach used in most fluid SPH simulations, forces in the elastic case are determined not from virtual pressure but by deviation of particle neighbourhoods from their initial configuration. The gradient of the displacement of the particles from their initial position by an external force is interpreted as a change in energy through calculation of strain, followed by calculation of the Cauchy stress. In a perfectly elastic case, the resultant forces cause the neighbourhood to return to its initial configuration once external forces are removed.

The density and the volume of the particle neighbourhood are pre-calculated, since the neighbourhood remains constant throughout the simulation, with the density computed in the SPH form

$$p_i = \sum_j m_j W\left(x_i^0 - x_j^{0\prime}, h\right)^{\prime} \tag{10.4}$$

$$'\widetilde{v}_i = \frac{m_i}{p_i} \tag{10.5}$$

where p is the particle density, $\widetilde{v}$ is the particle volume, x^0 is the position of the particle in its initial configuration and m is the particle mass, which is the same for all particles. The subscripts i and j refer to the current particle and its neighbour, respectively.

Because the strain energy is calculated based on any change in positions of the particles in a neighbourhood, local rotations will be misinterpreted as strain and mistakenly cause forces that prevent realistic rotation. This problem can be rectified by accounting for rotations during force calculations by rotating particle neighbourhoods back to their original orientation so that the "true strain" can be determined, performed in this case via a corotational approach.

The first step in the corotational approach used here is to determine the transformation matrix in an SPH form as proposed by Becker [29], which only considers particles within the neighbourhood. Their degree of influence is dependent on the smoothing kernel W

$$A_i = \sum_j m_j W\left(x_i^0 - x_j^{0\prime}, h\right)\left(\left(x_j - x_i\right)\left(x_j^0 - x_i^0\right)^T\right)^{\prime} \tag{10.6}$$

where the superscript T represents a matrix transpose.

The corotational approach requires the extraction of the rotation matrix of the deformation gradient of the particle and its neighbourhood from its initial state. In the majority of cases, polar decomposition is used, but in the case of inverted or degenerated neighbourhoods where the determinant of the deformation gradient is ≤ 0, which is likely to occur in collinear cases, then the always stable singular value decomposition (SVD) must be used [33]. Degenerated neighbourhoods can be verified by checking the matrix of the deformation gradient for 0 values in the diagonal. For brevity only the polar decomposition method is shown here.

In polar decomposition, the square root of a matrix multiplication must be found by computing the diagonalization of the symmetric matrix $A_i^T A_i$. This can be accomplished through approximately 5–10 sweeps of Jacobi rotations, yielding a diagonal matrix D_i and an orthonormal matrix V_i which contains the eigenvectors of $A_i^T A_i$ in its columns. The square root of $A_i^T A_i$ is then trivial as it is simply a matrix multiplications of the orthonormal matrix V_i and its transpose with the square root of D_i, which is computed as the square root of each element on the diagonal:

$$\sqrt{A_i^T A_i} = V_i \sqrt{D_i} V_i \tag{10.7}$$

If $S_i = \sqrt{A_i^T A_i}$, then the rotation matrix of particle i is determined by the equation

$$R_i = A_i S_i^{-1} \tag{10.8}$$

where the inverse S_i^{-1} is again trivially calculated as

$$S_i^{-1} = V_i \sqrt{D_i}^{-1} V_i \tag{10.9}$$

noting that the inverse of a diagonal matrix is just the inverse of each individual element.

Now that the rotational matrix has been found, it can be used to back-rotate the local deformation $\overline{u}_{ji}$ so that the "true strain" can be determined:

$$\overline{u}_{ji} = R_i^{-1}(x_j - x_i) - \left(x_j^0 - x_i^0\right) \tag{10.10}$$

noting that the inverse of an orthonormal matrix R is simply its transpose, thereby avoiding the computational overhead and notorious pitfalls inherent in matrix inversion. The gradient of the displacement field ∇u_i is similarly given an SPH approximation:

$$\nabla u_i = \sum_j \widetilde{v}_j \overline{u}_{ji} \nabla W\left(x_i^0 - x_j^0, h\right)^T \tag{10.11}$$

The calculation of strain E requires an appropriate choice of strain tensor, and in this case the non-linear Green-Saint-Venant tensor is used since it allows for more accurate handling of the large deformations one would expect with soft tissue:

$$E = \frac{\nabla u_i + \nabla u_i^T + \nabla u_i^T \nabla u_i}{2} \tag{10.12}$$

The Cauchy stress σ in this simple linear elastic case is defined as

$$\sigma = CE \tag{10.13}$$

$$C = \begin{matrix} \lambda + 2\mu & \lambda & \lambda & 0 & 0 & 0 \\ \lambda & \lambda + 2\mu & \lambda & 0 & 0 & 0 \\ \lambda & \lambda & \lambda + 2\mu & 0 & 0 & 0 \\ 0 & 0 & 0 & \mu & 0 & 0 \\ 0 & 0 & 0 & 0 & \mu & 0 \\ 0 & 0 & 0 & 0 & 0 & \mu \end{matrix} \tag{10.14}$$

$$\lambda = \frac{\varepsilon\nu}{(1+\nu)(1-2\nu)} \tag{10.15}$$

$$\mu = \frac{\varepsilon}{2(1+\nu)} \tag{10.16}$$

The Lamé constants ν and ε, respectively, are the Poisson ratio and Young modulus of the material. The Poisson ratio controls the volume conversation of the material with a value of 0.5 representing a perfectly incompressible object. The Poisson ratio for cartilage was chosen as 0.46, noting that the Poisson ratio of cartilage is likely to vary significantly from this value in a more rigorous formulation of the constitutive model. The Young modulus is varied throughout the simulation for testing, but cartilage is assumed to be in the range of 0.5 to 1.8 MPa for the current implementation [34].

The force exerted by particle j on particle i can now be determined as

$$f_{ji} = \widetilde{v}_j (I + \nabla u_i^T) \sigma_i d_{ij} \tag{10.17}$$

$$d_{ij} = \widetilde{v}_j \nabla W \left(x_i^0 - x_j^0, h \right) \tag{10.18}$$

where I is the identity matrix. For detailed derivations of eqs. 10.17 and 10.18, refer to the paper by Müller [35]. The force on particle i is then calculated in a symmetric way in accordance with Newton's 2nd law by using the respective particle rotation matrices to orient the forces in the correct directions

$$f_i = \sum_j \frac{-R_i f_{ji} + R_j f_{ij}}{2} \tag{10.19}$$

10.2.2 Rigid Boundary Collision Handling

Boundaries are treated using a modified virtual particle method [36]. Boundary particles are placed along the edges of any rigid boundary or non-penetrating object and are supported by two layers of virtual particles to reinforce the boundary particle neighbourhoods. The result is that any particle from the deformable solid that approaches within the smoothing length of a boundary will experience a repulsive force. The pressure enforces incompressibility so that penetration into the smoothing length of a boundary is insignificant and so acts like a rigid surface. Penetration is prevented through a simple implementation of Desbrun's spiky kernel using standard fluid SPH pressure force calculations [37]:

$$W_{spkyi}(r,h) = \frac{15}{\pi h^6}\left(h^2 - r^2\right) \quad 0 \leq r \leq h \tag{10.20}$$

which is equal to 0 for all other values of r.

10.2.3 Implementation

The simulation was programmed in C++ and, with the exception of the initialization stage, is run entirely in the CUDA framework on an Intel Core I7 PC with 3.1 GHz, 8 GB of RAM and an NVIDIA GeForce 670 m with 3GB of onboard memory. Visualization is accomplished by importing particle positions and stress values into Maya 2013.

Figure 10.1 outlines the algorithm followed during the simulation. During the initialization phase, particles are assigned initial positions and velocities by the host before *cudamemcpy* is invoked on each of the separate particle variables to transfer them to the GPU for use in CUDA kernels. Each particle variable (e.g. its stress tensor matrix) is assigned its own object of an array so that only the required data for each operation is sent to the GPU rather than being stored in a larger class. This reduces the amount of data sent to the GPU for each kernel and thereby raises the simulation limit on particle numbers. Particle variables are organized according to an index so that objects relevant to each particle correspond to that particle's index. Using the CUDA kernel for constructing the reference neighbourhood as an example, each particle index is assigned a thread, and calculations are performed for each object corresponding to that index for all particles. Data is operated on in blocks of 256 threads until all particle calculations are complete. All calculations outlined in the materials and methods section above are performed in this fashion. Functions were created that performed all necessary memory copy operations to and from the host and GPU, as well as all related allocation and memory freeing operations, to expedite their use and reduce the volume of code required. Error checking was performed after every CUDA kernel function using a custom macro.

The neighbour search is performed using a modified cell indexing method adapted to CUDA to run in parallel for each particle. For the majority of particles, it is only necessary to run the neighbour search algorithm once to determine their initial neighbourhood. Rigid boundary and collision detection requires a neighbour search in each frame for only the outer surface particles. The main program loop is separated into four distinct parts, each of which corresponds to a separate CUDA kernel. In the first kernel, the current neighbourhoods of the boundary particles are updated. In the second kernel, the particle rotation matrices are extracted from their deformations. The third kernel computes deforming particle strains, stresses and forces based on their deformations, and then a solution of the boundary forces is obtained. In the last kernel, symmetric forces are calculated followed by integration to obtain the new particle positions and velocities.

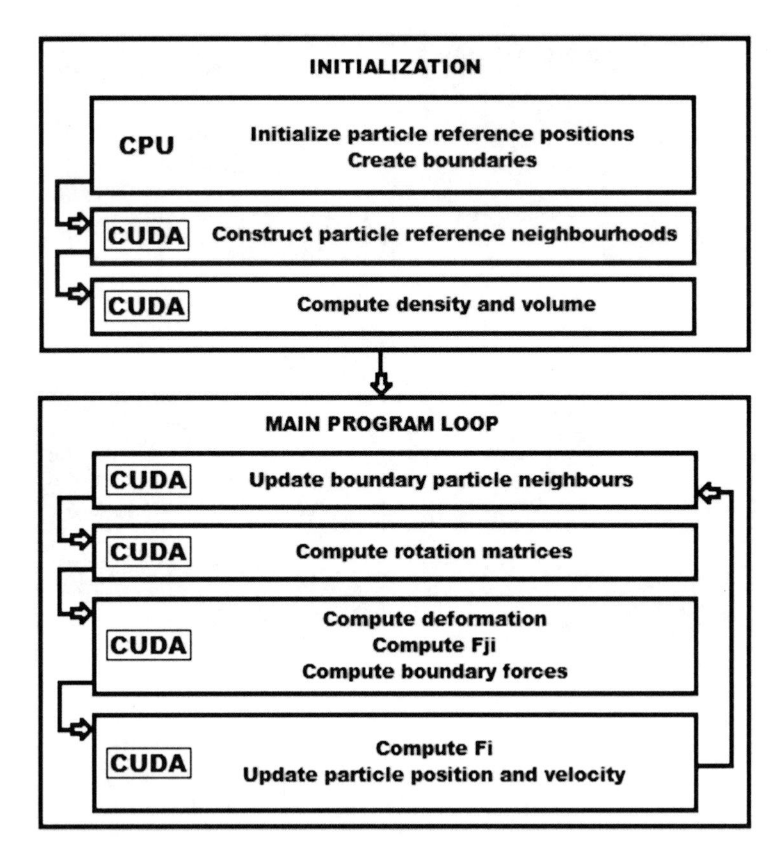

Fig 10.1 Elastic solid SPH algorithm

10.3 Results

Three tests were performed using the provided SPH procedure. Two of the tests—the solid rod and falling wedge—were similar to those shown in the work by Becker [29] and were designed to be rigorous validations of the formulation outlined in this chapter. The third scenario was a simple indenter test of a block of perfectly elastic solid intended to represent cartilage.

10.3.1 Solid Rod and Falling Wedge Tests

Poisson's ratio was held constant at 0.46 throughout both the solid rod and falling wedge tests, while Young's modulus was varied between 1,000 Pa and 1.5 MPa, the latter value representing the high range of what is to be used to represent cartilage in the acetabulum. The smoothing length was set to 0.1 m, and particles were arranged

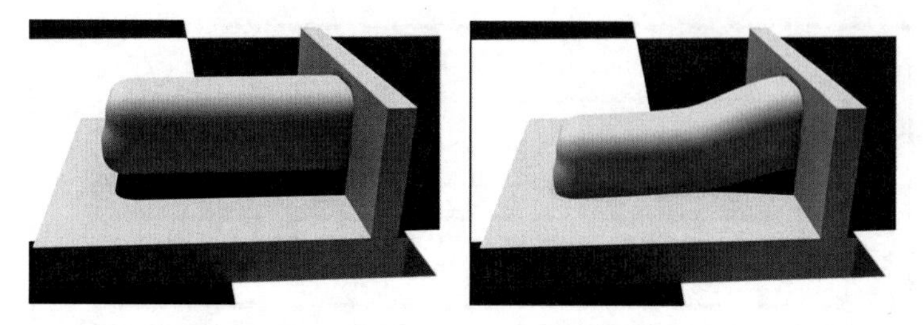

Fig 10.2 Solid rod elasticity test: $t = 0$ s and $t = 0.5$ s

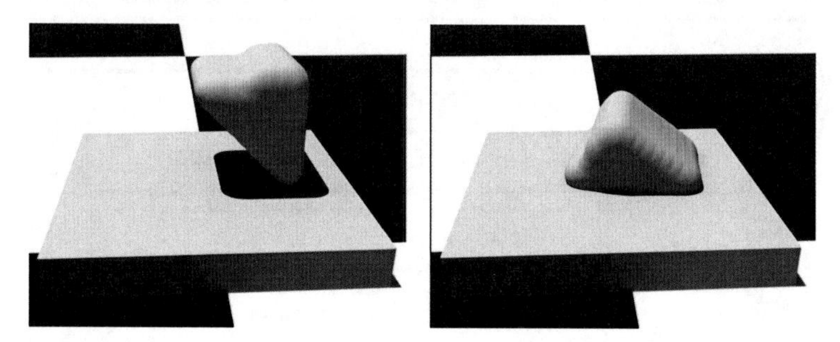

Fig 10.3 Falling wedge test: $t = 0$ s and $t = 0.5$ s

at 1/2 smoothing length interval spacing. It was found necessary to decrease the size of the time step with increasing Young's modulus to maintain stability, similar to the requirements of increasing stiffness in a fluid SPH simulation.

The first test was a model of a solid rod of SPH particles that were initialized in a box-like configuration parallel to a lower boundary (Fig. 10.2). A vertical boundary was erected, and the deformable particles alongside it were fixed in place. Gravity was applied to the model, and the particles reacted in a realistic manner by falling to rest on the lower boundary. At low Young's modulus values, the particles continued to oscillate in a jelly-like manner for some time afterwards, whereas at high Young's modulus values, the material was much stiffer and oscillations were significantly restrained, as would be expected.

The falling wedge test was a verification of the capacity of the formulation to accurately handle rotations. When the corotational method by Becker was employed, the wedge of particles fell a short distance under gravity to a lower horizontal boundary before tipping on to its side where it came to rest (Fig. 10.3). Without the corotational method, the wedge would not tip, but would instead bounce on its lower edge while remaining upright. This proved that the corotational method was allowing for proper rotation handling instead of introducing a false strain force that prevented the wedge from tipping over.

The results of tests with simulations of particle numbers from 1,000 to 60,000 from a computational speed perspective are shown in Fig. 10.4. As would be expected, frame rate decreased significantly with increasing number of particles

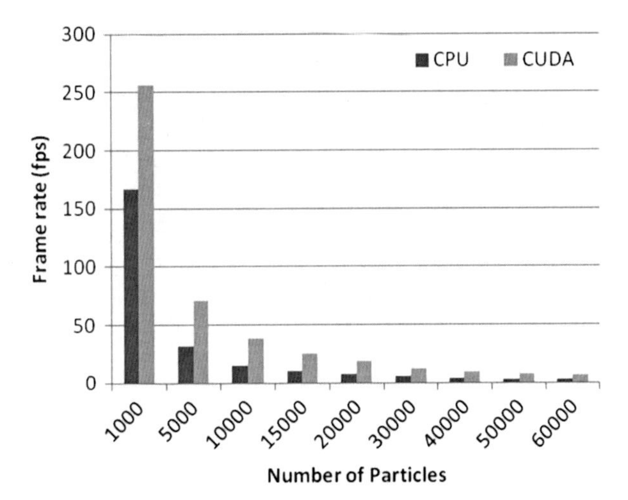

Fig 10.4 Frame rates of elastic solid tests with varying particle number

simulated. Interestingly, the computational speed gain of CUDA over pure CPU implementation was more prominent at higher particle numbers. With 1,000 particles, CUDA performed at 257 fps versus 167 fps for the CPU, 1.54 times as fast. With 60,000 particles, CUDA performed at 6.3 fps versus 2.6 fps for the CPU, 2.4 times as fast. This discrepancy is likely explained by the computational overhead required in memory transfer to the GPU in CUDA. At higher particle numbers, this overhead becomes a much smaller fraction of the total computational work performed during the simulation.

60,000 particles were determined to be the upper limit that could be simulated using the current framework. From a 2GB GPU to a 3GB GPU, there was found to be a linear increase in the number of particles that could be simulated, indicating that it was the onboard memory of the GPU that was the primary factor in this restriction. Reducing the data size of the particles and associated functions sent to the GPU was found to increase the particle limit. Should it be necessary to run a simulation using more than 60,000 particles, either the onboard memory would need to be increased, the size of the transferred data would need to be reduced, or a method of transferring from the host and operating on the data in "chunks" on the GPU would need to be investigated.

At first glance, it would appear in comparison to most fluid SPH formulations that the method presented in this chapter is not competitive in terms of computational speed, but that is not the case. The calculations required in an elastic simulation are much more complex and extensive than those required in a fluid simulation, and as such it is expected that the simulation would require greater lengths of time to perform a main program loop with similar particle numbers. It is this computational intensity rather than a fault in programming efficiency or in the method itself that prevents it from approaching the speeds of a typical SPH fluid program.

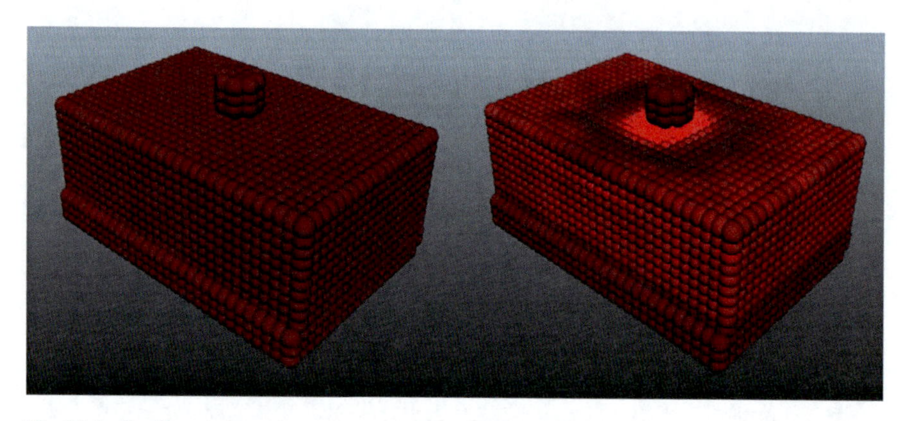

Fig 10.5 Cartilage indentation test: (**a**) initial condition (**b**) peak deformation. Brighter areas correspond to higher stress values

10.3.2 Cartilage Simulation

An indenter test was created to determine the viability of the method introduced in this chapter in its application to cartilage loading. The cartilage was simulated as a block composed of 6,800 particles resting upon a surface comprised of boundary particles (Fig. 10.5a). A cylindrical indenter similarly composed of boundary particles was incrementally lowered into the top surface of the cartilage at 0.3 m/s over a period of 0.065 s to induce a deformation of 4 %. The indenter position was maintained for 0.05 s before receding at the same rate.

For visual clarity, colourization of particles during the simulation was chosen as the average of the magnitudes of the three principle stresses. At peak deformation (Fig. 10.5b) the area of highest stress occurs directly beneath the indenter and is surrounded by a region of lower stress. It is possible that this lower stress region corresponds to negative strain, as is evident in the paper by Lu [38], but further research will need to be performed to verify this hypothesis. If the number of SPH particles used to simulate the block of cartilage were too low, stress concentrations would appear at the edges of the block rather than being distributed to the volume surrounding the indenter, so larger volumes are a necessity to obtain reasonable results.

In the case of the previous tests and those presented in the paper by Becker, high Young's modulus values in the MPa range tended to restrict any deformation in the simulated solid so that internal forces did not exceed a certain limit. By contrast, the current indentation test induces deformation on a solid with high Young's modulus values. The result is extremely high internal forces that tend to affect the stability of the test. For this reason, it was found necessary to restrict the Young's modulus value in this simulation to 0.5 MPa and to deformations of less than 5 %. The time step used was 0.00001 s, which was set to this low value because of stability requirements, since as in the case of most SPH simulations, higher stiffness requires a reduction in the time step size.

The second major obstacle found with the current method is the lack of stability for solids composed of highly dense particles. This is the case with cartilage, where 1–2 mm thickness is the norm. Because of this, it was found necessary to scale up the simulation by 100× and maintain the 0.05 m particle spacing used in the previous tests so that the cartilage block was in fact 1.5 m × 0.8 m × 0.5 m.

Conclusions and Future Work

Although the results presented in the cartilage test above are promising, more work needs to be done before SPH can be considered a viable alternative to finite element analysis of cartilage deformation. Stability presents the greatest challenge. The large scale of the simulation and the restricted maximum Young's modulus value imposed by the stability issues are unacceptable requirements for any simulation that strives for physical realism. More research is needed to arrive at a solution, but it is thought that an implicit solver or a position-based solver such as that presented by Macklin and Müller [39] may be worth considering instead of the leapfrog integration that is currently implemented. An implicit solver would presumably also allow for a larger time step.

While the simulation appeared to behave reasonably at the scale used in the indentation test, the resulting reaction forces on the applied indenter were several orders of magnitude higher than would be expected at the cartilage scale, and so were not presented here. However, considering the cartilage was modelled as purely elastic, the overall reaction force trend exhibited the expected rise, plateau and decline during the course of the simulation.

Of course, a perfectly elastic solid does not approach a realistic representation of cartilage behaviour under applied loads, so it will be necessary to include more complex constitutive models that attempt to reproduce physically accurate cartilage deformation. When this is accomplished, the next step would be to verify results obtained from the simulation against experimental testing or a similar finite element analysis, of which many such simulations exist. It is feasible that once the limitations in the current method are resolved, it could be applied towards the ultimate goal of a real-time, preoperative simulation of patient-specific acetabulum models derived from MRI data of FAI cases.

References

1. Martin, D.E., Tashman, S.: The biomechanics of femoroacetabular impingement. Oper. Tech. Orthop. **20**, 248–254 (2010)
2. Tannast, M., Goricki, D., Beck, M., Murphy, S.B., Siebenrock, K.A.: Hip damage occurs at the zone of femoroacetabular impingement. Clin. Orthop. Relat. Res. **466**, 273–280 (2008)
3. Krekel, P.R., Vochteloo, A.J.H., Bloem, R.M., Nelissen, R.G.: Femoroacetabular impingement and its implications on range of motion: a case report. J. Med. Case Rep. **5**, 143 (2011)

4. Asheesh, B., et al.: Surgical treatment of femoroacetabular impingement improves hip kinematics: a computer-assisted model. Am. J. Sports Med. **39**, 43S–49S (2011)
5. Terzopoulos, D., Platt, J., Barr, A., Fleischer, K.: Elastically deformable models. Com Graph **21**, 205–214 (1987)
6. Molino, N., Bridson, R., Teran, J., Fedkiw, R.: A crystalline, red green strategy for meshing highly deformable objects with tetrahedra. In: Proceedings of 12th IMR 103-114 (2003)
7. Maciel, A., Boulic, R., Thalmann, D.: Deformable tissue parameterized by properties of real biological tissue. Surg. Sim. Soft Tissue Model **2673**, 74–87 (2003)
8. Lloyd, B.A., Szekely, G., Harders, M.: Identification of spring parameters for deformable object simulation. IEEE Trans. Vis. Comp. Graph. **13**, 1081–1094 (2007)
9. James, D.L., Pai, D.K.: ArtDefo: accurate real time deformable objects. SIGGRAPH **1999**, 65–72 (1999)
10. MeieMollemans, W., Schutyser, F., Najmi, N., Maes, F., Suetens, P.: Predicting soft tissue deformations for a maxillofacial surgery planning system: from computational strategies to a complete clinical validation. Med. Image Anal. **11**, 282–301 (2007)
11. Cotin, S., Delingette, H., Ayache, N.: A hybrid elastic model allowing real-time cutting, deformations and force-feedback for surgery training and simulation. In: Proceedings on Computer Animation, pp, 70–81 (2000)
12. Niroomandi, S., Alfaro, I., Cueto, E., Chinesta, F.: Accounting for large deformations in real-time simulations of soft tissues based on reduced-order models. Comput. Methods Programs Biomed. **105**, 1–12 (2012)
13. Bro-Nielsen, M.: Finite element modeling in surgery simulation. Proc. IEEE **86**, 490–503 (1998)
14. Cotin, S., Delingette, H., Ayache, N.: Real-time elastic deformations of soft tissues for surgery simulation. IEEE Trans. Vis. Comput. Graph **5**, 62–73 (1999)
15. Meier, U., López, O., Monserrat, C., Juan, M.C., Alcañiz, M.: Real-time deformable models for surgery simulation: a survey. Comput. Methods Programs Biomed. **77**, 183–197 (2005)
16. Monaghan, J.J.: Smoothed Particle Hydrodynamics. Rep. Prog. Phys. **68**, 1703–1759 (2005)
17. Hieber, S.E., Koumoutsakos, P.: A Lagrangian particle method for the simulation of linear and nonlinear elastic models of soft tissue. J. Comput. Phys. **227**, 9195–9215 (2008)
18. Müller, M., Chentanez, N.: Solid simulation with oriented particles. ACM Trans. Graph. **30** (92), 1–9 (2011)
19. Gingold, R.A., Monaghan, J.J.: Smoothed particle hydrodynamics—theory and application to non-spherical stars. Mon. Not. R. Astron. Soc. **181**, 375–389 (1977)
20. Müller, M., Charypar, D., Gross, M.: Particle-Based Fluid Simulation for Interactive Applications. In: Eurograph/SIGGRAPH Symposium on Computer Animation, pp. 154–159 (2003)
21. Bao, K., Zhang, H., Zheng, L., Wu, E.: Pressure corrected SPH for fluid animation. Comput. Animat. Virtual Worlds **20**, 311–320 (2009)
22. Lenaerts, T., Adams, B., Dutré, P.: Porous Flow in Particle-Based Fluid Simulations. ACM Trans. Graph. **49**, 1–8 (2008)
23. Cleary, P.W., Das, R.: The potential for SPH modelling of solid deformation and fracture. In: IUTAM Symposium on Theoretical, Computational and Modelling Aspects of Inelastic Media, pp. 287–296 (2008)
24. Gray, J.P., Monaghan, J.J., Swift, R.P.: SPH elastic dynamics. Comp. Methods Appl. Mech. Eng. **190**, 6641–6662 (2001)
25. Qin, J., Pang, W.M., Nguyen, B.P., Ni, D., Chui, C.K.: Particle-based Simulation of blood flow and vessel wall interactions in virtual surgery. In: SolCT, pp. 128–133 (2010)
26. Mesit, J., Guha, R.K.: Experimenting with real time simulation parameters for fluid model of soft bodies. In: Proceedings of SpringSim, pp. 1–8 (2010)
27. Hieber, S.E., Walther, J.H., Koumoutsakos, P.: Remeshed smoothed particle hydrodynamics simulation of the mechanical behavior of human organs. Technol. Health Care **12**, 305–314 (2004)

28. Solenthaler, B., Schläfli, J., Pajarola, R.: A unified particle model for fluid-solid interactions. Comput. Animat. Virtual Worlds **18**, 69–82 (2007)
29. Becker, M., Ihmsen, M., Teschner, M.: Corotated SPH for deformable solids. In: Proceedings of the 5th Eurographics Conference on Natural Phenomena, pp. 27–34 (2009)
30. Mow, V.C., Holmes, M.H., Lai, M.W.: Fluid transport and mechanical properties of articular cartilage: a review. J. Biomech. **17**, 377–394 (1984)
31. Korhonen, R.K., et al.: Fibril reinforced poroelastic model predicts specifically mechanical behavior of normal, proteoglycan depleted and collagen degraded articular cartilage. J. Biomech. **36**, 1373–1379 (2003)
32. Wilson, W., Huyghe, J.M., van Donkelaar, C.C.: Depth-dependent compressive equilibrium properties of articular cartilage explained by its composition. Biomech. Model. Mechanobiol. **6**, 43–53 (2007)
33. Schmedding, R., Teschner, M.: Inversion handling for stable deformable modeling. Vis. Comp. **24**, 625–633 (2008)
34. Jin, H., Lewis, J.L.: Determination of Poisson's ratio of articular cartilage by indentation using different-sized indenters. J. Biomech. Eng. **126**, 138–145 (2004)
35. Müller, M. et al.: Point based animation of elastic, plastic and melting objects. In: Proceedings of SIGGRAPH Symposium on Computer Animation, pp. 141–151 (2004)
36. Liu, M.B., Liu, G.R.: Smoothed particle hydrodynamics (SPH): an overview and recent developments. Arch. Comput. Methods Eng. **17**, 25–76 (2010)
37. Desbrun, M., Gascuel, M.P.: Smoothed particles: A new paradigm for animating highly deformable bodies. In: Proceedings of EG Workshop on Animation and Simulation, pp. 61–76 (1996)
38. Lu, X.L., Wan, L.Q., Guo, X.E., Mow, V.C.: A linearized formulation of triphasic mixture theory for articular cartilage, and its application to indentation analysis. J. Biomech. **43**, 673–679 (2010)
39. Macklin, M., Müller, M.: Position based fluids. ACM Trans. Graph. **32**(104), 1–5 (2013)

Chapter 11
A GPU-Based Real-Time Algorithm for Virtual Viewpoint Rendering from Multi-video

Kyrylo Shegeda and Pierre Boulanger

Abstract In this chapter, we propose a novel GPU-based algorithm capable of generating free viewpoints from a network of fixed HD video cameras. This free viewpoint TV system consists of two main subsystems: a real-time depth estimation subsystem, which extracts a disparity map from a network of cameras, and a synthetic viewpoint generation subsystem that uses the disparity map to interpolate new views between the cameras. In this system, we use a space-sweep algorithm to estimate depth information, which is amiable to parallel implementation. The viewpoint generation subsystem generates new synthetic images from 3D vertices and renders them from an arbitrary viewpoint specified by the user. Both steps are computationally extensive, but the computations can be easily divided from each other and thus can be efficiently implemented in parallel using CUDA. The framework is tested using publicly available image sequences published by Microsoft. Experimental results are presented.

Keywords Real-time free viewpoint television • GPU-accelerated algorithms • CUDA

11.1 Introduction

Traditional videos are passive and two-dimensional in nature. Viewers can only observe video images from only one camera viewpoint. Recent technologies in video cameras, computer vision, and graphics have recently been used to create a new type of video delivery system called free viewpoint TV (FTV) [1]. Using FTV users can choose their viewpoints interactively from a network of cameras located at a remote site. By allowing people to choose arbitrary viewpoints, one can create a better immersive experience. To build such a system, there is a need for algorithms that can render arbitrary viewpoints from a discrete number of cameras in real time, which is not an easy task.

K. Shegeda (✉) • P. Boulanger
Computing Science Department, University of Alberta, Edmonton, AB, Canada
e-mail: shegeda@ualberta.ca; pierrb@ualberta.ca

Y. Cai, S. See (eds.), *GPU Computing and Applications*,
DOI 10.1007/978-981-287-134-3_11

Numerous systems geared towards solving this challenge can be found in the scientific literature. Roughly, most systems can be classified into two main categories: model-based rendering (MBR) and image-based rendering (IBR) techniques.

The first category includes methods that estimate the geometry of the scene by solving the correspondence problem and then use this information to generate new views using simple computer graphics techniques. Examples of such systems include shape-from-silhouette techniques where the scene is represented by a visual hull [2, 3], global multi-view stereo reconstruction techniques which use both photo-consistency and some additional information such as silhouette constraints or shape priors [4], surface-growing approaches that perform reconstruction from a set of reliable seed points and surface-growing algorithm that reconstructs a 3D map from the seed points [5], and view-dependent multi-view stereo reconstruction which obtains a separate reconstruction of a scene for each of the cameras and then merge them together [6]. In [7], the authors use the segmentation to extract the foreground and subsequently apply shape-from-silhouette methods to obtain a voxelized model that is then rendered and textured. The important aspect of this chapter is the fact that by utilizing the GLSL (OpenGL Shader Language), authors were able to speed up the rendering by a factor of 100. These types of methods have two main disadvantages: the computational cost of scene reconstruction is huge, and the virtual images produced are usually not photo-realistic.

The second category is based on techniques where a large number of images are used to keep light ray information. Three closely related methods were presented at approximately the same time: lumigraph [8], light field rendering [9], and ray space method [10]. These methods are based on describing how light rays travel through space. A good example of the quality of images generated by IBR approach is presented by Mori in [11]. In Mori's paper, the images are generated through warping precomputed depth maps for each camera and then post-processing (smoothing, boundary matting, and painting) the results to create an image without artifacts. Numerous versions of these methods were proposed, and some real-time implementation was developed, particularly, in [12] where new images are rendered in real time, but the algorithms assume that prior off-line depth-map estimation is performed beforehand. In [13] a plane sweeping algorithm together with shader programming techniques is used to create an algorithm that can process in real-time images of a size 320×240. The disadvantage of this method is that it is not able to handle occlusions properly. In [14], a new algorithm is proposed to generate new views using ray interpolation in parallel using 16 "client" processors under the command of a server machine. The system is able to generate new views at 16 fps rate for images of 640×480 pixels with a 4 cm baseline. In [15], the authors propose a method for free viewpoint generation in real time using the plane-sweep-based approach, where the scoring is based on a color variance between the cameras for each of the fragments of depth planes. The speed is achieved by utilizing the capabilities of GLSL. In [16], an algorithm that is based on a precomputed 3D proxy of a scene together with camera images is used. Shaders allow the method to generate realistic new views in real time for $1{,}024 \times 768$ images.

One can see that most of the previous papers were focused either on videos of small size or on some kind of precomputed off-line information, such as depth maps or scene geometry. Such an approach makes it impossible to use it when covering events online.

In this chapter, we address the problem of arbitrary view generation in real time. The goal is to develop a system that can generate arbitrary views from a network of cameras with known extrinsic and intrinsic parameters. Since the process of view generation is always computationally expensive, but in essence parallel, we decided to utilize general-purpose graphic processing units (GPUs) to achieve real-time performance. In the next sections, we will describe the proposed algorithms and the peculiarities of its implementation on a GPU. We will then present experimental results using standard image sequences from Microsoft and then conclude on the pros and cons of these algorithms.

11.2 Common Plane Sweeping Algorithm

The proposed algorithm is based on a relatively old idea presented by Collins in [17] called the *common plane sweeping algorithm*. Originally, the method was proposed to work as a way to match features that were obtained from images from multiple viewpoints. The general idea is to discretize the space in front of the camera using planes parallel to the plane of the camera and then project each of the features on all of the depth planes calculating the number of features from different cameras that happen to fall within a region of a specific size. Based on that processing, a decision is made whether those features correspond to each other.

We adapted this algorithm to work in real time for estimating depth information from a set of synchronized HD cameras. Let's assume that we have synchronized color images $\mathbf{I}_1, \mathbf{I}_2, \ldots, \mathbf{I}_N$ obtained from cameras 1,..., N with their corresponding intrinsic calibration projective matrices $\mathbf{A}_1, \mathbf{A}_2, \ldots, \mathbf{A}_N$ and extrinsic parameters defined by the rotation matrices $\mathbf{R}_1, \mathbf{R}_2, \ldots, \mathbf{R}_N$ and translation matrices $\mathbf{T}_1, \mathbf{T}_2, \ldots, \mathbf{T}_N$. The algorithm consists of two independent steps:

1. Estimation of depth maps for the cameras
2. Rendering a new viewpoint for a virtual camera $\mathbf{I}_V$ with given intrinsic and extrinsic matrices $\mathbf{A}_V$, $\mathbf{R}_V$, and $\mathbf{T}_V$

11.2.1 Depth-Map Estimation Algorithm

Let us assume that one wants to estimate the depth map for camera *j*. If one takes only the image of neighboring camera $\mathbf{I}_{j+1}$ and try to estimate *j*'s depth map $\mathbf{D}_j$, there will be conditions where some of the points will be occluded. To resolve this problem, we propose to look at two neighboring cameras $\mathbf{I}_{j-1}$ and $\mathbf{I}_{j+1}$ to estimate the

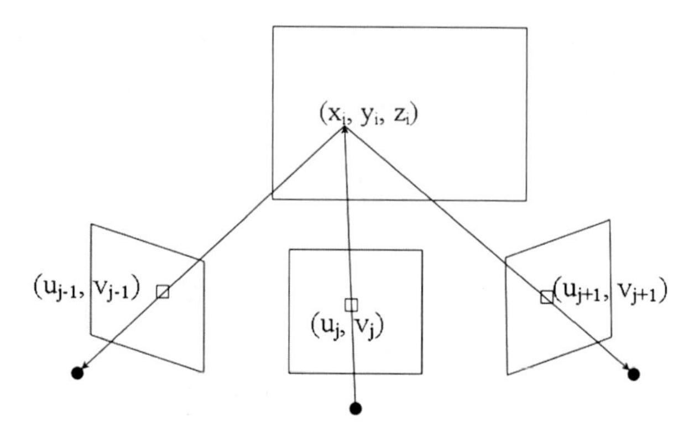

Fig. 11.1 Ray re-projection

depth map $\mathbf{D}_j$ for $\mathbf{I}_j$. Of course, it is possible that some of the points in $\mathbf{I}_j$ will be occluded in both $\mathbf{I}_{j\text{-}1}$ and $\mathbf{I}_{j+1}$, but one can assume that in most cases this will not happen if the baseline is small enough.

Let us sweep a single plane through space along the Z axis which is perpendicular to the camera sensor plane. As a result the plane always has equation $Z = z_i$, $i = 1 \ldots k$ where k is a specified number of depth levels that one wants to sweep through. Then the algorithm to compute the depth map from the two cameras is the following:

1. Cast rays from the center of the jth camera through every pixel of image $\mathbf{I}_j$, and record the intersection with the plane z_j, i.e., the ray that goes through pixel (u_j, v_j) intersects the plane z_j in the point with coordinates (x_i, y_i).
2. Re-project the point (x_i, y_i, z_i) back to the image planes of the neighboring cameras $\mathbf{I}_{j\text{-}1}$ and $\mathbf{I}_{j+1}$, with coordinates $(u^{i}_{j\text{-}1}, v^{i}_{j\text{-}1})$ as being the coordinates in an image plane of camera j-1 and $(u^{i}_{j+1}, v^{i}_{j+1})$ of camera $j + 1$ (see Fig. 11.1).
3. Quantify the similarity between pixels $\mathbf{I}_j(u_j, v_j)$ and $\mathbf{I}_{j\text{-}1}(u_{j\text{-}1}, v_{j\text{-}1})$ and $\mathbf{I}_j(u_j, v_j)$ and $\mathbf{I}_{j+1}(u_{j+1}, v_{j+1})$ using some similarity function $F(\mathbf{I}_t(u_t, v_t), \mathbf{I}_s(u_s, v_s))$. Add the two values obtained from the similarity function. This value is defined as the in-between camera consistency for the depth level z_i.
4. Repeat steps 1–3 for all the depth levels z_i where $i = 1, \ldots, k$.
5. Set the depth value of the pixel (u_j, v_j), $\mathbf{D}_j(u_j, v_j)$, to be the z_i corresponding the optimal in-between camera consistency.

The result of this algorithm is $\mathbf{D}_j$—the matrix of depth values for each of the pixels of jth camera. Note that the first and second steps of the algorithm can be done because the intrinsic and extrinsic parameters of all the cameras are known. Using the calibration information, one can easily calculate the camera projection matrix for any camera j using $\mathbf{C}_j = \mathbf{A}_j[\mathbf{R}_j; \mathbf{T}_j]$ which links a point in the 3D world coordinates to its counterpart in the sensor 2D coordinates. One can project a point (u_i, v_i) from virtual camera i to its 3D coordinates value with $Z = z_i$ where the x_i and y_i coordinates are determined by a simple line to plane intersection algorithm. Then one can re-project the 3D point back to camera j with 2D coordinates (u_j, v_j) using

$$\begin{bmatrix} u_j \\ v_j \\ 1 \end{bmatrix} = C_j \bullet \begin{bmatrix} x_i \\ y_i \\ z_i \\ 1 \end{bmatrix}. \tag{11.1}$$

The depth values obtained, using this algorithm, are actually in the coordinate system of the scene and not in the coordinate system of the camera j. The motivation for this unusual transformation will be explained in the view synthesis part of the chapter.

11.2.2 Pixel Similarity Function

One of the most simple and popular functions for quantifying pixel similarity is the sum of absolute differences (SADs). Smaller SAD values mean that pixels in neighboring cameras are similar. Because of its computational simplicity, we decided to use it for the GPU implementation except that in our implementation, we compare rectangular blocks of pixels that are located around the target pixel (block matching) in order to reduce noise.

The standard block-matching SAD algorithm is quite simple: for each pixel that has to be compared, the block is placed exactly in a rectangle, and the sum of the absolute difference of the corresponding pixels in blocks is calculated. This approach works great for rectified images. However, because cameras in our system are not rectified and they might be far from being parallel to each other, instead of re-projecting pixel in the center of SAD block and putting a rectangular block around them, we propose to use a modification of the block-matching algorithm called projective block matching.

11.2.3 Projective Block Matching

In projective block matching, we use a rectangular $(2M+1)*(2L+1)$ pixel grid around the pixel (u_j,v_j) to compute SAD. Most pixels in this grid should have the depth values close to each other. That is why in this scheme, the whole grid is projected to a given depth z_i and then re-projected back to neighboring cameras j-1 and $j+1$ (Fig. 11.2). Then the similarity function between neighboring $\mathbf{I}_{j-1}$ and reference $\mathbf{I}_j$ images is defined as $F(\mathbf{I}_{j-1}(\alpha_j,\beta_j),\mathbf{I}_j(u_j,v_j))$ where (α_j,β_j) is the re-projected 2D coordinate of the grid centered at (u_j,v_j) and can be found using

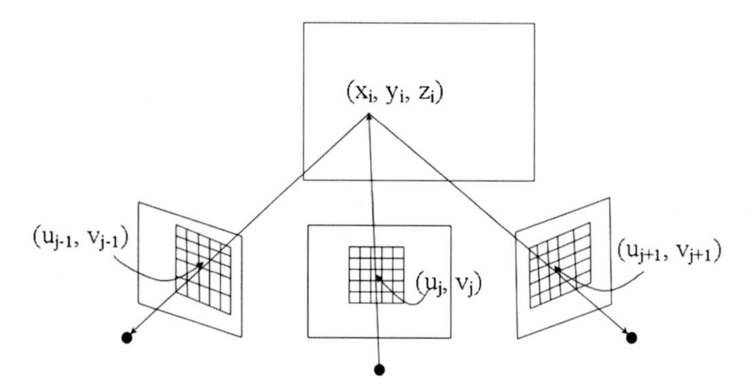

Fig. 11.2 Projective block matching

$$SAD = \sum_{k=-M}^{M} \sum_{p=-L}^{L} \left| I_{j-1}\left(\alpha_{j+k}, \beta_{j+p}\right) - I_j\left(u_{j+k}, v_{j+p}\right) \right|. \tag{11.2}$$

The advantage of projective block matching over conventional orthographic projection technique is due to the fact that shapes of objects in different cameras are different depending on the angle that the camera makes with the scene. The method has two main disadvantages:

1. The closer the angle between the cameras is equal to 90 degrees, the worse the results get. This can be explained by the fact that in such cases the rectangular block of pixels will be strongly distorted by perspective transform.
2. For large SAD blocks, the pixels closer to the edge of the block have a higher probability of having different depth than the central pixel.

11.2.4 Virtual Viewpoint Rendering

The next step is to generate an arbitrary viewpoint specified by the user. We propose to generate the new virtual viewpoint $\mathbf{I}_m$ from the two neighboring real cameras $\mathbf{I}_j$ and $\mathbf{I}_{j-1}$ in three steps:

1. Project the pixels from the neighboring cameras into 3D spaces using the depth-map information.
2. Re-project the 3D points onto the image plane of the virtual camera using one of the available 3D rendering APIs, e.g., DirectX/OpenGL.
3. Post-process the rendered image by filling void pixels created by occlusions, sampling, and low-textured regions using an interpolation technique.

The first step consists of creating 3D vertices that can be used by a rendering API. The number of vertices equals to the number of pixels in each of the cameras multiplied by the number of neighboring cameras used for the interpolation, which

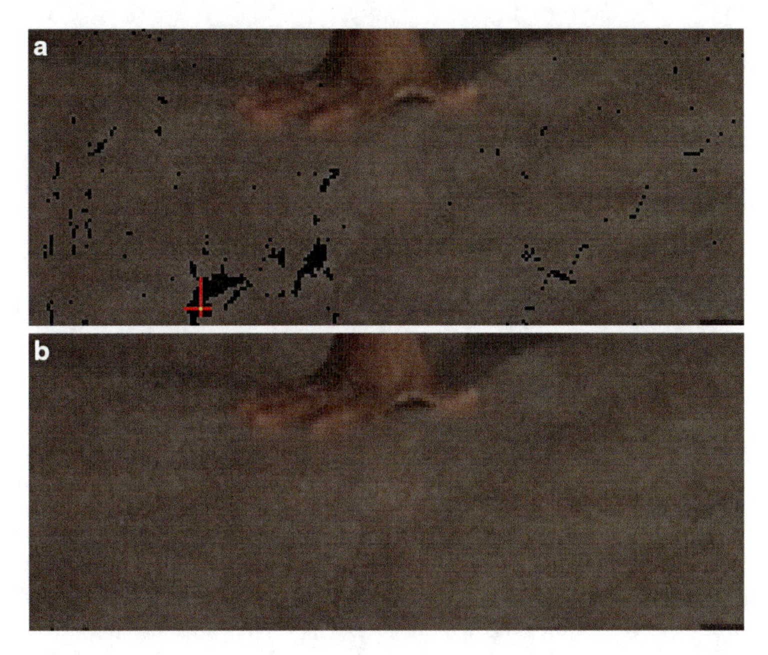

Fig. 11.3 (**a**) Generated view with void pixels (one of the pixels to be interpolated is in *yellow*, and the directions of search for pixels to be used in bilinear interpolation are in *red*) and (**b**) interpolated void pixels using the proposed algorithm

in our case is equal to two. In this scheme, there is 3D vertex associated with each pixel of the real camera. The coordinates of the vertices are calculated based on the computed depth map, with color being the color of a projected pixel. One could also project the pixels from all available cameras, which would allow us to represent the scene more accurately as it will reduce occlusions and increase image quality. However, the amount of computations would be prohibitive for a real-time implementation with current GPU. For this reason, we decided to use only two neighboring cameras.

Note that since one needs to recreate a scene from 3D vertices in order to be compatible with graphic cards, all depth maps must be in the same global coordinate system. The rendering API for the viewpoint rendering is using OpenGL. At this step, the data obtained from the previous step is transformed using OpenGL with the virtual rotation, translation, and projection matrices. As a result, we get a rendered picture of the 3D world, as if it was viewed from a camera located at a virtual viewpoint. Because of sampling and occlusions problems, the image will have unrendered pixels that will deteriorate the image quality. In order to solve this problem, a post-processing step is necessary where all unrendered pixels are filled using a linear interpolation technique.

The final step is done by interpolating each void pixel from the nearest valid pixels that have color information in four directions: left, right, top, and bottom (see Fig. 11.3). So if a pixel that needs to be interpolated is $\mathbf{I}_m(u_h,v_h)$ and the four nearest pixels are $\mathbf{I}_m(u_l,v_h)$, $\mathbf{I}_m(u_r,v_h)$, $\mathbf{I}_m(u_h,v_t)$, and $\mathbf{I}_m(u_h,v_b)$ in each direction, the void pixel will be filled using

$$\begin{aligned} I_m(u_h, v_h) = {} & \frac{v_b - v_t}{(u_r - u_t) + (v_b - v_t)} \times \\ & \left(\frac{(u_r - u_h) I_m(u_l, v_h)}{u_r - u_l} + \frac{(u_h - u_l) I_m(u_r, v_h)}{u_r - u_l} \right) + \\ & \frac{u_r - u_l}{(u_r - u_t) + (v_b - v_t)} \times \\ & \left(\frac{(v_b - v_h) I_m(u_h, v_t)}{v_b - v_t} + \frac{(v_h - v_t) I_m(u_h, v_b)}{v_b - v_t} \right). \end{aligned} \tag{11.3}$$

11.2.5 GPU-Accelerated Algorithm and Its Implementation

Unlike CPU that works at really high frequency to achieve high speed, GPUs have a parallel architecture composed of numerous streaming multiprocessors that work at a lower frequency allowing for lower power consumption and faster speed if the algorithm is parallelizable. In February 2007, NVIDIA introduced CUDA which made it possible to write general-purpose algorithms that run on GPUs in an efficient manner. From the programmers' viewpoint, CUDA looks like an extension of the C language. The code that is written in CUDA is running in threads that are grouped together in blocks of code. Threads in the same block share a fast memory region called shared memory. For the full CUDA specification, please refer to [18].

The algorithms presented in the chapter can be easily made parallel using CUDA. Thanks to CUDA graphics interoperability, one can do both depth estimation and viewpoint rendering without the need to process data on the CPU which reduces significantly the processing speed as CPU to GPU transfers are known to be very slow. Since getting a maximum performance on a GPU depends on many small things, we will go into the smallest details of the implementation.

Stam in [19] presented an efficient way of computing disparity maps of rectified images using CUDA. We modified his algorithm to deal with our context where rectification is not possible. The design specifications for our algorithm implementation are:

1. Avoid obscenely redundant computations—many computations performed for one pixel can also be used by neighbors.
2. Keep global memory coalesced.
3. Minimize global memory reads/writes.
4. Exploit texture hardware—textures in CUDA have the interpolation implemented on a hardware level.
5. Create enough threads and thread blocks to keep the streaming processors busy.

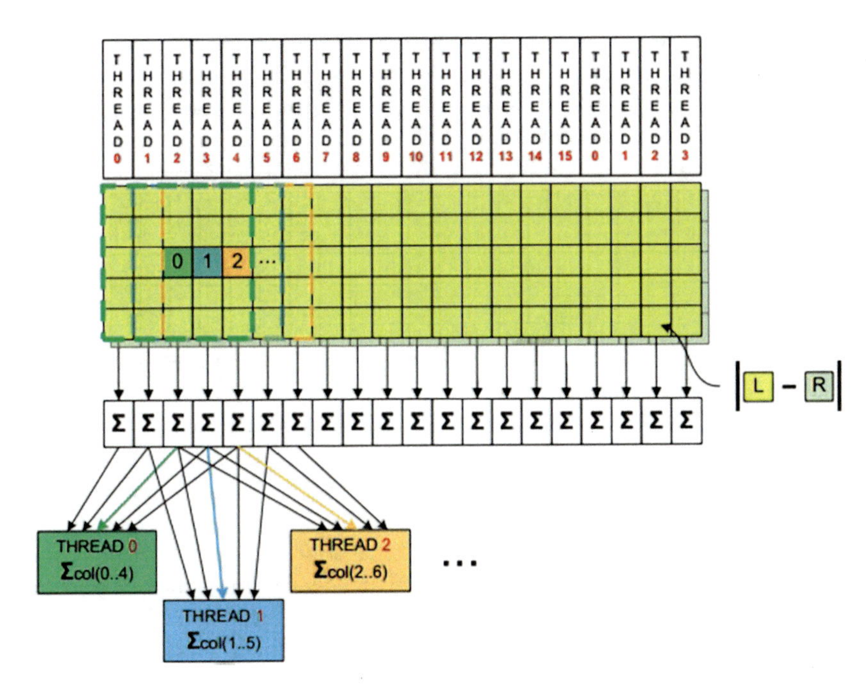

Fig. 11.4 Thread cooperation when calculating SAD by using shared memory [19]

During the step of SAD calculations, it is hard to keep the global memory access coalesced since after the re-projection of a 3D point to left and right cameras the access pattern to the memory storing information about the values of the pixels of the respective cameras is completely random. Hopefully texturing from CUDA arrays helps us to circumvent the requirement of coalesced memory access, as well as providing boundary clamping and hardware interpolation. This is really important considering the fact that after the re-projection step, the coordinates of the pixels in left/right image aren't going to be integers. That is why the images obtained from the cameras are stored in CUDA texture memory.

In the first part of the algorithm, each thread is dedicated to process part of a column of pixels of the camera for which the depth map is being estimated. Unlike most of the algorithms for depth estimation on a GPU, each thread computes SAD of column of pixels with a block size of $(2M+1)$ instead of individual pixel. It accumulates the SAD function between pixels in the reference image and its corresponding re-projected pixels in the neighboring images. The SAD functions for each column are stored in shared memory arrays. To illustrate thread cooperation for a 5×5 kernel, a 16-thread block size is shown in Fig. 11.4. Thus, we have one shared memory array that contains the sum of SAD functions between the reference and the left cameras and the reference and the right cameras. After the calculations for a column are done, each thread sums up neighboring column values within the block width $(2L+1)$. Once the first row of pixels has been processed by a thread block, a rolling window is used to speed up the calculations. Instead of repeating the SAD calculation and summation for each of the columns, the SAD function for the pixels in the first row is subtracted from the corresponding one in the accumulation arrays, and then the SAD value of the pixels in a new row is added. This reduces the amount of calculations that has to be done and more importantly reduces the need for reads and writes into memory. The whole process is repeated for every possible z_i. The trade-off between parallelization and sequential update of the column SAD functions depends on the number of stream processors that GPU has and can only be determined empirically.

Because the SAD blocks on the far left side and far right side of the block of threads require extra $2(L+1)/2$ SAD calculations, some threads will have to make extra calculations, while others will stay idle. In Fig. 11.4, these are threads 0–3 on the far right side.

Let's analyze the number of operations that this approach saves for us. Assuming that the height of the block is $2M+1$, to calculate the whole column of pixels each thread needs to make $2M+1$ reads from texture memory. It is actually $3*(2M+1)$ reads since we need to make an extra read for left and right cameras as well and the same number of writes to shared memory. At the same time, we need to re-project each pixel to global coordinate system and then back to the left camera and right camera, which takes at least $3*(2M+1)$ reads. In reality, it is going to be a bit bigger because re-projection consists of several arithmetic operations. Using the algorithm described previously, each thread for each new row of pixels will only need to make three reads from texture memory - to subtract the SAD of the top row of pixels and to add the new one and write to the shared memory, the number of operations that will be done at this step will be three as well. That gives a significant reduction in number of operations performed by GPU especially for block matching with large SAD blocks.

One thing regarding the use of shared memory has to be mentioned. Since the order of execution of the threads in a block of code is random, some of the threads may be far behind in number of performed instructions than others. That means that for the threads in the same block, before performing SAD accumulation step, we have to synchronize all the threads in order to get correct results. Hopefully, CUDA provides an easy barrier synchronization mechanism using **__syncthreads( )** command. This allows waiting until the shared data is filled for the current row of blocks of pixels.

One can further improve performance by using a trick that shows how even smallest changes to the code on a GPU can lead to a completely different behavior. The local variables defined inside the GPU code are stored in registers, which is the fastest memory type on GPUs. However, developers are not able to allocate the arrays of arbitrary size in CUDA registers; they have to use global memory instead. Interestingly if the size of the array is known in advance, then the allocation can be performed in a register memory. We decided to utilize this fact and store the height of the blocks of pixels that has to be processed by one thread using a preprocessor definition (**#define** directive in C/C++). This allows us to allocate an array responsible for handling the SAD of the whole column in a register memory, as the size of the array is known during the compilation time. As a result, we have a higher memory access speed and a reduced the number of operations. We do not need to recalculate the SAD for the top row of pixels now, as one can look it up in a corresponding row of an array. The test results have shown that this approach allows to increase the speed of the depth-map calculation twofold by adding the use of only one extra register per thread. This also shows how important the details of the implementation can be when it comes to general-purpose GPU programming.

Another interesting capability of the CUDA compiler is that it is able to unroll the loops of a known size. That is why, we decided to store the number of depth levels that are sweeping through in a preprocessor definition. That might seem to be impractical, since if one wants to use a different number of depth levels, then one would have to recompile the software, but if you consider the application domain of real-time FTV (concerts, sports events), then one will probably notice that these settings are known ahead of time, and they should not change during the event. We believe that this approach is a reasonable trade-off between having flexibility in the software and having a higher speed of computations.

Cameras' projection matrices are stored in constant memory, which provides cache for faster access. In order to further increase efficiency, we used CUDA intrinsic functions for multiplication and division. All calculations are performed in single precision.

For the image viewpoint interpolation, the processing is performed both using CUDA and OpenGL. The projection and post-processing steps are done using CUDA, whereas re-projection to a virtual plane is done using OpenGL. CUDA allows to use graphics interoperability which means that some of the OpenGL resources can be mapped into the address space of CUDA and vice versa. Particularly, vertex buffers and render buffer objects can be accessed from CUDA.

Using this scheme, the projection step is done by starting a single thread for each pixel that has to be projected into 3D with the following information: 3D coordinates (can easily be derived from the depth map), color of the pixel, and alpha value. The values are stored on the OpenGL 3D vertex buffer packed as three floats for coordinates and one integer representing color and alpha in base-256 system (since each of the color components is represented by the number between 0 and 255). As mentioned previously, each pixel has a separate 3D vertex. This allows us to avoid collisions when we are writing information to the depth buffer, making the projection step fast and simple. To generate the view in the virtual plane, we simply feed the view and projection matrices to OpenGL (see the next section for more details on how to do it) and render the virtual image into a frame buffer. The CUDA

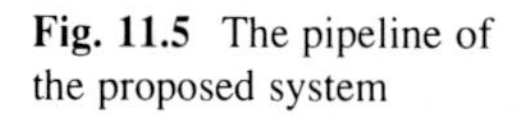

Fig. 11.5 The pipeline of the proposed system

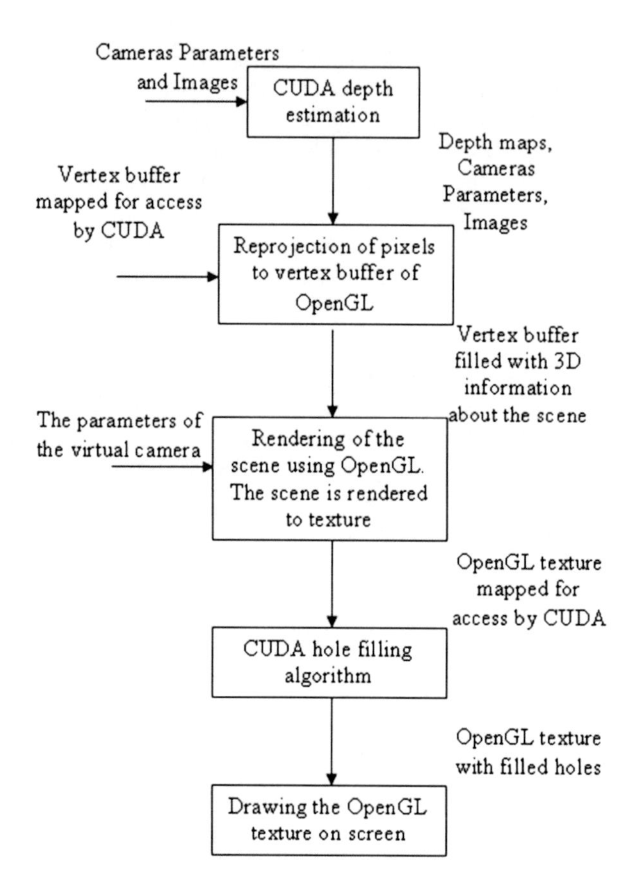

thread accesses this frame buffer once again during the post-processing step. If the pixel's alpha channel is not zero, then it does not need to be interpolated, and the thread is killed; otherwise, we perform the steps described in the previous section to write a new value into the frame buffer. The sequential search for the nonzero values is not the fastest operation, but since most hole sizes are small, it can be performed faster. Following hole filling, the resulting image is displayed to the user.

Note that because calculations of the depth map and view generation are independent across pixels, the approach is easily scalable by adding more GPUs. When scaling, each CPU thread is responsible for working with one GPU. Unfortunately, it does not scale linearly because often GPUs reside on the same bus and data transfer in parallel is currently impossible.

The overview of the whole pipeline is shown in Fig. 11.5. Note that all the computations are performed completely without the CPU being involved.

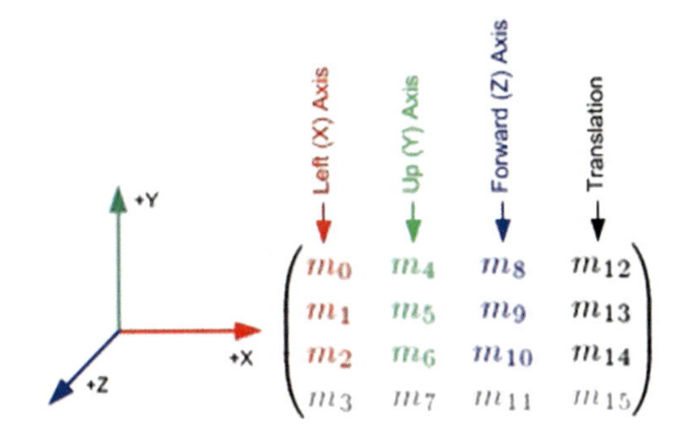

Fig. 11.6 The model–view matrix structure

11.2.6 Constructing OpenGL Model-View and Projection Matrices

To transform the 3D coordinates of objects to 2D coordinates on the screen, OpenGL uses two types of matrices: model-view matrix and projection matrix. In all the resources of OpenGL that we encountered, the construction of these matrices is performed through the OpenGL functions: **gluLookAt**, **glTranslatef**, **glRotatef**, and **glScalef** for the model-view matrix and **gluPespective** for the projection matrix. Since OpenGL does not provide a way to set up all the needed parameters of the matrices (e.g., coordinates of the principal point or the skew coefficient between the axis of the viewport), we decided to give more details on how to build these matrices by programming the intrinsic and extrinsic parameters of the virtual camera.

The model–view matrix, which is the combination of model and view matrices, transforms the coordinates from object space to eye space. Internally, it is represented as a 4×4 matrix (Fig. 11.6). Keeping in mind that in OpenGL the camera always faces Z eye coordinates, this matrix (**MV**) can easily be built from rotation (**R**) and translation (**T**) matrices of a virtual camera using

$$MV = \begin{pmatrix} R[0][0] & R[0][1] & -R[0][2] & T[0] \\ R[1][0] & R[1][1] & -R[1][2] & T[1] \\ -R[2][0] & -R[2][1] & R[2][2] & -T[2] \\ 0 & 0 & 0 & 1 \end{pmatrix}. \tag{11.4}$$

Projection matrix defines the viewing frustum as well as the projection of 3D scene onto the screen. It is also a 4x4 matrix. One can construct this matrix from the near and far Z clipping planes, the intrinsic parameters of the virtual camera (**A**), as well as the height (**h**) and the width (**w**) of the viewport using

$$\mathbf{P} = \begin{pmatrix} \frac{2 \cdot A[0][0]}{w} & \frac{2 \cdot A[0][1]}{w} & -\left(\frac{2 \cdot A[0][2]}{w} - 1\right) & 0 \\ 0 & \frac{2 \cdot A[1][1]}{h} & -\left(\frac{2 \cdot A[1][2]}{h} - 1\right) & 0 \\ 0 & 0 & \frac{\text{far} + \text{near}}{\text{near-far}} & \frac{2 \cdot \text{far} \times \text{near}}{\text{near-far}} \\ 0 & 0 & -1 & 0 \end{pmatrix}. \quad (11.5)$$

These matrices can be passed successfully to OpenGL using **glLoadMatrixd** function. The only thing you have to remember is that in OpenGL all the matrices are in column-major order, so the matrices have to be transposed.

11.3 Experimental Results

To test our framework, we decided to use the sequence "Breakdancing" published by Zitnick et al. [12] from Microsoft research. The sequence is 100 frame long with a camera resolution of 1,024 × 768 pixels to simulate HD format. In the original paper, the depth map was pre-calculated off-line: for each camera, the depth map spans 256 depth levels. In order to speed up the calculation process, we decided to sweep through every eighth depth plane. Thus, we sacrifice some of the quality to get the algorithm working in real time. OpenCV used the same approach in their calculations on a GPU of disparity for rectified views.

To get the full cycle of image rendering in real time, we used two Quadro FX 5800 GPUs. Each of them was dedicated to compute depth maps allowing us to achieve real-time speed for two cameras. Following this calculation, all depth information is flushed to one of the cards where the viewpoint interpolation steps are performed.

Quadro FX 5800 has 30 streaming multiprocessors (SM). That is why the number of SADs calculated by one thread is chosen to be 52, so that the total number of blocks of code that are spawned on a GPU is dividable by 30. The

Table 11.1 The output of the CUDA profiler on Quadro FX 5800

Grid size	[8 15 1]
Block size	[128 1 1]
Register ratio	1 (16,384/16,384) [32 registers per thread]
Shared memory ratio	0.25 (4,096/16,384) [648 bytes per block]
Active blocks per SM	4 (max active blocks per SM: 8)
Active threads per SM	512(max active threads per SM: 1,024)
Potential occupancy	0.5(16/32)
Achieved occupancy	0.5 (on 30 SMs)
Occupancy limiting factor	Registers

maximum number of active blocks of code per SM for this particular video card is equal to 8, whereas the maximum number of threads is equal to 1,024. Based on that, the number of threads in one block of code is chosen to be 128. When we ran a performance profiler, one can see that the maximum occupancy achieved was 0.5, with number of registers being the bottleneck (see Table 11.1). The problem is that the maximum number of registers across all the active threads on SM cannot be higher than 16,384. Therefore, to get a higher occupancy, we need to lower the number of registers in use (to get a 1.0 occupancy factor, the number of registers has to be no more than 16 per thread). However, a number of experiments have shown that lowering the number of registers by either limiting it during the compilation time or by physically removing some register variables from the code gives the maximum occupancy factor, but lowers the performance. This can be explained by the fact that using registers reduces the number of reads to global memory, which is slow, despite the fact that it limits the number of threads that are being executed concurrently on a multiprocessor.

The time breakdown spent to generate a virtual viewpoint is shown in Table 11.2. As mentioned previously, the processing speed does not scale linearly with additional GPUs. In our case by using two Quadro FX 5800, we were only able to achieve a 1.3 speedup compared to one GPU. This is due to the data transfer on the PCIe 2.0 (16 GBps) which is much slower than the internal bus of a GPU (~150 GBps). Newer version of the system will allow us to accelerate the transfer rate by a factor of two, as it will have a PCIe3 bus with a transfer rate of 32 GBps. In addition new GPU card like the Kepler has 3072 CUDA cores compared to the Quadro FX 5800 with 240 CUDA cores. Another thing is that the number of registers per thread in the new architecture is higher, so we expect to have a higher occupancy factor. Nevertheless, the overall frame rate we achieved was 29.7 frames per second, which is very promising.

An example of generated viewpoint is shown in Fig. 11.7c without missing pixel filling and in Fig. 11.7(d). The image was generated from two images shown in Fig. 11.7a, b for cameras 4 and 5. We also compared the reconstruction of a virtual camera 4 with the real one by generating a virtual view from cameras 3 and 5. To do that, we first calculated the weighted luminance channel L of both images. Then the peak signal-to-noise ratio (PSNR) value is computed between the real camera $\boldsymbol{L}$ and virtual camera $\hat{L}$.

Table 11.2 GPU time breakdown to generate a virtual viewpoint

Framework step	Elapsed time (ms)
Depth estimation	16
Projection to 3D	1.15
OpenGL rendering	1.10
Hole filling	2.4
Total	20.65

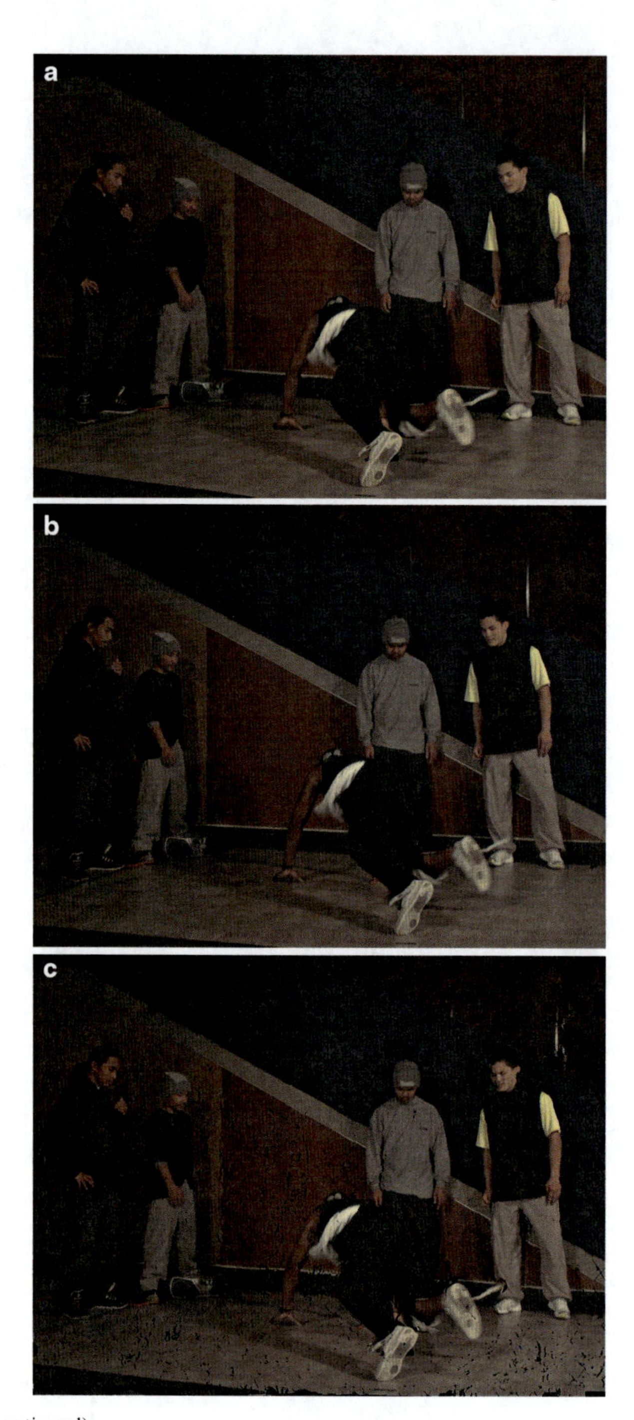

Fig. 11.7 (continued)

Fig. 11.7 (a) Camera 3 image, (b) camera 5 image, (c) virtual image rendered from cameras 3 and 5 without pixel filling, and (d) virtual image generated from cameras 3 and 5 with pixel filling

Table 11.3 PSNR average over 100 frames

Depth map	PSNR
Our algorithm	29
Zitnick et al.	33

To determine the effect of using a higher resolution depth map, we also tested our view generation algorithm with depth maps provided by Zitnick. The calculated PSNRs are presented in Table 11.3. Results show that a gain of 4 db in PSNR can be achieved by employing better depth resolution.

Conclusion

In this chapter, we proposed a novel GPU-based algorithm capable of generating arbitrary viewpoints from a network of HD video cameras in real time. The algorithms for depth estimation and view generation are independent from each other allowing them to be computed on different GPUs. The algorithm does not need view rectification and is capable of generating depth map from any virtual viewpoint if needed. The computation can also easily be scaled by adding more GPUs or more powerful GPU like the Kepler from NVIDIA. Since all computationally intensive operations are performed on a graphics processor, the CPU is free to do other tasks. The approach is directed towards utilization when covering events that need real-time computation such as football game or concert.

Future work will include the improvement of GPU code by using the features of the newly introduced Kepler architecture and CUDA 5, as well as the DMA access between the GPUs. We are also planning to test other

(continued)

similarity functions than SAD, which assumed that the cameras have exactly the same gain, that the surface is Lambertian, and that the illumination is uniform. To address this illumination invariance problem, we are planning to implement in the GPU the relative gradient functions proposed in [20]. We are also in the process of integrating into our video processing computer a new camera technology from Herodion Inc. This revolutionary technology is capable of streaming 12 pixel synchronized HD cameras on a single computer. Our main challenge will be the data transfer of this large video stream to the GPU memory using the DMA access mechanism.

References

1. Tanimoto, M.: Overview of free viewpoint television. Signal Processing: Image Communication **21**(6), 454–461 (2006), Special issue on multi-view image processing and its application in image-based rendering
2. Matusik, W., Buehler, C., Raskar, R., Gortler, S.J., McMillan, L.: Image-based visual hulls. In: Proceedings of the 27th annual conference on Computer graphics and interactive techniques, pp. 369–374. SIGGRAPH '00. ACM Press/Addison-Wesley Publishing Co, New York, NY. (2000)
3. Franco, J.S., Boyer, E.: Exact polyhedral visual hulls. In: British Machine Vision Conference (BMVC'03), vol 1, pp. 329–338. Norwich, Royaume-Uni (2003)
4. Starck, J., Hilton, A.: Surface capture for performance-based animation. IEEE Comput. Graph. Appl. **27**(3), 21–31 (2007)
5. Furukawa, Y., Ponce, J.: Accurate, dense, and robust multiview stereopsis. IEEE Trans. Pattern Anal. Mach. Intell. **32**(8), 1362–1376 (2010)
6. Starck, J., Hilton, A.: Virtual view synthesis of people from multiple view video sequences. Graph. Models **67**(6), 600–620 (2005)
7. Orman N., Kim H., Sakamoto R., Toriyama T., Kogure K., Lindeman R.: GPU-based optimization of a free-viewpoint video system. In: Proceedings of the 2008 symposium on Interactive 3D graphics and games (2008)
8. Gortler, S.J., Grzeszczuk, R., Szeliski, R., Cohen, M.F.: The lumigraph. In: Proceedings of the 23rd annual conference on Computer graphics and interactive techniques, pp. 43–54. SIGGRAPH '96. ACM, New York, NY (1996)
9. Levoy, M., Hanrahan, P.: Light field rendering. In: Proceedings of the 23rd annual conference on Computer graphics and interactive techniques, pp. 31–42. SIGGRAPH '96. ACM, New York, NY (1996)
10. Fujii, T., Kimoto, T., Tanimoto, M.: Ray-space coding for 3d visual communication. In: Picture Coding Symposium (PCS), vol. 2, pp. 447–451 (1996)
11. Mori, Y., Fukushima, N., Yendo, T., Fujii, T., Tanimoto, M.: View generation with 3d warping using depth information for ftv. Image Commun. **24**(1–2), 65–72 (2009)
12. Zitnick, C.L., Kang, S.B., Uyttendaele, M., Winder, S., Szeliski, R.: High-quality video view interpolation using a layered representation. ACM Trans. Graph. **23**(3), 600–608 (2004)
13. Yang, R., Welch, G., Bishop, G.: Real-time consensus-based scene reconstruction using commodity graphics hardware. In: Pacific Conference on Computer Graphics and Applications. **225–235** (2002)
14. Suzuki, K., Fukushima, N., Yendo, T., Tehrani, M., Fujii, T., Tanimoto, M.: Parallel processing method for realtime ftv. In: Picture Coding Symposium (PCS), pp. 330–333 (2010)

15. Nozick V., Saito H.: On-line free-viewpoint video: From single to multiple view rendering. Int. J. Autom. Comput. **5**, 257–267 (2008)
16. Starck J., Kilner J., Hilton A.: A free-viewpoint renderer. J. Graph. GPU Game Tools **14** 57–72 (2009)
17. Collins, R.: A space-sweep approach to true multi-image matching. In: Computer Vision and Pattern Recognition. IEEE Computer Society Conference on proceedings CVPR '96, pp. 358–363 (1996)
18. NVIDIA Corporation: NVIDIA CUDA C programming guide (2012) Version 5.0.
19. Stam, J.: Stereo imaging with cuda. (2008)
20. Zhou, X., Boulanger, P.: Radiometric invariant stereo matching based on relative gradients. In: 19th IEEE International Conference on Image Processing (ICIP), pp. 2989–2992 (2012)

Chapter 12
A Middleware Framework for Programmable Multi-GPU-Based Big Data Applications

Ettikan K. Karuppiah, Yong Keh Kok, and Keeratpal Singh

Abstract Current application of GPU processors for parallel computing tasks shows excellent results in terms of speedups compared to CPU processors. However, there is no existing middleware framework that enables automatic distribution of data and processing across heterogeneous computing resources for structured and unstructured Big Data applications. Thus, we propose a middleware framework for "Big Data" analytics that provides mechanisms for automatic data segmentation, distribution, execution, information retrieval across multiple cards (CPU and GPU) and machines, a modular design for easy addition of new GPU kernels at both analytic and processing layer, and information presentation. The architecture and components of the framework such as multi-card data distribution and execution, data structures for efficient memory access, algorithms for parallel GPU computation, and results for various test configurations are shown. Our results show proposed middleware framework, providing alternative and cheaper HPC solution to users. Data cleansing algorithms on GPU show a speedup of over two orders of magnitude compared to the same operation done in MySQL on a multi-core machine. Our framework is also capable of processing more than 120 million of health data within 11 s.

Keywords GPGPU • CUDA • GPU • Architecture • Big Data • High-performance computing • Middleware framework

12.1 Introduction

NVIDIA CUDA-enabled GPGPU (general purpose graphic processing unit) has made its name by being part of world super computers to enable high-performance computation. Thus, GPGPUs are widely accepted and becoming common for many high-performance computing applications. GPGPUs are used for both specific and general purpose applications either running in large-scale system or desktop PCs.

E.K. Karuppiah (✉) • Y.K. Kok (✉) • K. Singh (✉)
MIMOS Berhad, Kuala Lumpur, Malaysia
e-mail: ettikan.karuppiah@mimos.my; kk.yong@mimos.my; keeratpal.singh@mimos.my

Y. Cai, S. See (eds.), *GPU Computing and Applications*,
DOI 10.1007/978-981-287-134-3_12

The design of PC pluggable GPGPU cards provides new programmable computing paradigm as the cost-effective solution. High-performance parallel applications and algorithms can be designed and developed, utilizing both CPU and GPU processors capabilities. However, the environment including middleware, framework, applications, and supporting tools must be capable of supporting parallel computing and execution, otherwise serial performance will be the outcome.

Big Data processing certainly has become imminent for enterprises that wish to process large amount of data which mainly comes from the social network, semantic web, sensor networks, geo-based service information, patient information, and employee or transaction-based applications. These areas observe quick growth of large data which needs either timely analytics or batched processing. Thus, the challenge is to analyze and mine these big data in order to effectively exploit the information to improve efficiency and quality of service for consumers and producers alike. However, the computing capabilities of current multi-core microprocessors are unable to meet the data mining requirements to effectively mine the data on time, thus needing parallel acceleration hardware such as GPUs [1] to accelerate the data mining. Even though high-performance computing solutions are available today for the above processing usage, the cost is still relatively high for general deployment and usage. For example, Netezza-, Teradata-, and Vertica-based systems are computationally fast and cater for terabytes of data processing in milliseconds but not affordable for small and medium enterprises. On the other hand, MapReduce framework-based applications such as Apache Hadoop and Drill which are free and stable are suitable for large-scale data processing. GPU-based Big Data processing system complements the above MapReduce-based solution. In our proposed Big Data and BI (business intelligence) solution framework, we have positioned GPU in two different layers, namely, analytics and processing, as illustrated in Fig. 12.1. These positioning provides flexibility to application-specific analytics algorithm coupled with data processing algorithms. For example, edit distance algorithms (analytics component) which are written in CUDA/GPU are tightly coupled with other generic data processing (processing component) component such as sorting, searching, etc., providing high-performance solutions. Thus, we believe GPU-based solution will coexist with other MapReduce systems as a complementing solution. The implementation section will demonstrate an example of this combination.

GPUs are massively parallel multi-threaded multi-core processors that allow large amounts of data to be processed in parallel to speed up computation time. The single instruction multiple threads (SIMT) architecture of the NVIDIA GPU places it between the single instruction multiple data (SIMD) architecture for vector processing and the simultaneous multithreading (SMT) architecture for hardware multithreading in terms of flexibility and efficiency. Current benchmarking shows that GPUs can execute up to a few orders of magnitude faster than CPUs for certain types of algorithms [2] and a large set of work has been done in order to leverage on the GPU computing capabilities [3, 4].

Since GPUs are treated as a coprocessor with its own architecture, applications must be designed to reflect the two-processor nature of the system. As such, data

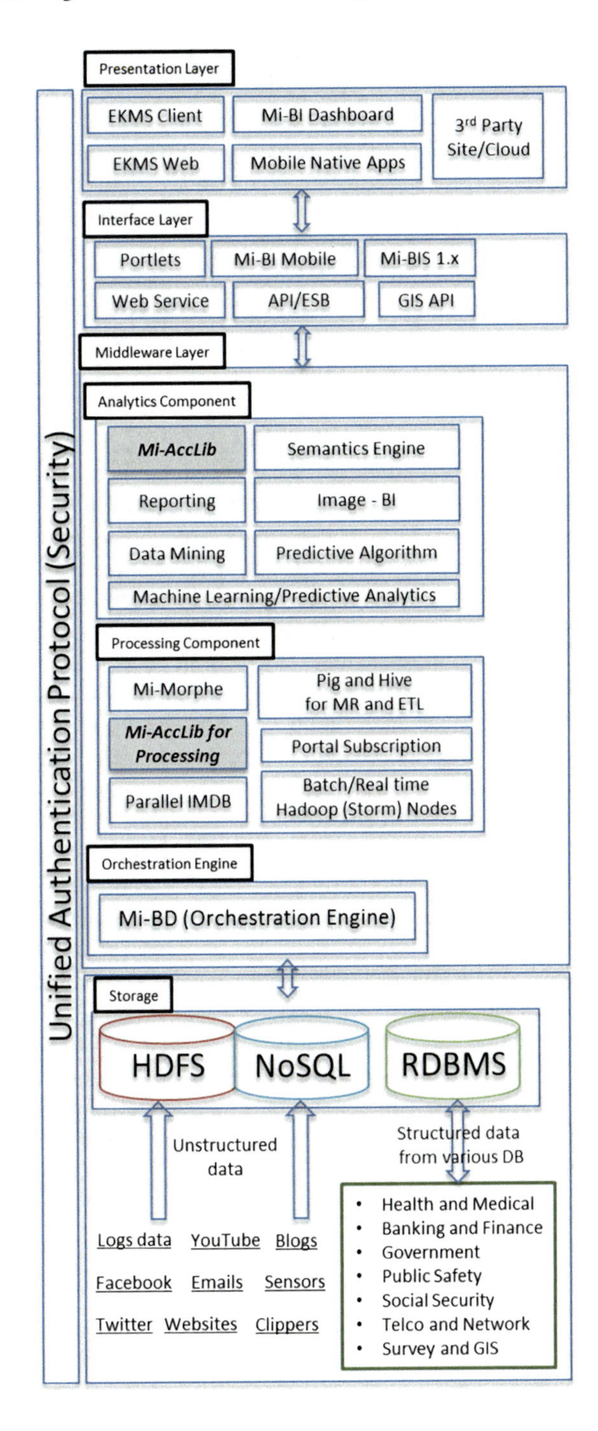

Fig. 12.1 Architecture of middleware framework

needs to be transferred from host (PC) processor to GPUs (device) for processing. Even though there are performance gains by using GPUs, functional-specific algorithms and application-specific algorithms exploiting GPU architecture need to be designed for optimum data processing. Thus, we have designed a set of library suites named as "MIMOS Accelerator Libraries" (Mi-AccLib) for various domain-specific (analytics component) and generic applications (processing component). These algorithms are categorized into different groups, namely, "Common Mi-AccLib," "Finance Mi-AccLib," "Text/String Mi-AccLib," "DB Mi-AccLib," etc. These libraries are designed and developed using Mi-AccLib framework such that the code can run seamlessly on different processor (GPU and multi-core for now) architectures exploiting underlying parallelization capabilities. The processed information in turn is fetched and displayed at presentation layer facilitated by interface layer (refer to Fig. 12.1).

In order to exploit current GPU computing capabilities for Mi-AccLib, we have to take into consideration the characteristics of the GPU and how it can cooperate with the CPU. One such consideration is the disparity of the computation capabilities between versions of the NVIDIA GPU cards. As such, a chunking and load balancing mechanism that splits and distributes data to different GPU cards in the system based on their computing capability has to be developed. Secondly, the I/O delays due to moving data to and from the hard disk, RAM, and GPU cards need to be considered when designing the framework in order to ensure that the overall system performance (multi-core CPU and GPU) is actually better by at least an order of magnitude compared to a multi-core CPU alone. Otherwise, there is no justification for multi-architecture development.

To meet these requirements, we designed the Mi-AccLib framework for multiple GPU support along with CPU synchronization. Our initial goal was to exploit GPUs for text-based processing and analytics work. In order to evaluate our middleware framework, we implemented one search and one sort algorithm for text processing on our framework and demonstrate how we can utilize these algorithms for data cleansing application. We then evaluate these algorithms against multi-core GPU versions.

Section 12.2 describes related works, while Sect. 12.3 details our Big Data framework and system architecture embracing our Mi-AccLib libraries via analytic and processing components. We outline our implementation in Sect. 12.4 and show the results of our algorithms on various different GPU cards in Sect. 12.5. Finally, we conclude in the last section.

12.2 Related Work

The MapReduce [5, 6] framework for distributed computing has been widely adopted in large-scale data processing. Mars [4] applies a flexible parallel programming in managing tasks partitioning and data distribution by using a GPU, which is an accelerated run-time system. However, Mars only works by distributing the data set over streaming processors on a single GPU card. MapCG [7] provides source code level portability between CPU and GPU for a high-level programming model. Nevertheless, this implementation has scarified the usage of shared memory and constant memory in GPUs due to the compiler support issues. There are more MapReduce on GPUs implementation [8–10], yet, these systems have faced the overhead issues on data transfer and kernel launching issues. The proposed Big Data middleware framework using Mi-AccLib has a more macro-level data distribution orientation that works by chunking data into multiple GPU cards.

GPUMiner [1] is a parallel data mining framework for using GPUs for data mining work. It is composed of three parts – a storage and buffer management module, a visualization module, and a mining module. GPUMiner utilizes DirectX for visualization and CUDA for the data mining module. Chidchanok [11] works on an experimental framework for searching large Resource Description Framework (RDF) and performing the semantic query using JCUDA.[1] It takes advantage of GPUs parallel thread and block for retrieving, joining, and finding operations to the corresponding RDF graph. While the focus of these systems are on data mining, Mi-AccLib components utilize the GPU for a wide range of string processing functions including, but not limited to, data mining, analytics, and in-memory database like operations.

OpenAcc is a standard for the directives and programming model which has been developed by CAPS, Cray, The Portland Group, and NVIDIA [12]. There are two commercialized directive compilers integrating with NVIDIA NVCC compilers, which are CAPS (HMPP) and PGI (PGCC) [13] compilers. Directive-based high-level programming model is a simple and portable method to parallelize loops in C code. This intermediate high-level code is compiled by the NVCC compiler. Subsequently, it converts to a CUDA assembler source (PTX[2]) and optimizes the defined code. Then, it generates the final CUDA binary (a .cubin file). Ghosh, Liao, Calandra, and Chapman [14] evaluated the GPU directive compilers, which resulted 1.5× to 1.8× improvement in performance for both ISO and TTI kernels in single GPU against multi-core CPU by using OpenMP. In addition, they concluded this reduce efforts in code optimization with pragmas. Directive approach is complementary to Mi-AccLib rather than a competitor since it works on a different level of parallelization.

OpenCL [15] is a platform-independent standard for programming heterogeneous systems. OpenCL programs are compiled just in time for execution and can be used together with Mi-AccLib or other run-time libraries. These works [16–18] experienced a performance penalty on the NVIDIA GPU, due to the OpenCL abstraction layer. Thus, we have disabled OpenCL support as it is not optimized

[1] It is a CUDA binding for the Java language, which exploiting the features of NVIDIA GPU computing from Java-based applications.

[2] Parallel Thread Execution (PTX) is a pseudo-assembly language used in NVIDIA CUDA programming environment.

for GPUs at the moment, and real gains on GPUs can only be seen through optimized code as there are additional overheads from data movement.

CUDA [19, 20] or Compute Unified Device Architecture is NVIDIA's parallel computing architecture for their GPU cards. It is an intuitive and scalable programming model which is an extension of C [21]. Additionally, it provides the entire GPU platform accessing for developers. This architecture unifies the devices of CPU and GPU by performing a heterogeneous computation system [22]. It has rapidly evolved and scaling parallel performance since 2007. There are sets of libraries that are mostly for non-graphics-related processing. Mi-AccLib is built on CUDA for the GPU computation parts.

12.3 Middleware Framework Design

The proposed Big Data application framework in this chapter comprises of presentation layer, interface layer, middleware layer, and storage component. Middleware layer is further decomposed to analytics component, processing component, and orchestration engine. Mi-AccLib libraries are positioned at both analytics component and processing component. The **presentation layer** includes business intelligence (BI) dashboard and the **interface layer** includes MIMOS business intelligence suite (Mi-BIS 1.x API). Meanwhile, off-the-shelf technology will be used for the **storage layer**.

The following subsection describes these frame layers and middleware components in details, followed by specifically focusing on GPU-based solutions.

12.3.1 Big Data Needs

Digital data explosion has exceeded the petabytes and entered into the zettabyte era, based on IDC Digital Universe Study 2011 [23] as shown in Fig. 12.2. As of year 2011, as a society, we have generated and consumed ~1.8 zettabytes of data. But the question is, was the data analyzed for useful information in a timely manner for instantaneous usage? The ultimate value of a big data implementation will be judged based on one or more of these three criterions:

- Able to provide more useful information
- Able to improve the reliability of the information
- Able to improve the timeliness of the response

Thus, a Big Data application framework which meets the above three criteria is inevitable to provide reliable, useful, and timely information, enabling quick response by the data owner. Otherwise, Big Data is worthless.

Meeting the growing demand for Big Data processing, large-scale parallel processing for data mining and analytics has sparked innovative solutions both in

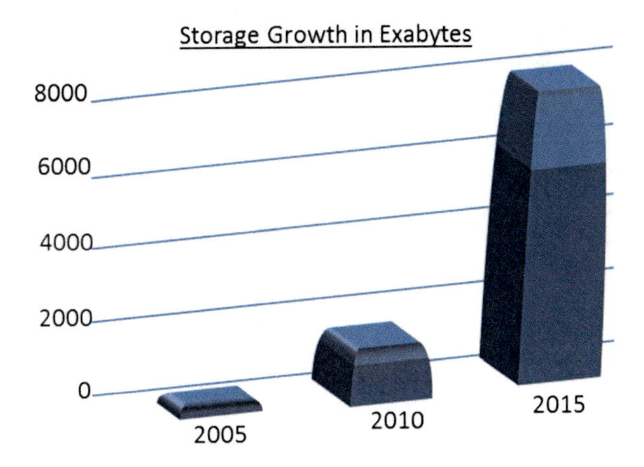

Fig. 12.2 Digital data growth in terms of storage size with forecasted data

commercial and scientific domain. Some of the commercial applications (e.g., Netezza, Teradata, Vertica) are computationally fast and cater for terabytes of data processing in milliseconds, however, relatively expensive to be used by small and medium enterprises. On the other hand, generally scientific communities rely on the MapReduce framework like applications such as Apache Hadoop which is free and stable for large-scale data processing. Following this open-source success, many applications are designed and developed in a parallel manner. Usage of parallel computing hardware such as GPGPU and Intel MIC (Many Integrated Core) coupled with parallel computing capability aware middleware/ application can provide another less expensive approach of Big Data processing.

Companies like Intel and NVIDIA are on track to realize many-core and multi-core parallel computing hardware with increasing number of parallel cores. Intel MIC equips with 60 cores and 244 threads for hyper-threading. NVidia Tesla K20c offers 14 SMX (streaming multiprocessor extension) with 2496 CUDA cores. Both Intel and NVIDIA claim their processor is much faster compared to the others. The fact is that this competition is important for total paradigm shift in hardware enables parallel computing.

For example, Kepler architecture of Tesla series of NVIDIA GPU promises higher 1.3 teraflops (double precision), while Intel MIC Xeon Phi provides 1.2 teraflops with both having 6 GB memory for big data analytics. Streaming functionality enables seamless data movement between CPU and GPGPU for ultrahigh speed data processing. Leveraging the hardware technological capabilities, MIMOS is building various solutions including Mi-AccLib to enable ultra-speed big data processing with data and process parallelism. MIMOS Mi-AccLib libraries reside at both analytics component and processing component of the middleware, being part of the overall building block of the MIMOS Big Data (Mi-BD) processing solution.

12.3.2 Presentation Layer

The presentation layer is responsible to provide precise, concise, and simple visualization capability for the processed big data, enabling the user to make sense of the data in an informative manner. MIMOS business intelligence and Big Data framework provides client, web, dashboard, and native mobile solutions for easy representation of data for various audiences with visually comprehensible format. Usually the information is presented in chart and graph forms. It also allows easy integration of other 3rd-party software tools.

12.3.3 Interface Layer

This layer provides the interfacing between the middleware layer and the presentation layer. It caters for various types of application programming interface (API) toward presenting on various types of user interfaces, such as through business intelligence dashboards, where the orchestration engine interacts with Mi-BIS 1.x to output results through tabular, graphical, and charts display. At the same time, portlet, web service, ESB (enterprise service bus), and Mi-Mobile BIS outputs result on third-party Web sites, third-party cloud, and MIMOS web EKMS (MIMOS interactive dashboard) and also on native Windows applications and native Android/IOS applications for smartphones and tablets. GIS API is responsible for producing mapping results on web, MIMOS web, native client, and mobile display.

12.3.4 Middleware Layer

Middleware layer consists of orchestration engine, analytic component, and processing component. The orchestration engine is responsible to orchestrate the entire process from acquiring/ingesting data contained in the storage; processing data; providing the required analytic library requested by the user interface, in order to produce the desired results, which are mapped via API of the interface layer; and finally presenting results through web, mobile, native client, and cloud connections.

The analytics component, which would be utilized to handle selected types of analytics, algorithm, statistical analysis, and prediction depending on the required user needs, currently, consists of:

- Mi-Acclib (GPGPU libraries)
- Reporting and OLAP (online analytical processing) for business intelligence (BI)
- Data mining libraries
- Machine learning and predictive analytics

- Semantic engine
- Image BI for video/image analytics
- Predictive algorithm suites

The processing component, which handles data processing such as extracting, transforming, and loading (ETL) data from databases, before passing to analytics components or passing data directly to the API layer as per instructions from the orchestration engine, consists of:

- Mi-Morphe (MIMOS ETL tool)
- Mi-Acclib for fast processing
- Parallel in memory DB
- Pig and Hive for big data ETL
- Portal subscription
- Batch (Hadoop) and real-time (Storm [24] and Impala [25]) big data processing

Processed data before analysis and after analysis are stored in the storage components in the form of data warehouse for BI API to utilize. The orchestration engine will determine, based on user selection from the presentation layer or predefined configuration, which analytics model and processing model to be utilized. The orchestration engine will ingest incoming data (structured or unstructured) and pass the result to interface layer which will be mapped to the display channels, such as MIMOS dashboard (EKMS), BI dashboard, or native mobile application.

12.3.5 Orchestration Engine (with Example of Use Case)

As mentioned above, the orchestration engine plays a significant part to process, analyze, and send the processed data to the respective display channels via the appropriate interface API layers. Through the API on the interface layer, the orchestration engine could also call hybrid technologies. For example, orchestration engine may access libraries with various algorithms for preconfigured purposes such as utilizing GPGPU libraries to edit distance algorithm and using Pig with Hadoop to perform MapReduce function and processing structured data from RDBMS (relational data bases management system) data while grouping the unstructured result within the same system to achieve the aggregated task.

The orchestration engine could very well be used for different scenarios. The following paragraph explains the implementation of Mi-BIS for video analytics (or image analytics) as an example, where the purpose is to identify the various types of events such as "event detected," "face detected," and "motion detected" per given camera location for given time stamp information as reported by the video analytics system. This data when analyzed on real time requires real-time processing speed, volume, and complexity of aggregating from other structured database tables such as listing the names of guards in charge during the occurrence of certain type of events for a period of 1 year pertaining to the camera locations. This would help the security companies or public safety organization to place their staff at strategic locations within their planned coverage area for patrolling, based on predictive analytics. Thus, probability of similar occurrence could be observed within certain time frame. In terms of implementation, the sources of video files from all the cameras being monitored are stored in HDFS of Hadoop nodes. Mi-BD (MIMOS Big Data) will sqoop in the files based on schedule time with the various types of events detected and performs ETL using Pig/Hive. Hadoop is used here for batch MapReduce processing, where the results would be stored in the storage component for later date. When a business user logs in through the display channel of Mi-BIS dashboard, the user could construct the required dimensions to view the report chart such as events, day, month, year, personnel names, and camera locations from the various data sources (structured or unstructured as stored in the storage components). Mi-BIS presentation/display layer will communicate with Mi-BD via the interface layer; Mi-BIS 1.x. Mi-BD will determine if this request is for real time or could utilize any previously stored preprocessed data. If real-time big data (from large video files of many camera sources) is required, then Storm would be used. In this example, the batch Hadoop ETL data that was pre-produced earlier is utilized by the BI dashboard in order to view the relationship in a bar chart (or tabular format) between time of year, type of events, camera locations, and guards in charge. The user can view an image and an 8-s video clip of the event (6 s before and 2 s after the event) when each individual event is clicked. If the user decides to open any statistical analysis tools, then Mi-BD, the orchestration engine, would be responsible to call machine learning/predictive analysis libraries or Mi-Acclib to drill down on the statistics of occurrences and pass the result back to the presentation layer via API from the interface layer.

12.3.6 Storage

The storage component consists of storage for structured and unstructured data. RDBMS such as SQL, MySQL, and Postgres supports the storage of structured data, while HDFS (Hadoop Distributed File System) and NonSQL such as Mongo DB or Cassandra could support unstructured data such as live feed from Twitter, Facebook, and log files. The storage component also store processed and analyzed data in the storage or as data warehouse in order to be used by API to display results in the various display channels and devices. In the next section, Mi-Acclib as an analytics component with GPGPU libraries is further elaborated.

12.3.7 Mi-AccLib and Analytics Component

We designed the Mi-AccLib framework to be modular in order for it to be extensible. Mi-AccLib is divided into two layers, which are an application-specific layer (analytic component) and a functional algorithm layer (processing component) as shown in Fig. 12.3. The Mi-AccLib framework is built to run on top of different processor architectures. One of the challenges of such a system is the need to support libraries written on different languages for different architectures. However, we focus on our work with GPU in this chapter.

The library interface wrapper layer provides a common interface for users to utilize functions that have been implemented for various hardware processors and coprocessors. For example, a search function can be used on either a GPU card or a multi-core CPU card or on both as the user requires. The functions exposed to the users at this layer share a common format as shown in Fig. 12.4.

The function takes in a variable number of parameters. All these parameters provide users with a fine-grained level of control when executing tasks. However, they can also leave the parameters to the default value for the framework to determine the best parameters for performance based on the available resources, function profiling, and data size.

The application-specific libraries (analytic component) are a set of basic functions that have been linked together to perform a certain task. An example of such a

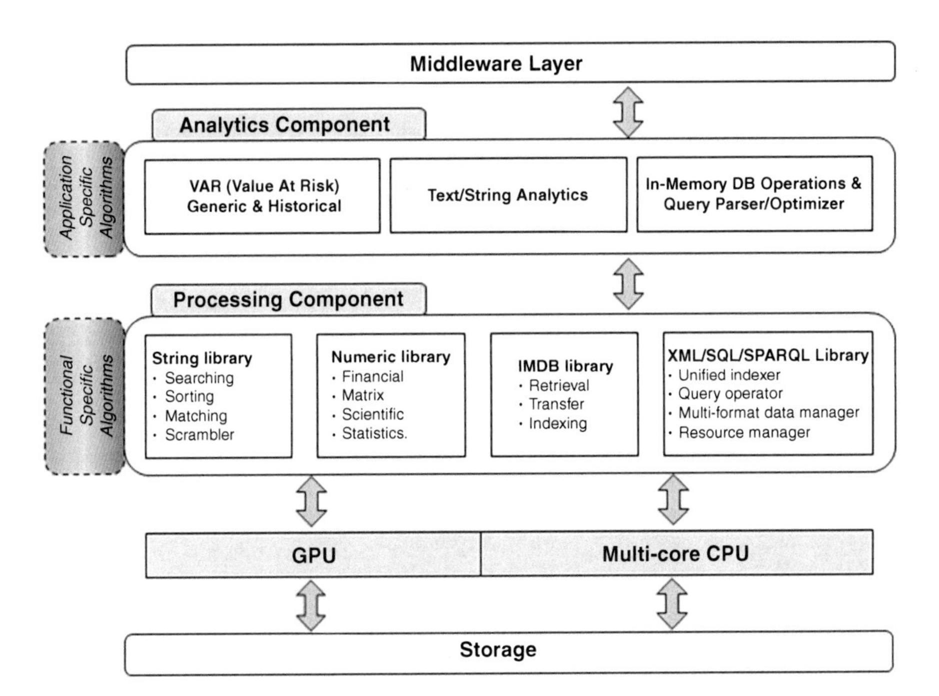

Fig. 12.3 Mi-AccLib framework architecture

```
miacclibError_m MiAccLib::MiLoad(
     char*                        filename,       /// specifies the input data file path
     miacclibComputePlatform_m    platform,       /// specifies the computing architecture
     migpuaccMode_m               gpumode,        /// specifies the GPU execution model
     unsigned int*                retTotalRows,   /// returns total of rows
     unsigned long long*          retTotalBytes)  /// returns total of bytes
{
     ...
}

miacclibError_m MiAccLib::MiExactKeywordSearch(
     unsigned char*   keyword,            /// specifies keyword to match
     unsigned char*   retResult)          /// returns results
{
     ...
}

miacclibError_m MiAccLib::MiEditDistanceKeywordSearch(
     unsigned char*   keyword,            /// specifies keyword to match
     unsigned char*   approval_level,     /// specifies edit distance approval level
     unsigned char*   retResult)          /// returns results
{
     ...
}
```

Fig. 12.4 Format of function

task is a financial calculation of value at risk (VAR) [26]. The VAR application-specific library (finance Mi-AccLib) takes in a set of data and first preprocesses the data for sorting using a preprocessing kernel (processing component) (e.g., changing floating point numbers to unsigned integers), sorts the preprocessed data using a sorting kernel, sends the sorted data to a percentile kernel to obtain the result, and converts the final result back into a floating point value.

The functional algorithm libraries (processing component) are a collection of kernels that perform tasks in similar areas and are the basic building blocks for the library framework. For example, the string processing library contains a set of different search algorithms that are exposed through the library wrapper interface. Each search algorithm gives a different set of performance that users can try to use for different application purposes. Sets of kernels from different functional algorithm libraries can also be integrated to perform a larger task by the application-specific libraries (analytic component).

In order to achieve an overall improvement of the whole system rather than emphasizing just faster execution of parts of the workflow, a holistic view needs to be taken during system architecture design. This is especially true when considering latency for I/O and load balancing for data distribution to GPUs of varying capabilities.

The first step is to decide which parts of the workflow are more suitable to be executed on the CPU or GPU. Typically, functions that require a lot of calculations that can be parallelized or have a lot of uncorrelated data to be processed are suitable for execution on the GPU. For example, matching data from columns of two different tables is very suitable for GPU processing as each entry of a column can be compared independently from any other entry. This allows the system to leverage on the parallel nature of the GPUs.

The second step is to determine how to order the data transfer and processing so that latency can be hidden. Mi-AccLib provides methods to split data into chunks that are readily transferrable to multiple GPU cards based on their available memory and computing power profiles. This ensures the data is distributed and processed optimally so that delays are kept to a minimum. Data chunking also needs to be done in a manner that allows the data structure to be preserved so that each chunk can be processed individually. The data in the chunks will normally need to be converted into a structure of arrays for faster processing. This is identical to the normal column format in databases, so data will need to be transposed when copied to the GPU.

12.4 Implementation

We have currently implemented a few different functional algorithm libraries into our framework. We will discuss some of the implementation details and challenges in this section.

One of the biggest limitations to using GPU cards for text processing is the large amounts of data that must be moved through the PCI-e bus to and from the GPU and also the reading and writing of data from the hard disk, which is five times or more slower than the PCI-e bus. Streaming is the method commonly used to hide or minimize the data transfer latency, where data is sent and received in small chunks using direct memory access (DMA) methods in the background, while data processing is performed on the chunks already received by the GPU.

To enhance the parallel processing of data and memory transfer, profiling of kernels needs to be performed beforehand to determine the duration of the kernel execution time. Based on the kernel execution time, the size of the data chunks used can be determined using the PCI-e data transfer speed as well as the kernel execution time. On initial observation, it may seem that chunking data to the minimum size chunk would seem like the logical choice to minimize overall delay as that will give the most overlap between data transfer and data processing. However, the transfers of many small chunks give rise to additional overhead time between data transfers. For example, the transfer of a single chunk size of 64 MB gives the highest data transfer throughput, but multiple transfers of 64 MB chunks incur a 33 % overhead on that of a single chunk transfer as we observed.

Getting outputs from kernel execution on the GPU is another trade-off issue that has to be considered during implementation. For example, an algorithm returning multiple search results in an array needs to ensure that the global variable that serves as the index of the array is not accessed simultaneously by multiple threads. Since there is no concept of critical sections in CUDA, a mutex must be implemented using the "atomicCAS" instruction to allow CUDA cores to lock the variable for reading and incrementing before releasing it for other cores.

While this works well if there are only a few results to be returned, searches with many results will cause many threads to be executed serially for this section. Besides this, atomic instructions in CUDA are slower than other instructions since they need to access global memory every time a read and write is performed, and a timeout may occur if too many threads in the same warp try to lock the same global variable.

A trade-off using additional memory can be done by allocating an array of bytes or bits equivalent to the size of search data. Each thread can mark the equivalent byte or bit in the result array without the need for a mutex. However, this requires a much larger memory allocation for the output and a larger delay when moving the results back to the CPU.

As an alternative HPC solution to GPU, we have implemented Big Data applications using the proposed middleware framework. In this experiment, we have incorporated a sample scenario using the presentation layer (Mi-BI dashboard), via Mi-BiS 1.x (API and connectors), middleware, storage, and the orchestration engine with the MapReduce functions managing the nodes. We have configured seven virtual machines with one master node (8 cores) and six worker nodes (4 cores each) running on a few of HP DL380p G8 servers installed with Apache Hadoop, Cloudera's Hadoop, and Impala. We have also installed Postgres on another same model of HP server with 8 GB RAM with 4 cores and another high-end HP machine with 96 GB RAM and 48 cores. At the same time, we have installed a GPU card, NVIDIA Tesla K20c, on a DELL Precision T5500 workstation with Windows Server 2008 R2 Enterprise SP1 64-bit operating system, running on Intel Xeon E5630@2.53 GHz processor, with 12GB RAM, 1 TB SATA hard drive (7,200 rpm). The Kepler GK110 GPU card provides 2496 CUDA cores. Next, we have written scripts within the orchestration engine to import ~120 million records from four significant tables of inpatient record table, state code names table, disease code names, and age groups from the hospital database, originally residing in Postgres database of our high-end physical server, into the storage of the Hadoop clusters which is in HDFS (Hadoop distributed file storage) format. Through our libraries in the middleware framework, our ETL processing component uses Hive to extract, cleanse, transform, and load the dataset to be accessed for the new data warehouse. Mi-BiS 1.x uses the cleansed data from the warehouse using connectors (API that had been developed for the interface), given by the business user, when logged on through BI dashboard. For example, the user could request for information such as the type of disease and age group distribution on a pie chart or view the trend of the selected disease for the duration of several years. Mi-BIS dashboard creates the reports for the types of query (or SQL select statements) for the different API/connectors. The processing time has been recorded with at least five trials for each of the different setup. The results of the average processing time are shown in Table 12.1.

The orchestration engine is responsible to interlink the storage layer toward the self-service report creation from the Mi-BIS dashboard or running the real-time analytics via the dashboard. Upon identification of large data request, the predefined orchestration engine will use HDFS and run the search either in Impala, GPU, or by combining the queries as hybrid parallel process and presenting the output to the Mi-BIS dashboard.

Table 12.1 Comparison of processing time for types of search query using RDBMS, Big Data Hadoop/Impala nodes, and GPU parallel DB

No	Description of search query vs. average processing time (seconds)	SQL (8GB/ 4Core)	SQL (96GB/ 48Core)	Hadoop-Hive	Impala-Hive	GPU-parallel DB
1	Selecting sum from one column of 120 million records	1,466.7 s	218.7 s	347.6 s	3.7 s	0.3 s
2	Selecting a name column, counting the name and ordering by top 10 names	7,901 s	1612s	505 s	64.2 s	NA
3	Selecting state code, years from hospital patient records with one disease code selected, group by years and state code, order by years and state code	1,464.7 s	103.6 s	383.5 s	3.5 s	3 s
4	Selecting state, years, disease name from hospital patient records where one disease name type is selected and joining disease code with disease name and state code with state names; grouping by years, states, and disease names; ordering by years and state code	1,688.7 s	102.7 s	N/A	2.9 s	1.6 s
5	Inserting the results of selecting state code, years from hospital patient records with ALL disease type, group by years and state code, order by years and state code	Failed	7,878 s	557.3 s	10.1 s	N/A
6	Selecting state, years, disease name from hospital patient records where three disease name types are selected and joining disease type with disease name and state code with state names; grouping by years, states, and disease names; ordering by years, states, and disease names	1,893s	704 s	N/A	3.7 s	6.3 s

12.5 Results

Two of the most important features of text processing sorting and matching of processing component and edit distance of analytics component are explored. In this section, we discuss the results from our implementations of the string matching in our Mi-AccLib framework, while results of edit distance operations are illustrated for various configurations. Finally, we present the results of experiment based on the different setups using the middleware framework.

We implemented a string matching algorithm, which matches all the characters in a keyword to a string from a large text file. This search is O(n) in complexity,

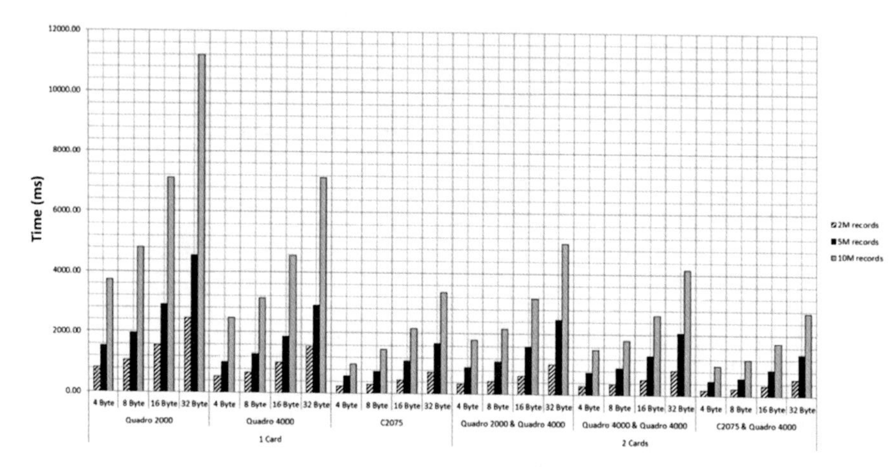

Fig. 12.5 Performance of single and dual GPU cards for string matching

where n is the number of characters to be searched. To speed up matching, we copy chunks of the text into the shared memory of the GPU, which is much faster than accessing global memory. Then, each thread in a warp searches for the keyword at different points of the text. The same search is repeated for other chunks of memory at other streaming multiprocessors throughout the GPU. The results of the search are returned either in array size of the search text or in arrays of integers pointing to the positions of the characters in the search text, as detailed in the previous section.

From our results in Fig. 12.5, we can see that the algorithm scales well according to the number of CUDA cores and shading processor speeds of the GPUs. For example, the Tesla C2075 GPU is three times as fast as the Quadro 2000 due to having almost three times as many CUDA cores as well as having a much higher memory bandwidth between the global memory and the shared memory. This allows it to complete memory-intensive jobs, such as matching and sorting, much faster than the Quadro 2000.

When we distribute the matching load between two cards, the throughput for the searches is slightly lower than the sum of the throughput of each card individually. The reason for this is mainly due to the overhead of distributing data to two separate cards on the same PCI-e bus. For example, performing a search for a 4-byte keyword from two million characters takes an average of 275.85 ms on a C2075 and 545.84 ms on a Quadro 4000, but distributed matching on a Quadro 4000 and C2075 takes only 198.80 ms. This gives us a throughput of 7.25 MB/s and 3.66 MB/s for the C2075 and Quadro 4000, respectively, and a combined throughput of 10.91 MB/s. However, the distributed matching on both cards gives a throughput of 10.06 MB/s.

We also implemented Levenshtein distance (edit distance) matching on CPU and GPU. This parallel version of edit distance feature is part of the analytic component in Mi-AccLib. The following explanation describes how application-specific analytic component and functional-specific processing components are

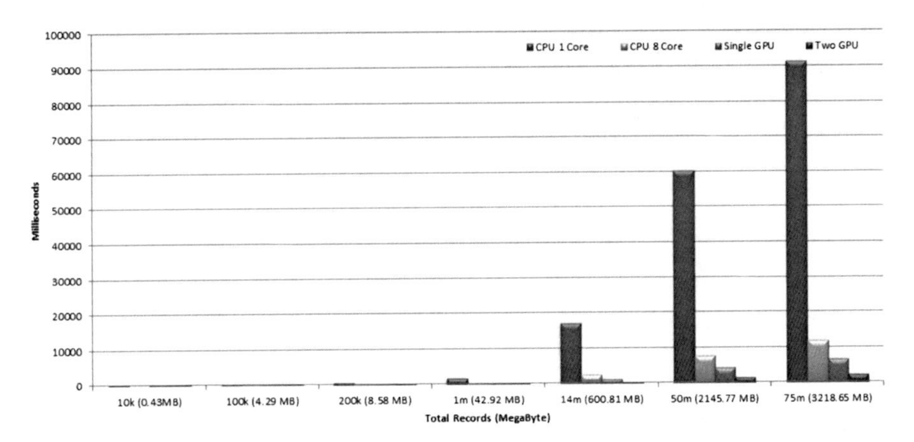

Fig. 12.6 Edit distance result for CPU and GPUs

used for data cleansing example using the middleware framework. It is certain these results prove GPU-based solution is an alternative to existing MapReduce-like application for Big Data processing.

It is defined to be the smallest number of edit operation (insertions, deletions, and substitutions) required to change one string into another [27]. Figure 12.6 shows one-to-many matching of execution time versus total records for the edit distance algorithm by utilizing CPU cores and GPUs (C2075). Single GPU and dual GPUs outperform the CPU multiple cores for processing much larger size. The speedup (CPU time/GPU time) on single CPU cores is 13.9×, and the 8 cores CPU is 1.78× for processing 75 million of records (3.14 GB). There is more speedup by utilizing dual GPUs, as utilizing the single CPU core is as high as 38.34×, and the 8 cores CPU is 4.92×. By comparing the speedup of single and dual GPUs processing, there is ~2.7× for the size of records from 14 to 75 million of records.

An application that we put together using sort and search was a data cleansing application project. In this project, we compared the national registration identification (NRIC) numbers from a database of 14 million records, which we call database A, against a clean set with 13 million records, which we call database B. Once the number matches, the kernel compares the names associated with the identification numbers at each table to confirm the match.

For this application, we first extracted the data from the two databases. Then we developed a kernel that first performs a sort on database B using the identification numbers as keywords. The identification numbers vary in length based on whether they are old, new, or army identification numbers. Then, for each record in database A, we search through database B for a match. We use a binary search algorithm to perform the search, and the brute force algorithm to perform the match. The binary search algorithm takes an average of O(logN − 1) to find if there is a match. After all the records from database A have been iterated through, we use the results to perform a brute force name matching from database A to database B. The total time taken using the GPU kernel for sorting and matching alone is below 15 s compared to the time of over 12 h on an Intel Xeon Quad Core E5620 running MySQL. The total time including data extraction from the databases and preprocessing was below 5 min for the GPU processing.

The results are shown as in Table 12.1 for six types of searches for the different environment setups using our middleware framework including the Mi-BIS presentation layer.

The average time for each query type was analyzed for Postgres database with 8 GB server and 96 GB server, Hadoop with Hive (7 nodes), Impala with Hive (7 nodes), and GPU server processing in order to compare the processing time (in seconds). The result shows the performance comparison of various setups for real-time analytics processing of Big Data in the health sector, using Mi-BIS presentation dashboard to analyze ~120 millions of records in HDFS and Postgres (RDBMS) servers. Some of the results are reported as NA (not available) because the work is still in progress.

GPU parallel DB processing takes the shortest time to process ~120 million of records and the cost is also cheaper than implementing 7 nodes of Hadoop or SQL on Postgres (medium and high-performance fully tuned Postgres database server). SQL was not able to compute on big data especially for real-time analytics. It even failed for insertion of output data in the same database. Hadoop with Hive is not suitable for real-time processing and would only be useful for batch processing of big data. Impala-Hive is as beneficial as GPU for general queries and could be used to complement in the hybrid parallel processing approach especially led by the orchestration engine with predefined rules within the scripts, developed for the middleware layer. Impala-Hive is faster compared to GPU parallel DB when there are multiple tables to be joined and with huge strings operations to be performed.

Conclusion

We have presented a middleware framework for big data processing using data cleansing as an example application. It is certain the above results prove GPU-based solution is an alternative to existing MapReduce-like application for Big Data processing. Our layered middleware framework approach with GPU capable analytic and processing components has facilitated seamless integration of our Mi-AccLib. It allows users to exploit the powers of the GPU by providing the ability for efficient work distribution across multiple GPUs with regard to I/O access and load balancing. Using the Mi-AccLib framework, we implemented and tested radix sort and string matching algorithms on single and multiple GPU cards as part of processing component. On the other hand, the edit distance algorithm as part of analytic component used underlying processing component functionalities for application-specific needs. Our results show a significant improvement by using two GPU cards over single GPU cards and single GPU cards over multi-core CPUs for text data sorting, matching, and cleansing. The performance of the GPU

(continued)

implementation for data cleansing shows a speedup of over two orders of magnitude over the same operation done in MySQL on a multi-core machine. The proposed middleware framework can perform real-time analytical queries using the hybrid Impala and GPU libraries of ~120 million records for the selected hospital database, within less than 11 s.

Acknowledgement This research was done under joint lab of "NVIDIA-HP-MIMOS GPU R&D and Solution Center." This is the first GPU solution center in Southeast Asia established in October 2012. Funding for the work came from MOSTI, Malaysia. The authors would like to thank Prof. Simon See and Pradeep Gupta from NVIDIA for their support.

References

1. Fang, W., et al.: Parallel data mining on graphics processors. Technical Report (2008)
2. Gregg, C., Hazelwood, K.: Where is the data? Why you cannot debate CPU vs. GPU performance without the answer. In: IEEE International Symposium on Performance Analysis of Systems and Software (ISPASS), pp. 134–144 (2011). doi:10.1109/ISPASS.2011.5762730
3. Bakkum, P., Skadron, K.: Accelerating SQL database operations on a GPU with CUD. In: Proceedings of the 3rd Workshop on General-Purpose Computation on Graphics Processing Units, pp. 94–103. New York, NY: ACM (2010). ISBN: 978-1-60558-935-0, doi:10.1145/1735688.1735706
4. He, B., et al.: Mars: a MapReduce framework on graphics processors. In: Proceedings of the 17th international conference on Parallel architectures and compilation techniques, pp. 260–269. New York, NY: ACM (2008). ISBN: 978-1-60558-282-5, doi:10.1145/1454115.1454152.
5. Dean, J., Ghemawat, S. MapReduce: simplified data processing on large clusters. Communications of the ACM, vol. 51, pp. 107–113. New York, NY: ACM (2008). ISSN: 0001-0782, doi:10.1145/1327452.1327492
6. Wolfe Gordon, A., Lu, P.: Elastic phoenix: Malleable MapReduce for shared-memory systems. In: Altman, E., Shi, W. (eds.) Network and Parallel Computing, vol. 6985, pp. 1–16. Springer, Heidelberg (2011)
7. Hong, C., et al.: MapCG: writing parallel program portable between CPU and GPU. In: Proceedings of the 19th international conference on Parallel architectures and compilation techniques, pp. 217–226. New York, NY: ACM (2010). ISBN: 978-1-4503-0178-7, doi:10.1145/1854273.1854303
8. Shirahata, K., Sato, H., Matsuoka, S.: Hybrid map task scheduling for GPU-based heterogeneous clusters. In: IEEE Second International Conference on Cloud Computing Technology and Science (CloudCom), pp. 733–740 (2010). doi:10.1109/CloudCom.2010.55
9. Stuart, J. A., Owens, J. D.: Multi-GPU MapReduce on GPU clusters. IEEE Computer Society. In: Proceedings of the 2011 I.E. International Parallel & Distributed Processing Symposium, pp. 1068–1079. Washington, DC. (2011). ISBN: 978-0-7695-4385-7, doi:10.1109/IPDPS.2011.102
10. Catanzaro, B., Sundaram, N., Keutzer, K.: A map reduce framework for programming graphics processors. In: Third Workshop on Software Tools for MultiCore Systems (STMCS) (2008)

11. Choksuchat, C., Chantrapornchai, C.: Experimental framework for searching large RDF on GPUs based on key-value storage. In: 10th International Joint Conference on Computer Science and Software Engineering (JCSSE), pp. 171–176 (2013). doi:10.1109/JCSSE.2013.6567340
12. NVIDIA Corporation. OpenACC. https://developer.nvidia.com/openacc (2011). Accessed 4 Aug 2013
13. Wolfe, M.: Implementing the PGI accelerator model. New York, NY: ACM. In: Proceedings of the 3rd Workshop on General-Purpose Computation on Graphics Processing Units, pp. 43–50 (2010). ISBN: 978-1-60558-935-0, doi:10.1145/1735688.1735697
14. Ghosh, S., et al.: Experiences with OpenMP, PGI, HMPP and OpenACC directives on ISO/TTI Kernels. In: High Performance Computing, Networking, Storage and Analysis (SCC), 2012 SC Companion, pp. 691–700 (2012). doi:10.1109/SC.Companion.2012.95
15. Munshi, A.: The OpenCL specification. Khronos OpenCL Working Group. Technical Report (2009)
16. Torres, Y., Gonzalez-Escribano, A., Llanos, D.R.: Using fermi architecture knowledge to speed up CUDA and OpenCL programs. In: IEEE 10th International Symposium on Parallel and Distributed Processing with Applications (ISPA), pp. 617–624 (2012). doi:10.1109/ISPA.2012.92
17. Wezowicz, M., Taufer, M.: On the cost of a general GPU framework: the strange case of CUDA 4.0 vs. CUDA 5.0. In: High Performance Computing, Networking, Storage and Analysis (SCC), SC Companion, pp. 1535–1536 (2012). doi:10.1109/SC.Companion.2012.310
18. Shen, J., et al.: Performance traps in OpenCL for CPUs. In: 21st Euromicro International Conference on Parallel, Distributed and Network-Based Processing (PDP), pp. 38–45 (2013). ISSN: 1066-6192, doi:10.1109/PDP.2013.16
19. NVIDIA Corporation.: CUDA C Programming Guide. s.l. NVIDIA Corporation (2012)
20. Sanders, J., Kandrot, E.: CUDA by example: an introduction to general-purpose GPU programming. Addison-Wesley Professional. (2010). ISBN: 0131387685
21. Wilt, N.: CUDA handbook: a comprehensive guide to GPU programming. Addison-Wesley Professional, (2013). ISBN: 0321809467
22. Kirk, D.B., Hwu, W-m.W.: Programming massively parallel processors, second edition: a hands-on approach. Morgan Kaufmann, Burlington, MA (2012). ISBN: 0124159923
23. Hollis, C.: IDC digital universe study: Big data is here, now what? http://chucksblog.emc.com/chucks_blog/2011/06/2011-idc-digital-universe-study-big-data-is-here-now-what.html (2011). Accessed 18 July 2013
24. Storm—Distributed and fault-tolerant realtime computation. http://storm-project.net/ (2011). Accessed 10 Aug 2013
25. Impala—The platform for big data. http://www.cloudera.com/ (2013). Accessed 10 Aug 2013
26. Holton, G.A.: Value at risk: theory and practice. Academic Press, Amsterdam (2003). ISBN: 0123540100
27. Navarro, G. A guided tour to approximate string matching. ACM computing surveys, vol. 33, pp. 31–88. New York, NY: ACM. (2001). ISSN: 0360-0300, doi:10.1145/375360.375365

Chapter 13
On the Efficient Implementation of a Real-Time Kd-Tree Construction Algorithm

Byungjoon Chang, Woong Seo, and Insung Ihm

Abstract The kd tree is one of the most commonly used spatial data structures for a variety of graphics applications because of its reliably high-acceleration performance. Several years ago, Zhou et al. devised an effective kd-tree construction algorithm that runs entirely on a GPU. In this chapter, we present improved GPU programming techniques for implementing the algorithm more efficiently on current GPUs. One of the major ideas is to reduce the number of necessary kernel functions by replacing the essential, segmented-scan, and reduction computations by simpler per-block atomic operations, thereby alleviating the overheads from multiple synchronous kernel calls. Combined with the efficient implementation of intrablock scan and reduction, using recently introduced intrinsic functions, these changes achieve remarkable performance enhancement to the kd-tree construction process. Through an example of real-time ray tracing for dynamic scenes of nontrivial complexity, we demonstrate that the proposed GPU techniques can be exploited effectively for various real-time applications.

Keywords Real-time ray tracing • Kd-tree construction • GPU computing • CUDA • Scan and reduction operations

13.1 Background and Our Contribution

For many important applications in computer graphics, such as ray tracing and those relying on particle-based computations, adopting a proper acceleration structure will affect their run-time performance greatly. Among the variety of spatial data structures, the kd tree is frequently used because of its reliably high-acceleration performance. Compared to other techniques such as grids and bounding volume hierarchies, its relatively higher construction cost has been

B. Chang
Digital Media & Communications R&D Center, Samsung Electronics, Suwon-si, Gyeonggi-do, South Korea
e-mail: bj81.chang@samsung.com

W. Seo • I. Ihm (✉)
Department of Computer Science and Engineering, Sogang University, Seoul, South Korea
e-mail: wng0620@sogang.ac.kr; ihm@sogang.ac.kr

Y. Cai, S. See (eds.), *GPU Computing and Applications*,
DOI 10.1007/978-981-287-134-3_13

regarded as a drawback, despite efforts to develop an optimized algorithm (e.g., [1]), which has often restricted the use of the kd tree for real-time applications.

Recently, much effort has gone into accelerating kd-tree construction, particularly by developing effective parallel algorithms on modern CPUs and GPUs. Shevtsov et al. [2] and Zhou et al. [3] presented parallel construction algorithms for the CPU and GPU, respectively, in which, instead of applying a precise surface area heuristic (SAH) metric, median-splitting schemes were used to build the upper levels of the trees to enable effective parallelization on the respective processors. To alleviate memory usage issues, Hou et al. improved Zhou et al.'s method by modifying the kd-tree construction order [4]. In another approach, Choi et al. [5] and Wu et al. [6] attempted to build better kd trees for the CPU and GPU, respectively, by applying the accurate SAH metric to the entire tree structure. As pointed out in [5], the approximate approaches taken in [2,3] may often lead to kd trees of somewhat degraded quality, which would influence the kd-tree performance adversely. However, for interactive applications such as the real-time ray tracing of dynamic scenes, where the kd tree must be rebuilt for every frame after ray tracing the scene, it is important to adopt an effective kd-tree construction scheme that achieves a balance between tree-construction efficiency and run-time acceleration performance.

In this chapter, we present enhanced CUDA programming techniques for implementing the GPU method of Zhou et al. [3]. While their detailed algorithm, proposed several years ago, is still effective, current GPU designs enable it to be implemented more efficiently. In developing this CUDA implementation, we aim to enhance the GPU performance, particularly by minimizing the overheads caused by multiple synchronous kernel calls. For this, the essential, segmented-scan, and reduction computations are replaced by simpler per-block atomic operations. Coupled with an efficient implementation of intrablock scan and reduction, based on recently introduced intrinsic functions of the CUDA API, our methods achieve significant performance improvements in the kd-tree construction process. Via experiments on ray tracing for dynamic scenes of nontrivial complexity, we demonstrate that the proposed GPU techniques can be applied effectively to various real-time applications.

13.2 Optimizations for the Large-Node Stage

In Zhou et al.'s method, the upper levels of the kd tree were constructed using a node-splitting scheme that comprised spatial median splitting and empty space maximizing. In particular, based on the observation that the assumptions made in the SAH may often be inaccurate for large nodes, this stage of computation, called the *large-node stage*, simply selects the spatial median of the longest axis of the axis-aligned bounding box (AABB) of a node as its split position. For efficient parallel implementation on a GPU, all triangles in each large node are grouped into *chunks* of fixed size (i.e., 256), parallelizing the computation over the triangles in the chunks. (Note that the triangles and chunks are mapped to the threads and blocks, respectively, in the CUDA implementation.)

13.2.1 *Triangle Sorting with Respect to Splitting Planes*

The large-node stage iterates the node-splitting process until no large node is left. In Algorithm 2 [3], the most time-consuming parts of each iteration are the fourth and fifth steps, corresponding to lines 24–34 and 35–40, respectively, where the triangles for each large node are first sorted with respect to the splitting plane, and the triangle numbers of the resulting two child nodes are then counted. In this subsection, we present two different approaches to implementing these two steps on a GPU. We then analyze their performance in the section on experimental results.

13.2.1.1 Implementation Using Standard Data-Parallel Primitives

As was done in [3], the first implementation relies on standard data-parallel primitives such as (segmented) scan and reduction but uses a slightly different algorithm, which is computationally as efficient as the original one. The topmost part of Fig. 13.1 shows a situation where triangles in each large node are sorted into two child nodes. Here, we allocate two lists statically, *active list* and *next list*, to the global memory of the GPU to buffer the triangle indices. (Note that the triangle indices are grouped into chunks of size 256, as shown in the dashed boxes, which are then packed into the triangle index lists.)

For each triangle in a large node, mapped to a CUDA thread, the key issue is how to efficiently calculate its address(es) in parallel in the new triangle index list *next list*, whose production is complicated because of the simultaneous subdivisions of the large nodes in the current list *active list*. For this, a kernel is first executed over every thread block corresponding to a chunk of triangles, classifying each triangle against the respective splitting plane and generating two bit-flag sequences of size 256 per chunk *triangle bit flags*. Then, for each of these, an exclusive scan is performed using the shared memory of the GPU, resulting in the local *triangle offset* sequences. In addition, the kernel counts the number of triangles in each bit-flag sequence by simple addition and places this number in an array in the global memory. (Note that, for the example in Fig. 13.1, the two triangle counts of 201 and 75 are written to the array marked **[A]** as a result of execution over the first chunk of *node 0*.)

The next kernel then starts performing an inclusive *segmented* scan over this array, storing the scanned result in another array, marked **[B]** in the example figure, where each child node now comprises a segment in the sequence. After this scan, a per-element subtraction is carried out in parallel between these two arrays to build another *chunk offset* sequence that stores the displacement of the first triangle in each chunk within a new child node. In the subsequent third kernel, an exclusive

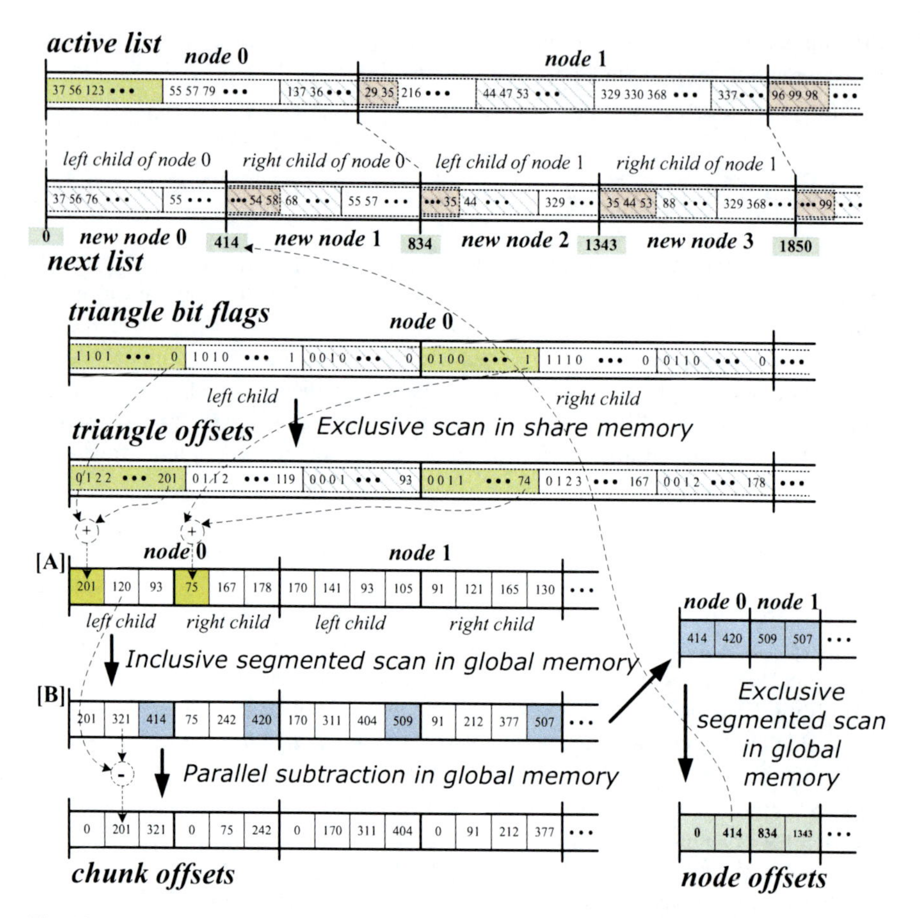

Fig. 13.1 Parallel triangle sorting over splitting planes: using standard segmented-scan primitives

segmented scan is carried out over the sequence of numbers formed by the last element of each child node in the scanned array **[B]**, whose resulting *node offsets* indicate the offsets of the first triangles of the new nodes within the new triangle index list. Finally, a fourth kernel is executed over the thread blocks of triangles in the *triangle bit flags* array, where, for a triangle whose bit flag is on, its triangle index is stored in the appropriate place in the new triangle index list, whose address can be calculated using the *node offsets*, *chunk offsets*, and *triangle offsets*.

13.2.1.2 Implementation Using Atomic Operations

The triangle-sorting technique described in the previous subsection requires a segmented scan to be carried out twice on the data sequences stored in the global memory and can easily be implemented using the data-parallel primitive functions provided by the CUDPP library [7], for example. Although very effective, such an approach forces the run-time execution to be split into a sequence of synchronous kernel calls, whose overheads will impact the run-time performance adversely.

To address this, observe that a side effect of using a standard segmented-scan method is that the relative order of triangle indices within a large node made of multiple chunks is retained in the respective child nodes. Such a property is important when the order of elements is essential, as in a radix sort algorithm, for example. However, retaining the strict order is unnecessary in the kd-tree construction algorithm because the order of triangles within a kd tree's leaf node is not critical in the later ray-tracing stage. This observation allows us to implement the triangle-sorting computation by using a single faster-running kernel and replacing the segmented-scan operations with simpler per-chunk atomic operations that are supported by the CUDA API.

In the new implementation, the memory configuration for the triangle index lists is slightly different, as shown in Fig. 13.2. For the ith large node with n_i triangles in the current *active list*, $2n_i$ elements, n_i per child node, are consecutively allocated to the *next list*. In addition, an array of integer-valued *chunk offset* counts, all initially set to zero, is allocated in the global memory, each of whose elements corresponds to a child node, i.e., a new node in the *next list*. As before, these *atomic* variables are intended to hold the displacements of the first triangles in the chunks within a new child node, although the order between chunks may no longer be preserved because of the use of the atomic operation.

For each chunk of triangle indices in the current list, the new kernel repeats the same computation until the triangle numbers are calculated in the array **[A]**. A representative thread then carries out two atomic additions, respectively fetching the local offsets, one for each child node, from the corresponding atomic variables and simultaneously adding the triangle counts to them, through which we will know where to start storing the sorted triangle indices in the child nodes. Then, once per child node, each thread checks the corresponding bit flag in the *triangle bit flag array*, and, if set to *on*, puts its triangle index in the proper place in the next triangle index list, whose location can easily be deduced from the fetched offset and the offset in the *triangle offset* array.

In this implementation, the two segmented scans over the arrays in the global memory have been replaced by two atomic add operations per thread block. While the computation time is already reduced markedly by this change, two per-block scans, one for each child, must still be carried out per chunk to compute the triangle offsets. While such scans can be performed effectively in the shared memory by using a standard scan method [8], recent GPUs offer useful intrinsic operations, such as __ballot() for warp voting, __popc() for bit counting, and __shfl() for warp shuffling, that can enable an efficient implementation of the per-block scan [9]. Therefore, to achieve a further performance enhancement, our implementation uses the __ballot() and __popc() functions for an intra-warp scan [10] and the __shfl_up() function for an inter-warp scan. (Details of our CUDA implementation of the kernel function are described in the Appendix.)

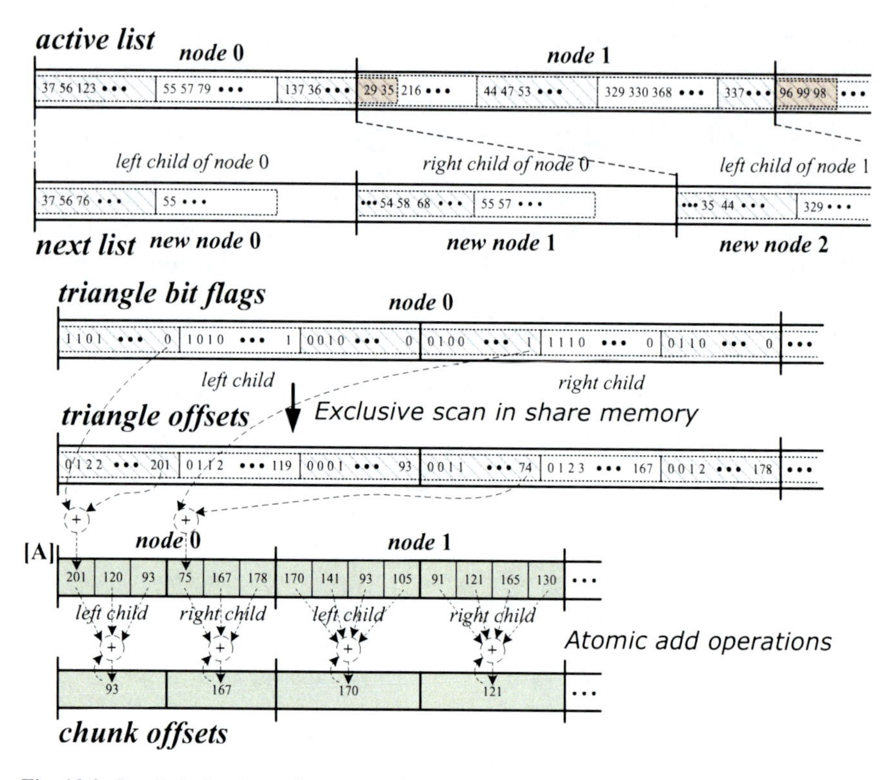

Fig. 13.2 Parallel triangle sorting over splitting planes: using atomic add operations

13.2.2 AABB Computations for Active Large Nodes

Another time-consuming part of the large-node stage is the second step (lines 9–14 of Algorithm 2), in which the AABB of all triangles in each node is calculated. The optimization techniques described in the previous subsection can also be applied to this AABB computation. The standard reduction in the shared memory for computing per-chunk bounding boxes can be implemented more efficiently on the GPU by a simple modification of the scan implementation using the intrinsic shuffle function __shfl_up(). Then, via three pairs of atomic *min* and *max* operations, the result of each chunk reduction is written in parallel to the location in the global memory that corresponds to the large node to which the chunk belongs. Although such atomic operations are still regarded as expensive on current GPUs, we observe that our single-kernel implementation based on atomic operations runs significantly faster on the GPU than the original implementation, which needed to perform segmented reductions six times.

13.3 Optimizations for the Small-Node Stage

After all large nodes are split into nodes whose triangle numbers do not exceed 64, the *small-node stage* starts. Because sufficient nodes are available, the computation in this stage is parallelized over nodes instead of triangles, evaluating the precise SAH metric to find the best splitting plane for each small node. The key to the efficient implementation of this stage is exploiting a preprocessed data structure that facilitates the iterative node-splitting process. For each initial small node, called the *small root node*, up to 384 (=64 (triangles) * 3 (x-, y-, z-axes) * 2 (min/max)) splitting-plane candidates are first collected from triangles in the node. Then, for each candidate, two 8-byte bit masks are generated to represent the triangle sets contained in both sides. To represent this information, 20 bytes of memory per node is necessary, including the 4 bytes used to store the location of the splitting plane, implying that up to 7,680 (=20 * 384) bytes of memory may be necessary for each small root node. It is important to choose an appropriate memory layout for the representation because the nontrivial amount of data will be accessed in parallel during the small-node stage. Although several different configurations are possible, we observed that the combination of a 4-byte access from the global memory for the splitting plane location and another 16-byte access from the texture memory for the triangle sets incurred the lowest memory latency on the GPU tested. (Our analysis of the generated PTX code showed that 16 bytes of data were fetched from texture memory even for a 4-byte access command.)

With this representation, the SAH cost evaluation and triangle sorting in the subsequent node-splitting step can be performed efficiently using simple bitwise operations. In this process, a parallel bit-counting operation is carried out very frequently to obtain the numbers of triangles in the child nodes. Whereas the method presented in [11] was used in the original description of Zhou et al.'s algorithm, we find that the __popc() intrinsic function accelerates the counting process significantly, as will be shown in the next section. Furthermore, we can also accelerate the intrablock scan, using the same intrinsic functions as for the triangle-sorting computation, which improves the performance slightly.

13.4 Experimental Results

To measure the performance improvement achieved by the optimization techniques presented here, we first implemented the kd-tree construction algorithm of Zhou et al. on an NVIDIA GeForce GTX 680 GPU, effectively as described in the original paper. In doing this, we used the scan and reduction techniques described in [8] for both *intra-chunk* and *segmented* scan and reduction. Here, the CUDPP primitive functions [7] were utilized for the segmented data. Furthermore, the parallel bit-counting operation needed in the small-node stage was implemented as proposed in [11]. Starting with this original implementation, we applied the

Table 13.1 Applied optimization techniques

	Stage	Operations	Our implementation
[A]	LNS1	Two segmented scans	Two atomic operations per block
[B]	LNS1	Intrablock scans	__ballot()/__popc()[a]
			__shfl_up()[b]
[C]	LNS2	Six seg. reductions	Six atomic operations per block
[D]	LNS2	Intrablock reductions	__shfl_up()
[E]	SNS	Intrablock scans	__popc()
			Same as [B]

LNS1 and LNS2 denote the computation for triangle sorting (Sect. 13.2.1) and AABB computation (Sec. 13.2.2), respectively, in the large-node stage, while SNS denotes the small-node stage
[a]Intra-warp scan of binary numbers
[b]Inter-warp scan

optimization techniques described above one at a time, in the order given in Table 13.1, and measured their impact on the timing performance. Note that the order of triangle indices in the leaf nodes of the produced kd trees may be different because of the simultaneous atomic operations performed in our method. To check for any effects on rendering performance, we also measured the time to render a 1024 * 1024 image by full ray tracing with shading, textures, reflection, and shadows. To experiment with dynamic scenes of nontrivial complexity, we synthesized some test scenes from commonly used scenes, made available by the Utah 3D Animation Repository, Joachim Helenklaken, and Marko Dabrovic: "Sponza with *i* Runners" (SR*i*) and "Kitchen with *i* Runners" (KR*i*), for $i = 1, 2, 3$. These scenes comprise $66{,}454 + i * 78{,}029$ triangles and $101{,}015 + i * 78{,}029$ triangles, respectively. (See Fig. 13.3.)

For seven representative scenes, Table 13.2 gives the stage-by-stage reduction in the kd-tree construction time, achieved as a result of the application of the series of optimization techniques described above. It is clear that the replacement of segmented scan and reduction by per-block atomic operations produced significant improvements despite atomic operations still being regarded as costly in the current CUDA architecture. (See the changes from "Original" to "[A]" and "[A]–[B]" to "[A]–[C]".) A major reason is that the operations explained in Sects. 13.2.1 and 13.2.2, respectively, were able to be executed more efficiently on the GPU using fewer numbers of kernels, as clearly indicated in the "Kernel calls" row, which markedly reduced the overheads from multiple synchronous kernel calls. (Note that a single kernel was sufficient for the triangle-sorting process, while four plus those necessary for the two segmented-scan calls were needed, which was repeated per each iteration.) Also, by exploiting the intrinsic functions offered by the more recent CUDA compute capability, we could reduce the computation cost for the intrablock scan and reduction further.

As can be verified from the ray-tracing time "*R*" in Table 13.3, the modifications to the original kd-tree construction algorithm did not incur any noticeable degradation in the quality of generated trees except a few cases, despite the different

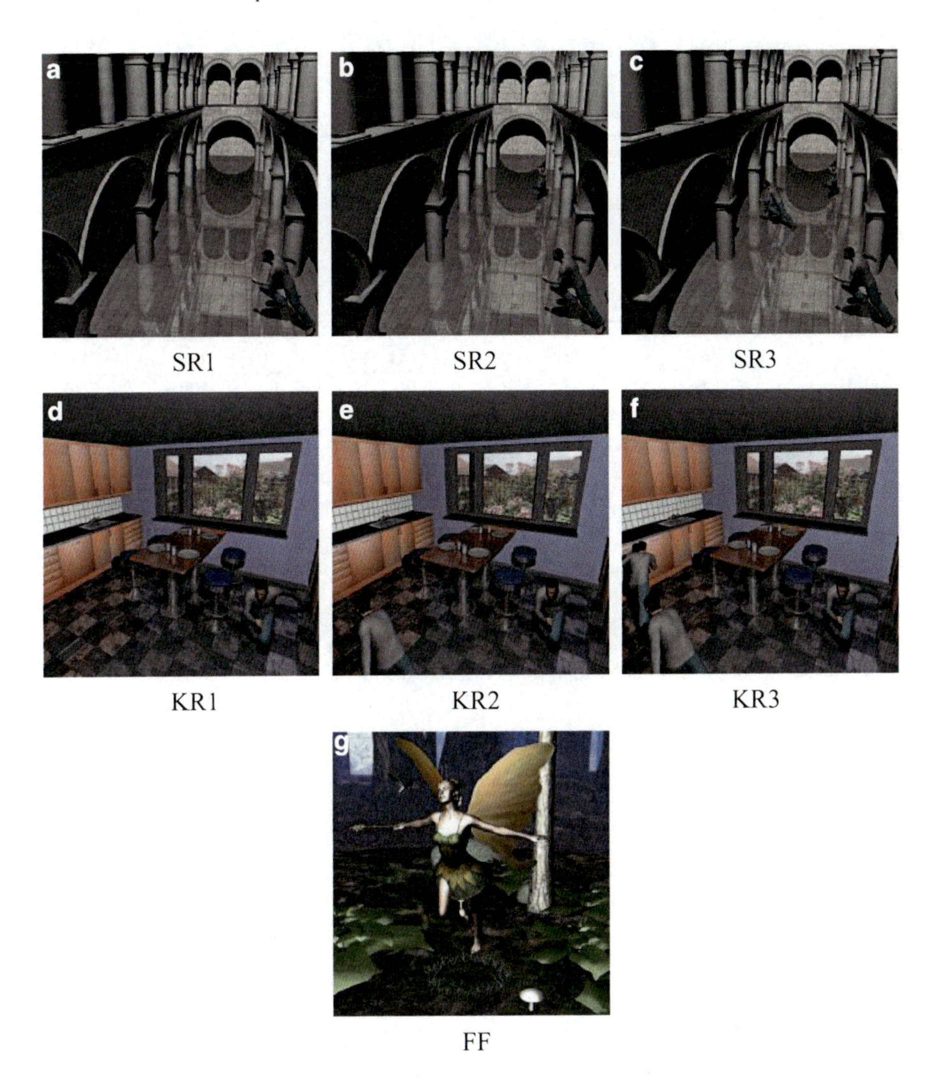

Fig. 13.3 Test scenes. The numbers of triangles in these scenes are 144,483(SR1), 222,512(SR2), 300,541(SR3), 179,044(KR1), 257,073(KR2), 335,102(KR3), and 174,117(FF)

orders of triangle indices stored in the leaf nodes. As a result, we were able to accelerate the process of interactive ray tracing of nontrivial dynamic scenes on the GPU effectively. (This is shown by the "T" values in Table 13.3.)

Table 13.2 Performance of the kd-tree construction (Rendering time (in ms)) and the numbers of kernel calls)

	SR1	SR2	SR3	KR1	KR2	KR3	FF
Original	120.9 (46.3)	146.2 (59.1)	181.8 (77.7)	130.9 (47.1)	149.7 (60.0)	185.5 (77.8)	107.4 (42.1)
[A]	88.6 (34.8)	93.0 (49.9)	109.7 (64.2)	91.8 (35.1)	109.1 (50.0)	118.9 (64.9)	76.2 (36.8)
[A]–[B]	76.3 (34.1)	89.2 (49.0)	107.4 (62.8)	85.7 (34.3)	94.7 (48.6)	104.4 (62.6)	67.1 (36.2)
[A]–[C]	52.3 (29.9)	67.5 (44.7)	79.8 (59.0)	62.3 (31.1)	67.6 (44.9)	90.6 (59.3)	58.5 (34.0)
[A]–[D]	51.6 (29.2)	66.4 (43.3)	78.1 (56.9)	60.1 (29.8)	67.1 (43.6)	88.6 (57.4)	54.5 (32.8)
Ours(all)	48.5 (26.7)	64.3 (39.1)	74.1 (52.4)	48.8 (27.4)	64.0 (40.6)	72.5 (52.8)	48.1 (30.4)
Kernel calls	1,031 /221	1,041 /229	1,069 /232	977 /214	995 /219	1,034 /223	681 /153

In this table we provide the total time spent on the kd-tree construction, averaged for the given animation sequences, where the number in parentheses represents the timing obtained by summing each kernel's execution time. Also, the averaged numbers of CUDA kernel calls made by the original and our implementations are compared, in which the extra calls within the CUDPP functions for the original implementation were not counted here

Table 13.3 Performance of the kd-tree construction (Rendering time (in ms))

		SR1	SR2	SR3	KR1	KR2	KR3	FF
Orig.	*R*	92.5	83.9	95.4	90.0	83.5	88.7	92.1
	T	213.4	230.1	277.2	220.9	233.2	274.2	199.5
Ours	*R*	90.1	92.7	93.5	88.2	94.3	100.5	103.3
	T	138.6	157.0	167.6	137.0	158.3	173.0	157.8

In this table the average ray-tracing time (R) and the average total time (T), which includes both construction and ray tracing, are given. FF denotes the "Fairy Forest" scene comprising 174,117 triangles

Concluding Remarks

In this chapter, we have presented efficient GPU programming techniques for implementing the well-known kd-tree construction algorithm [3] and demonstrated its effectiveness through several examples. With current GPUs, executing a CUDA kernel is still a relatively expensive operation, and thus it is important to make an effort to minimize the number of kernel calls made. As shown in the result section, our method was shown to be very successive in building kd trees using much fewer numbers of kernels, which resulted in a markedly more efficient GPU implementation. We believe that the ideas presented are also relevant to the development of applications that use other hierarchical spatial data structures.

Acknowledgments This work was supported by the National Research Foundation of Korea (NRF) grant funded by the Korean government (MOE) (No. 2012R1A1A2008958).

Appendix: A Single-Kernel Implementation for the Triangle-Sorting Process (Sect. 13.2.1.2)

```
/* This kernel corresponds to the fourth and fifth steps of the large node stage
described in [11]. */

__global__ void MedianSplitChunk (float *TriAABB, int *ChunkNodeIDs,
                                  int *NodeTriOffsets, int *NodeTriNums,
                                  int *ChunkStartIndices, int *ActiveNodeList,
                                  char *NodeSplitAxes, float *NodeSplitPoss,
                                  int *ChunkOffsets, int*NextNodeList) {

    __shared__ volatile int LChildTriOffsets2[9], RChildTriOffsets2[9];
    __shared__ int LChildNodesID, RChildNodesID, LOffset, ROffset;

    int LaneID = threadIdx.x & 0x0000001f;
    int NodeID = ChunkNodeIDs[CurBlockIndex];
    int CurBlockIndex = blockIdx.x
    int TriNum = NodeTriNums[NodeID], TriOffset = NodeTriOffsets[NodeID];
    int LChildTriOffsets, RChildTriOffsets;

    if (threadIdx.x < 9)
        LChildTriOffsets2[threadIdx.x] = RChildTriOffsets2[threadIdx.x] = 0;

    int TriIndex, StartPos = ChunkStartIndices[CurBlockIndex];
    int CurPos = StartPos + threadIdx.x;

    /* Classify the current triangle w.r.t. splitting plane. */
    unsigned int LChildTriBitFlag = 0, RChildTriBitFlag = 0;

    if (CurPos < TriNum) {
        /* The last chunk may have fewer than 256 triangles. */
        int SplitAxis = NodeSplitAxes[NodeID];
        float SplitPos = NodeSplitPoss[NodeID];

        TriIndex = ActiveNodeList[TriOffset+ CurID];
        float MinPos = TriAABB[TriIndex + SplitAxis * TRI OFFSET];
        float MaxPos = TriAABB[TriIndex + (SplitAxis + 3) * TRI OFFSET];
        LChildTriBitFlag = (MinPos < SplitPos);
        RChildTriBitFlag = (MinPos >= SplitPos);
        if (LChildTriBitFlag)
                RChildTriBitFlag = (SplitPos < MaxPos);
    }

    /* Perform intra-warp scan. */
    unsigned int LeftMask = ballot(LChildTriBitFlag), LaneMaskLT = 0;
```

```
unsigned int RightMask = ballot(RChildTriBitFlag), LaneMaskLE = 0;

asm("mov.u32 %0, %%lanemask lt;" : "=r"(LaneMaskLT));
asm("mov.u32 %0, %%lanemask le;" : "=r"(LaneMaskLE));

LChildTriOffsets = popc(LeftMask & LaneMaskLT);
RChildTriOffsets = popc(RightMask & LaneMaskLT);

if (LaneID == 31) {
    LChildTriOffsets2[(threadIdx.x >> 5) + 1] = popc(LeftMask & LaneMaskLE);
    RChildTriOffsets2[(threadIdx.x >> 5) + 1] = popc(RightMask & LaneMaskLE);
}
syncthreads();

/* Perform inter-warp scan. */
float Scan8[2];
if (threadIdx.x < 8) {
    Scan8[0] = LChildTriOffsets2[threadIdx.x + 1];
    Scan8[1] = RChildTriOffsets2[threadIdx.x + 1];

    for (int i = 1; i <= 4; i *= 2) {
        float n0 = shfl up(Scan8[0], i, 8);
        float n1 = shfl up(Scan8[1], i, 8);

        if (LaneID >= i) {
            Scan8[0] += n0;
            Scan8[1] += n1;
        }
    }
}

if (threadIdx.x < 8) {
    LChildTriOffsets2[threadIdx.x + 1] = Scan8[0];
    RChildTriOffsets2[threadIdx.x + 1] = Scan8[1];
}

/* Fetch start positions for the current chunk. */
if (threadIdx.x == 0) {
    LChildNodesID = 2*NodeID; RChildNodesID = 2*NodeID + 1;

    LOffset = atomicAdd (&ChunkOffsets[LChildNodesID], LChildTriOffsets2[8]);
    ROffset = atomicAdd (&ChunkOffsets[RChildNodesID], RChildTriOffsets2[8]);
}
syncthreads();

LChildTriOffsets += LChildTriOffsets2[(threadIdx.x >> 5)];
```

```
        RChildTriOffsets += RChildTriOffsets2[(threadIdx.x >> 5)];

        if (LChildTriBitFlag != 0)
            NextNodeList[LOffset + LChildTriOffsets] = TriIndex;

        if (RChildTriBitFlag != 0)
            NextNodeList[ROffset + RChildTriOffsets] = TriIndex;
}
```

References

1. Wald, I., Havran, V.: On building fast kd-trees for ray tracing, and on doing that in O(Nlog N). In: Proceedings of the EEE Symposium on Interactive Ray Tracing, pp. 61–69 (2006)
2. Shevtsov, M., Soupikov, A.: Highly parallel fast Kd-tree construction for interactive ray tracing of dynamic scenes. Comp Graph Forum (Proceedings of Eurographics) **26**:395–404 (2007)
3. Zhou, K., Hou, Q., Wang, R., Guo, B.: Real-time KD-tree construction on graphics hardware. ACM Trans. Graph. **27**, 1–11 (2008)
4. Hou, Q., Sun, X., Zhou, K., Lauterbach, C., Manocha, D.: Memory-scalable GPU spatial hierarchy construction. IEEE Trans. Vis. Comput. Graph. **17**, 466–474 (2011)
5. Choi, B., Komuravelli, R., Lu, V., Sung, H., Bocchino, R., Adve, S., Hart, J.: Parallel SAH k-D tree construction. In: Proceedings of High-Performance Graphics (HPG'10), pp. 77–86 (2010)
6. Wu, Z., Zhao, F., Liu, X.: SAH KD-tree construction on GPU. In: Proceedings of High Performance Graph (HPG'11), pp. 71–78 (2011)
7. CUDPP Google Group.: CUDA data parallel primitives library release 2.0. http://code.google.com/p/cudpp/ (2011). Accessed 1 June 2013
8. Sengupta, S., Harris, M., Garland, M., Owens, J.: Efficient parallel scan algorithms for many-core GPUs. In: Scientific Computing with Multicore and Accelerators, Taylor & Francis, pp. 413–442 (2011)
9. NVIDIA.: CUDA C programming guide: design guide (PG-02829-001 v5.0) (2012)
10. Skjellum, A., Whittaker, D., Bangalore, P.: Ballot counting for optimal binary prefix sum. In: Presented in the GPU Technology Conference 2010 (2010)
11. Manku, G.: Fast bit counting routines. http://cpptruths.googlecode.com/svn/trunk/c/bitcount.c (2002). Accessed 1 June 2013

Chapter 14
Fast Approximate k-Nearest Neighbours Search Using GPGPU

Niko Lukač and Borut Žalik

Abstract The k-nearest neighbours (k-NN) search is one of the most critical non-parametric methods used in data retrieval and similarity tasks. Over recent years, fast k-NN processing for large amount of high-dimensional data is increasingly demanded. Locality-sensitive hashing is a viable solution for computing fast approximate nearest neighbours (ANN) with reasonable accuracy. This chapter presents a novel parallelisation of the locality-sensitive hashing method using GPGPU, where the multi-probe variant is considered. The method was implemented using CUDA platform for constructing a k-ANN graph. It was compared to the state-of-the-art CPU-based k-ANN and two GPU-based k-NN methods on large and multidimensional data set. The experimental results showed that the proposed method has a speed-up of 30× or higher, in comparison to the CPU-based approximate method, whilst retaining a high recall rate.

14.1 Introduction

Nearest neighbour search (NNS) is one of the oldest and most widely applied algorithms in computer science. In terms of computational geometry, it is defined as a problem of finding the closest point p to a query point q within the set of points $S \in \mathrm{R}^{\mathrm{D}}$ and is defined as [1]

$$p = \arg\min_{x \in S} \ d(q,x); x \neq q, \tag{14.1}$$

where d is a distance function. The k-nearest neighbour (k-NN) problem is an extension of NNS, where the k-nearest neighbours $p_1, p_2, \ldots, p_k$ are considered for q, where $d(p_1, q) \leq d(p_2, q) \leq, \ldots, d(p_k, q)$. Obviously, the problem has in the worst case quadratic computational complexity when considering all points for query. Namely, the brute-force algorithm calculates the distances between all points and

N. Lukač (✉) • B. Žalik
Faculty of electrical engineering and computer science, University of Maribor, Smetanova ulica, 17, SI-2000 Maribor, Slovenia

Y. Cai, S. See (eds.), *GPU Computing and Applications*,
DOI 10.1007/978-981-287-134-3_14

sorts them in ascending order. k-NN is widely applied in a broad range of applications in data retrieval and similarity assessment, where performance and accuracy play an important role. Hence, various methods have been developed to speed up the k-NN search by data partitioning using spatial indexing structures [2]. These approaches have an expected time complexity of O(n log(n)) when considering n points. Unfortunately, they are inefficient for high-dimensional data, due to the curse of dimensionality phenomena [3], where the time complexity of spatial indexing methods increases significantly as the dimensionality increases. Additionally, high-dimensional data (i.e. $D > 20$) is very sparse, and the Euclidean distance between the furthest and nearest points can be considerably small.

One possible solution is based on dimension reduction over data, which can be performed by the prior application of data-partitioning k-NN search in order to retain high performance. Such approaches perform approximate nearest neighbour search (ANN), where the exact k-NN is not always guaranteed. This compromise is feasible when considering large multidimensional data (e.g. image and audio features) [4]. During the last decade, one of the more popular choices for ANN is the use of hashing functions that retain spatial locality: if two points are close in R^D, then there is a high chance they are close within the hashed domain, which generally consists of buckets. Various hashing methods have been developed, such as: cosine similarity [5], spectral hashing [6] and locality-sensitive hashing (LSH) [4, 7, 8].

Another way to solve this problem is by parallelisation of the time-consuming brute-force k-NN method. This substantially decreases the runtime and is even faster than any CPU-based approximate method. However, as the data sets become even larger in size and dimensionality, a parallel k-ANN method would perform reasonably better than parallel exact k-NN methods. This chapter proposes a novel parallel version of LSH-based k-ANN using general-purpose computing on a graphics processor unit (GPGPU). The LSH family that is based on p-stable distributions is considered for parallel hashing, since the hash calculation is easily parallelisable as no synchronisation between the points is required. Then the multi-probe LSH query algorithm is used in parallel over all points in order to speed up the query process. The query is by default less parallelizable, since there is no coalesced memory access. Therefore, an efficient approach has been developed by using fast parallel sorting of buckets' indices prior k-ANN query. The multi-probe query was efficiently parallelized by using new approximate scoring criteria. Furthermore, the method takes advantage of the skip-list data structure for faster update of the k-ANN results.

This chapter is organised as follows: Sect. 14.2 discusses the related works on parallel k-NN computation, GPGPU-based k-NN methods, and LSH-based methods on GPU. A brief theoretical overview of LSH and MLSH is provided in the Sect. 14.3, followed by the description of the proposed GPGPU-based method. Section 14.4 presents the results from the experiments, where the proposed method was implemented using NVIDIA's Compute Unified Device Architecture (CUDA) [9] and compared to the CPU-based LSH approach, as well as two GPU-based k-NN methods. Final section concludes this chapter.

14.2 Related Work

Fast k-NN computation has been a long studied problem in computer science, where several parallelisations have been proposed over the past decade [10–12]. With the increasing technological advancements of the GPUs, new methods based on GPGPU have been developed for fast k-NN computation. Garcia et al. [13] proposed one of the first GPU-based parallelisation of the brute-force k-NN approach, where they reported a 120× speed-up in comparison to the CPU-based implementation. Qiu et al. [13] proposed a parallel NNS for calculating the iterative closest point when solving the 3D registration problem. They used an array-based kd-tree for accelerating the k-NN search on GPU. Liang et al. [14] proposed CUKNN, a parallel k-NN implementation on CUDA. Their approach is based on a local k-NN calculation for each block of threads, then merging them in order to obtain a global k-NN. They also took advantage of CUDA's streaming capabilities. Leite et al. [15] proposed an efficient parallel scheme for nearest neighbour search in 3D point clouds on GPU. Their approach subdivides the data into cells in order to gain spatial locality and efficient parallelisation. Yeh et al. [16] developed an efficient GPU-based k-NN search using kd-tree, where they performed fast parallel radix sort for calculating the median values in kd-tree construction. Garcia et al. [17] proposed a newer GPU-based k-NN approach, by using the cuBLAS (CUDA Basic Linear Algebra Subroutines) library in order to efficiently calculate a parallel distance matrix for faster brute-force k-NN parallelisation. Barrientos et al. [18] improved nearest neighbour computation on GPU by using parallel lists of clusters (LC) and SS-index strategies in order to perform fast range search and k-NN, respectively. Arefin et al. [19] have recently proposed fast and scalable k-NN computation using GPUs, where the distance matrix is divided into smaller chunks in order to parallelise distance calculations and k-NN search over these sub-matrices. Sismanis et al. [20] have recently proposed a parallel k-NN implementation by using truncated bitonic sort in order to speed up the query computation.

Since the emerging of LSH a few years ago, only a few parallel k-ANN methods using LSH have been developed. Haghani et al. [21, 22] proposed a distributed version of LSH based on p-stable distributions for large-scale structured peer-to-peer networks. Their approach uses dual-level mapping from D-dimensional space to peer identifier space. Pan et al. [23] presented one of the first known GPU-based implementation of LSH for the motion planning application. Their method is based on parallel radix sort, difference, and prefix-sum operations over created buckets, in order to speed up the parallel query by knowing the size and starting position of each bucket. The authors used the parallel cuckoo hashing approach in order to quickly find the location of a bucket for a given query point. The more recent GPGPU-based approach by Pan and Manocha [24] is a novel k-NN parallelisation, where the bi-level LSH is used coupled with a parallel RP-tree indexing structure. Their hash tables are based on parallel cuckoo hashing and Morton curves. They accelerate the query process by using the clustered-sorting of output buckets.

14.3 Parallel Multi-probe LSH

The first subsection briefly introduces LSH and MLSH k-ANN approaches, whilst in the second subsection, the proposed GPGPU-based method for parallelisation of MLSH for constructing k-ANN graph is presented by using CUDA platform.

14.3.1 Locality-Sensitive Hashing

LSH as introduced by Indyk and Motwani [7] is an efficient randomised method for finding the ANN by performing probabilistic dimensional reduction. It has been successfully used in various applications, see [4] for an overview. The basic idea is that close neighbours within the R^D Euclidean space have a high probability of being close within the LSH-hashed space U. Therefore, for any two points $q_1, q_2 \in R^D$ within distance r, the following two premises hold for a locality-sensitive hash function $h: R^D \rightarrow U$ [7]:

- If $d(q_1,q_2) \leq r$, then there is a greater probability than P_1 that their hashed values are also close $[h(q_1) == h(q_2)] \geq P_1$.
- If $d(q_1,q_2) > cr$, then there is a lower probability than P_2 that their hashed values are also close $[h(q_1) == h(q_2)] \leq P_2$.

Generally, the constant c is considered greater than 1; hence, nearby points have a higher probability of being close in U than the far apart points (i.e. $P_1 > P_2$). LSH works in two steps, namely, initialisation and query. In the initialisation, the points are hashed into locality-sensitive buckets. The query step is then used to search the k-ANN of a given query point q by calculating the distances $d(q, x)\ \forall\ [h(q) == h(x)]$. The LSH-based k-ANN query is considerably faster than the brute-force k-NN query, due to smaller search space, where only the distances between the points are calculated that are within the same buckets.

The popular LSH family based on the p-stable distributions proposed by Datar et al. [25] is considered throughout this chapter. The basic idea is as follows: if the points within R^D that are near by using the l_p distance are projected onto a one-dimensional line, there is a high chance that they are also near on that given line. The hash function they proposed is defined as:

$$h^{a,b}(q) = \left\lfloor \frac{a \cdot q + b}{w} \right\rfloor, \tag{14.2}$$

where a is a random vector in R^D that belongs to p-stable distribution (e.g. if $p = 2$, then Gaussian distribution is considered), whilst b is a uniformly chosen random value in the [0, w] range. w is the bucket width that defines the resolution of the quantisation. If w is too large, too many points fall into the same bucket, whilst the opposite happens if w is too small. The former considerably reduces the method's

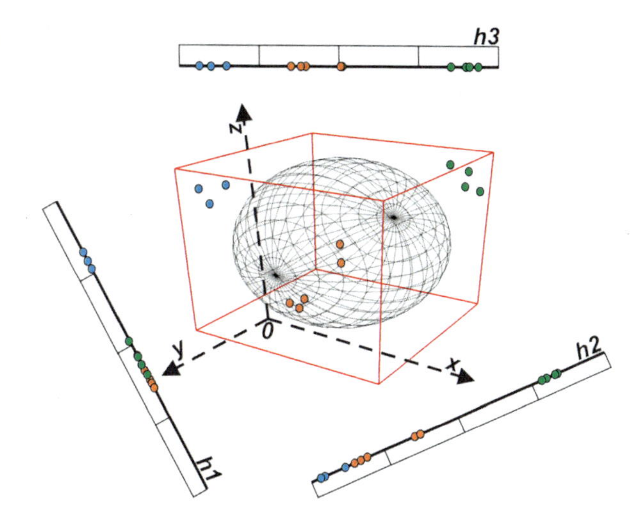

Fig. 14.1 Illustration of the LSH in R^3, where the points' projections are shown for three hash functions

efficiency, whilst the later affects the accuracy. Hence, LSH is sensitive to the choice of the hash function's parameters, which can be calculated using statistical knowledge derived from the data (see Slaney et al. [8]). After the points are hashed into buckets by using $h^{a,b}(q)$, a linear search is performed between the points within the same buckets in order to find the ANN. In practice, one hashing function does not suffice (i.e. unfortunate projection error), as shown in Fig. 14.1.

Therefore, K hash functions are considered [26] in order to increase the recall of the true nearest neighbours. Moreover, L hash tables are constructed in order to decrease the collision probability of false neighbours. Hence, the accuracy of LSH is significantly improved as the amount of hash tables increases. However, this also increases the algorithm's runtime. This represents the time-quality trade-off of LSH. In this chapter, the E2LSH [27] approach for hash tables' construction is considered, where the buckets obtained with per-table hash functions are quantised into a single bucket. They defined the hash-table function as [27]

$$H^{r,K}(q) = \left\lfloor \sum_{j=1}^{K} \left(|h_j(q)| r \mod M \right) \right\rfloor \mod c, \tag{14.3}$$

where M is a large prime close to the maximum number supported on a given architecture in order to avoid integer overflow and r is a smoothing parameter—a uniformly randomly chosen value between [0, c], where c denotes the maximum number of buckets (i.e. compressed hashing). The given hash-table construction approach and the reduced domain space of the buckets are highly suitable for parallelisation on GPU, due to memory constraints.

Lv et al. [28] proposed MLSH, which nowadays is one of the most popular improvements to the standard LSH. Given the nature of the LSH, there is a certain probability that the true k-NN can be located within the neighbouring buckets of a given bucket where the query point is hashed to (see Fig. 14.2).

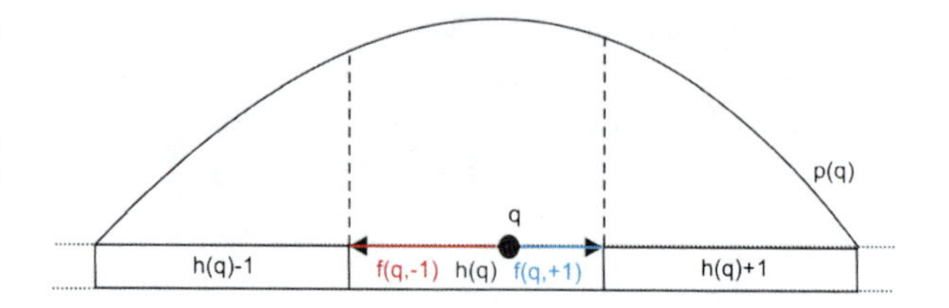

Fig. 14.2 Illustration of the probability density function $p(q)$ of a hashed point q, where there is a chance that the neighbouring buckets of $h(q)$ contain true k-NN

The hashing of q by using L hash tables can be denoted by the L-dimensional vector $v = (H_1(q),\ldots,H_L(q))$. MLSH is based on the perturbation vector $\Delta = (\delta_1(q),\ldots, \delta_L(q))$ that defines which neighbouring buckets need to be probed by considering $v + \Delta = (H_1(q) + \delta_1(q),..,H_L(q) + \delta_L(q))$. In this chapter, the set $\delta = \{-1, 0, +1\}$ is considered, as these buckets are most probable candidates when w is large enough. In order to determine the best Δ, the probability of each bucket $h_j(q) + \delta_i$ containing k-NN of q has to be taken into account. Therefore, the distances from q to neighbouring buckets $h_j(q) + \delta_i$ for $\delta_i = \{-1, +1\}$ are calculated (see Fig. 14.2):

$$f_j(q, -1) = \left(a_j \cdot q + b_j\right) - h_j(q)w, \tag{14.4}$$

$$f_j(q, +1) = w - f_j(q, -1). \tag{14.5}$$

Since the hash-table construction is based on E2LSH [27], it is necessary to calculate the distances to neighbouring buckets for each hash table $H_i(q) + \delta_i$. Although these distances may be less accurate, due to the quantised nature of $H_i(q)$, a viable approximation is proposed:

$$g_i(q, \delta_i) = \frac{1}{K}\sum_{j=1}^{K} f_j(q, \delta_i). \tag{14.6}$$

This is based on the premise that if q tends to be close to $h_j(q) + \delta_i$ for all K hashing functions within $H_i(q)$, then q is also close to $H_i(q) + \delta_i$. After the calculation of $g_i(q, \delta_i)$ for each $H_i(q)$, the best Δ can be estimated. MLSH uses query-directed probing in order to reduce the dimension of Δ by not probing neighbouring buckets of each hash function. This chapter proposes a simplified probing, due to the quantised nature of $H_i(q)$. The threshold $\tau \in [0, 5]$ is defined as an input parameter, where only the neighbouring buckets that satisfy the following approximate scoring criteria are used:

$$\Psi(q, \delta_i) = \begin{cases} 1 & \tau - g_i(q, \delta_i) > 0 \\ 0 & else \end{cases}. \tag{14.7}$$

Although these scoring criteria are less accurate than the probing that is originally used with MLSH, the increase of hash tables should improve the accuracy. In comparison to the standard LSH, the number of hash functions and tables is still lower, which significantly improves the initialisation speed and decreases the memory consumption.

14.3.2 Parallel MLSH Using CUDA

In the initial phase, the D-dimensional data is copied into 1D global memory residing on the GPU, and the L hash tables are constructed on the CPU. They are transferred to GPU's read-only constant memory, in order to allow fast parallel hash-table retrieval in a broadcast fashion. Then the parallel kernel of the proposed GPU-based MLSH method is executed on the given data in order to compute k-ANN graph. Since the LSH hashing can be performed independently for each point for the amount of L hash tables, the parallelisation of the initialisation (i.e. hashing) is highly suitable. As mentioned earlier, the output consists of L buckets' indices for each hashed point. This would require $4Ln$ bytes of additional memory, since it is considered that 2^{32}-1 suffices as the maximum amount of unique buckets per hash table. This is unacceptable in practice, due to the limited amount of memory available on the GPU. Hence, the proposed approach performs parallel hashing per one hash table at a time, whilst also executing the parallel k-ANN query. As shown in Sect. 14.4, the total LSH construction time is only a fraction of the total runtime. The total amount of memory for storing the output buckets indices for all points then remains at $4n$ bytes through the entire runtime. Other additional required memory is at $2(4kn)$ bytes, in order to store the output indices and distances of k-ANN for each point. Both of these reside in the GPU global memory. The host mapped pinned memory can be used, in case there is insufficient global memory.

After all the points have been hashed by i-th hashing table, the proposed k-ANN parallel query is executed, as shown in the Fig. 14.3. In order to execute a fast LSH-based query, given point distances are calculated to other points that were hashed into the same buckets. Since the point ordering is different than the resulting per-point bucket indices, this would introduce non-coalesced memory access and would significantly slow down the query process. Therefore, the buckets' indices are sorted in ascending order by using fast parallel radix sort proposed by Merrill and Grimshaw [29]. This is performed on the key-value $(H_i(q_j), q_j)$ basis for each point q_j. Therefore, the points are also sorted depending on the order of the buckets. Thus, a point-bucket locality is established, where close points within the same buckets are also close in memory. This allows coalesced memory access, whilst additional $4n$ bytes of memory are required for storing the points' indices. Of course, sorting the data can be quite demanding. However, since only buckets' integer indices need to be compared, this is performed very fast. As shown in the results within Sect. 14.4, the number of buckets is only a fraction of n. The parallel k-ANN query is then performed, where the distance computations are done between the points within the same bucket. As shown in the example in Fig. 14.3, the resulting output memory for storing k-ANN results is also coalesced, where

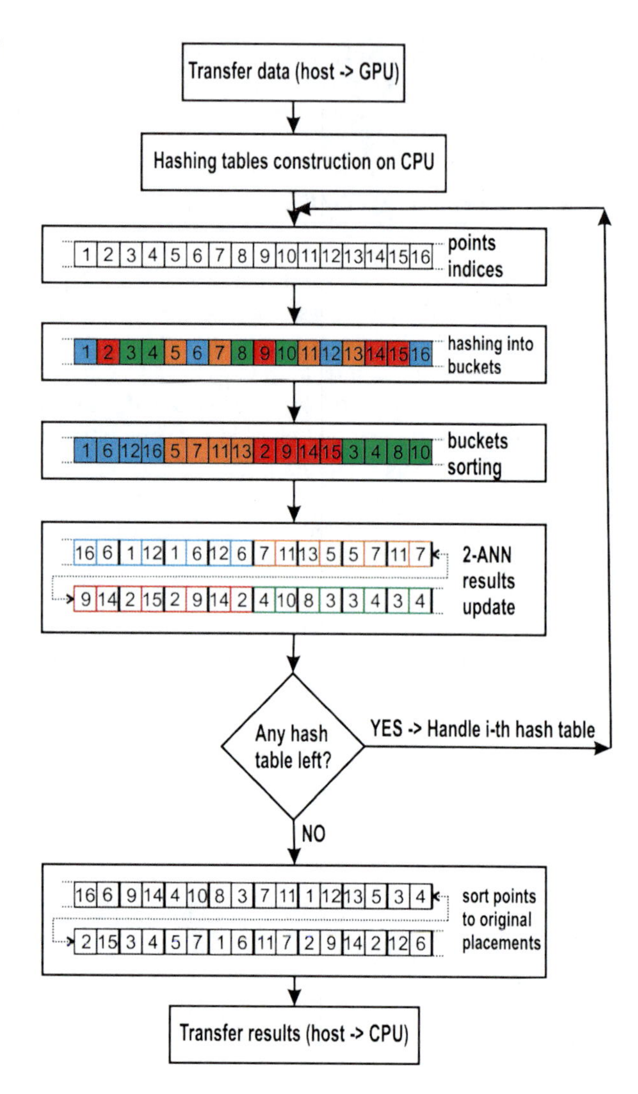

Fig. 14.3 Flow diagram of the proposed parallel LSH-based method, whilst also illustrating an example of 2-ANN

k slots are reserved for each point in order to store the indices of its k-ANN. Before the results are transferred back to host memory, they are sorted based on original placements of the points indices.

Considering CUDA's logical parallel execution model, the threads are grouped into blocks, and these are grouped into grids. CUDA is generally considered as STMD (Single Thread Multiple Data), since all threads execute the same parallel kernel code over different data. CUDA-supported GPUs have hardware consisting of stream multiprocessors (SMs) that are further composed of stream processors. The block size is set as a multiple of the size of a warp (e.g. 16 or 32 threads)—the SMs' core scheduling unit. When a given warp is delayed by a memory IO, several cycles of computation can be used in the meantime for executing workload from another warp. Therefore a block size that is a multiple of a warp size is the best choice. In order to hide latency, and maximise occupancy, each thread handles one point from the sorted points' data set by their corresponding bucket keys.

Figure 14.4 shows the pseudocode of the proposed GPGPU method. A given thread is responsible for calculating the distances of the assigned point to other points within the same bucket. Since the buckets are not memory-parallel with the blocks, the thread needs to probe neighbouring points within the 1D memory (left and right) in order to check whether they have the same bucket index. In order to speed up this initial probing, the buckets' sizes and positions within the sorted data set are calculated prior the k-ANN query is being performed (see Fig. 14.4 lines 12–13). The sizes are calculated by using a parallel histogram method [30], whilst the positions are calculated using a simple parallel check if the next point in the memory has a different bucket index, which corresponds to the beginning of the new bucket. Both of these operations require minute computing time, as shown in the next section. The calculated buckets' sizes and positions are then copied from the device's global memory to its constant memory (line 14). This requires in total $2c$ entries ($8c$ bytes) in constant memory, which is feasible due to the expected low value of c (number of buckets). During the k-ANN query (line 15), for each thread, it is then known which points reside in the same bucket. The distances to these neighbouring points are computed sequentially per each thread. When a given thread responsible for point q computes the distance to point p, the stored distance of the k-th neighbour of q is compared to $d(q, p)$. If the distance is higher, then the k-ANN result has to be updated, where the index and the distance of p are stored as an appropriate k-th neighbour of q. Such checking of the k-th neighbour is feasible, since the k-ANN results are always ordered in ascending order.

In order to speed up the k-ANN update (i.e. finding the k for p), a deterministic skip-list (DSL) [31] data structure is used to efficiently store the k-ANN in a sorted list. The DSL has an expected time complexity of log(k) for the insertion of the new element into the sorted set of k-ANN. By deterministically comparing $d(q, p)$ with the stored distance of the z_0-th neighbour, where $z_0 \in [0, k]$, then consequently it is known whether the given point p is located within the interval $\alpha_0 = [0, z_0]$ or $\beta_0 = [z_0, k]$. z_0 is located at the 0-th level of the DSL, where multiple levels can be defined (e.g. z_1 neighbour that splits the parent interval α_0 or β_0), as shown in Fig. 14.5. DSL does not require any additional memory for storing the results, as the levels are defined deterministically a priori, and allows very fast updating of k-ANN. This is desired especially in cases where k is large, since multiple calculated distances within k-ANN are unrequired for checking. This provides a flexible and fast alternative than sorting the k-ANN or by using a binary or heap-tree array-based structure [31]. Once the new neighbour is inserted into the DSL, the distances are compared in linear order at the last level, until $d(q, p)$ is larger than a given already calculated distance of a u-th neighbour or $d(q, p)$ is the new closest neighbour. The new point is stored as the $(u+1)$-th or 1st neighbour, and the remaining $[(u+2), k]$ or $[1, k]$ neighbours are repositioned for one place forward, where the old k-th neighbour is erased.

```
1: Input: D-dim dataset S,L,K,w,c,τ
2: transfer S to GPU memory
3: generate L hash tables on host
4: initialize GPU constant memory (n,L,K,w,c,τ)
5: allocate GPU memory for knn_dist,knn,bucket_sizes,bucket_pos,
                           pts_ind,pts_buckets,pts_scores
6: kernel_initialize(knn_dist, FLT_MAX)
7: kernel_initialize(pts_ind)
8: for i=1 to L do
9:      transfer i-th hash table to GPU constant memory
10:     kernel_LSH_hashing(pts_buckets,S,pts_scores)
11:     kernel_sort_buckets(pts_buckets,pts_ind,S,knn_dist,knn,pts_scores)
12:     kernel_calc_buckets_pos(pts_buckets,bucket_pos)
13:     kernel_calc_buckets_size(pts_buckets,bucket_sizes)
14:     transfer bucket_pos and bucket_sizes to GPU constant memory
15:     kernel_querry_aknn(pts_buckets,pts_ind,S,knn_dist,knn,pts_scores)
16: done
17: kernel_sort_pts_indices(pts_buckets,S,knn_dist,knn)
18: transfer knn_dist and knn to host memory
```

Fig. 14.4 Pseudocode of the proposed GPGPU method

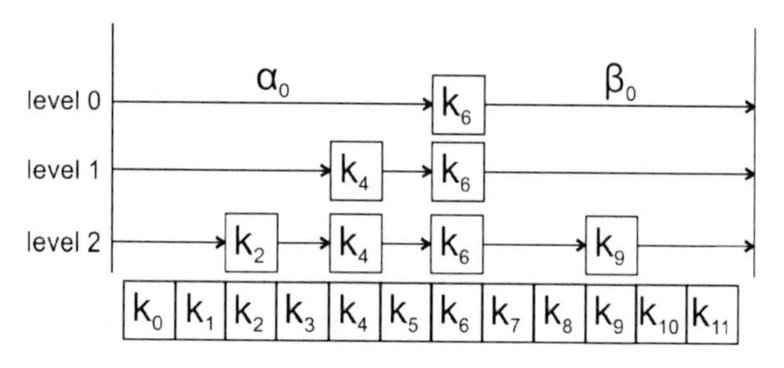

Fig. 14.5 Illustration of the DSL data structure for efficient storage of the sorted k-ANN result for a given query point. In this example, $z_0 = k_6$ and $z_1 = k_4$

The proposed parallel MLSH is a straightforward upgrade of the previously proposed parallel LSH method. Each point contains two calculated scores $\Psi(q, \delta_i)$ for the left and right buckets, respectively. These are calculated on the fly during the hashing phase (see Fig. 14.4 line 10) and additionally require $2n$ bytes of memory. During the k-ANN query phase, the thread probes additional points from the neighbouring buckets based on the MLSH score of the assigned point. In order to speed up this process, the left bucket's starting position and the right bucket's size are used.

14.4 Results

For the input data, the TEXMEX repository was used (see http://corpus-texmex.irisa.fr/) [32]. It consists of 1,000,000 128-dimensional points representing SIFT (Scale-invariant feature transform) descriptors for images. The NVIDIA TESLA C2050 GPU that uses the compute capability 2.0 was used for the experiments. At first, the timing of each main step in the proposed method was extensively analysed by using NVIDIA's visual profiler [33], as shown in Fig. 14.6. As can be seen, the

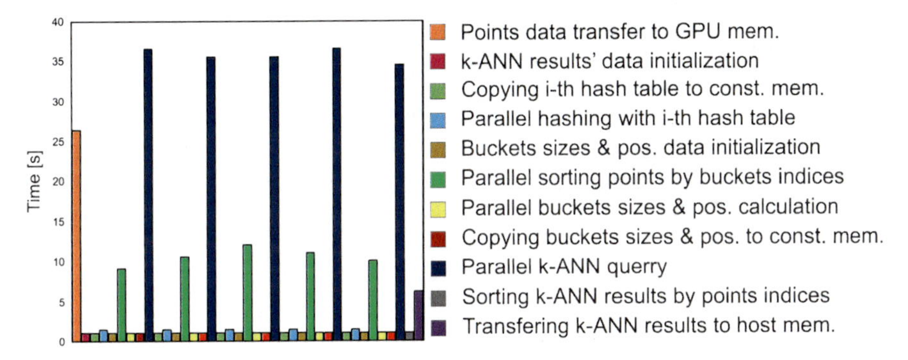

Fig. 14.6 Analysis of the timings for the main steps within parallel MLSH runtime, where the MLSH parameters were set at $L = 5$ and $K = 10$. The query was performed over the entire data set of $n = 1{,}000{,}000$ points, where $k = 100$

hash computation occupies only a fraction of the total time, whilst the k-ANN query is the most computationally intensive task. The radix sorting of data by bucket indices is the second most expensive operation. However, without this sorting, the query would have non-coalesced memory access, which would substantially decrease the search performance. The CUDA cache configuration was set to emphasise L1 caching over shared memory. The visual profiler reported of over 90 % L1 cache hit-rate during the k-ANN query, which is one of the most important performance factors. When measuring the total runtime, the data transfer from the host memory to the GPU, and vice versa, was also considered.

The method was not directly implemented on CPU, since several operations would be redundant due to architectural differences. Therefore, the proposed method was compared with the state-of-the-art CPU-optimised MLSH method (namely, LSHKIT; see http://lshkit.sourceforge.net/) [34] using Intel i7-950 3.07 Ghz CPU, in terms of runtime and speed-up, as shown in Fig. 14.7. Furthermore, the method was compared with the state-of-the-art GPGPU-based exact k-NN methods by Garcia et al. [13] (see http://vincentfpgarcia.github.io/kNN-CUDA/) and Sismanis et al. [20] (see http://autogpu.ee.auth.gr/doku.php?id=software). This was done whilst increasing the size of the input data (see Fig. 14.7a) with $k = 100$, or by increasing the number of nearest neighbours (see Fig. 14.7b), where the number of points was constantly at 100,000. The number of query points was set to the same as the number of input points (i.e. constructing a k-NN graph). The method proposed by Garcia et al. [17] was only tested on data with sizes lower than 50,000, since the method in its current version does not support bigger data sets. As expected, all the GPU-based methods are faster than the CPU-based k-ANN method, when the data set is reasonably high enough (i.e. $n \geq 1{,}000$). However, the proposed GPU-based k-ANN method is faster than the exact k-NN methods on GPU when the data sets become large enough. Moreover, the exact k-NN methods use matrices multiplications for calculating the distance matrix, which consumes considerably more memory than the proposed method.

The speed-up in comparison to the state-of-the-art CPU-based k-ANN when considering increasing input was up to 30× and increasing with the amount of input

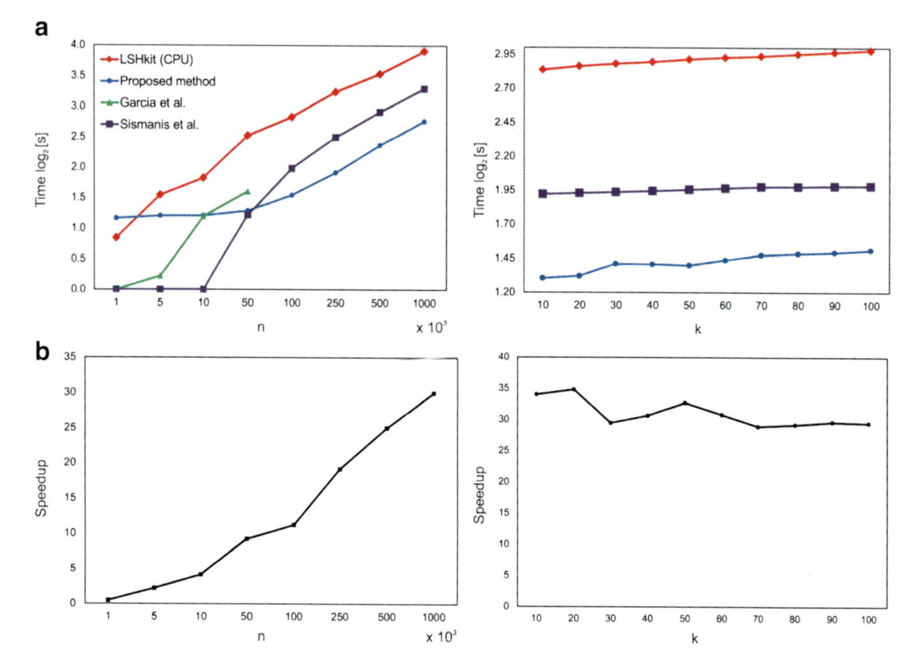

Fig. 14.7 Comparison between proposed GPU-based MLSH with CPU-based MLSH and two GPU-based k-NN methods, in terms of (**a**) timing and (**b**) speed-up, as the input data increases (*left-hand side*) or the amount of nearest neighbours increases (*right-hand side*)

data. When considering the increase of k, the speed-up was converging towards $30\times$. The MLSH parameters were initialised at $L=2$ and $K=10$. The number of buckets was initialized to $c=100$, which then increased for 250 after each increase of n. The number of used hash tables L was gradually increased by 2 as n and k increased during the experiments. This resulted in average recall of above 90 %. In the proposed parallel MLSH, the additional parameter for scoring was set as $\tau=0.25$. The number of threads per block was set at 256, in order to achieve full occupancy.

Conclusion

This chapter proposed a new GPGPU-based method for fast parallel k-ANN computation over multidimensional data sets. The nature of the considered LSH method allows adequate parallelisation on the GPU, where points are independently hashed into buckets, whilst simple scoring criteria are used in multi-probe LSH in order to speed up the calculation. The k-ANN is updated in parallel for each query point, where a deterministic skip-list data structure is used to hold the k-ANN neighbours' indices and the distances for each query point. The experimental results have shown that the proposed method is

(continued)

substantially faster than the CPU-based multi-probe LSH. For future work, the method could be extended to store the hashing results (i.e. construction phase), in order to be viable for fast incremental and streaming applications.

Acknowledgements Thanks to the TEXMEX Research Team for publicly providing the SIFT data set for research purposes. This work was supported by Slovenian Research Agency under research contracts 1000-13-0552, P2-0041, and J2-5479.

References

1. Cover, T.M., Hart, P.E.: Nearest neighbour pattern classification. IEEE Trans. Inf. Theory **13** (1), 21–27 (1967)
2. Samet, H.: Foundations of MultiDimensional and Metric Data Structures. Morgan Kaufmann, San Francisco, CA (2006)
3. Beyer, K., Goldstein, J., Ramakrishnan, R., Shaft, U.: When is "nearest neighbor" meaningful?. In: Proceedings Database Theory (ICDT'99), pp. 217–235 (1999)
4. Andoni, A., Indyk, P.: Near-optimal hashing algorithms for approximate nearest neighbor in high dimensions. In: Proceedings of the 47th Annual IEEE Symposium on Foundations of Computer Science, pp. 459–468 (2006)
5. Charikar, M.S.: Similarity estimation techniques from rounding algorithms. In: Proceedings of the thiry-fourth annual ACM symposium on Theory of computing, pp. 380–388 (2002)
6. Weiss, Y., Torralba, A., Fergus, R.: Spectral hashing. In: Advances in neural information processing systems (NIPS'08), pp. 1753–1760 (2008)
7. Indyk, P., Motwani, R.: Approximate nearest neighbors: towards removing the curse of dimensionality. In Proceedings of the thirtieth annual ACM symposium on Theory of computing, pp. 604–613 (1998)
8. Slaney, M., Lifshits, Y., He, J.: Optimal parameters for locality-sensitive hashing. Proc. IEEE **100**(9):2604–2623 (2012)
9. NVIDIA Corporation.: NVIDIA CUDA C Programming Guide (Version 4.2). http://docs.nvidia.com/cuda/cuda-c-programming-guide/index.html (2013)
10. Gao, Y., Chen, L., Chen, G., Chen, C. Efficient parallel processing for K-nearest-neighbor search in spatial databases. In: Computational Science and Its Applications-(ICCSA 2006), pp. 39–48 (2006)
11. Aparício, G., Blanquer, I., Hernández, V.: A parallel implementation of the k nearest neighbours classifier in three levels: Threads, MPI processes and the grid. In: High Performance Computing for Computational Science (VECPAR 2006) pp. 225–235 (2007)
12. Zhang, C., Li, F., Jestes, J.: Efficient parallel kNN joins for large data in MapReduce. In: Proceedings of the 15th International Conference on Extending Database Technology, pp. 38–49 (2012)
13. Garcia, V., Debreuve, E., Barlaud, M.: Fast k nearest neighbor search using GPU. In: Proceedings of the IEEE Computer Society Conference on Computer Vision and Pattern Recognition Workshops (CVPRW'08), pp. 1–6 (2008)
14. Liang, S., Wang, C., Liu, Y., Jian, L.: CUKNN: a parallel implementation of K-nearest neighbor on CUDA-enabled GPU. In: Proceedings of the IEEE Youth Conference on Information, Computing and Telecommunication (YC-ICT'09), pp. 415–418 (2009)
15. Leite, P., Teixeira, J.M.X.N., de Farias, T.S.M.C., Teichrieb, V., Kelner, J.: Massively parallel nearest neighbor queries for dynamic point clouds on the GPU. In: Proceedings of the IEEE

21st International Symposium on Computer Architecture and High Performance Computing (SBAC-PAD'09), pp. 19–25, (2009).
16. Yeh, T.T., Chen, T.Y., Chen, Y.C., Shih, W.K.: Efficient parallel algorithm for nonlinear dimensionality reduction on GPU. In: Proceedings of the 2010 I.E. International Conference on Granular Computing, pp. 592–597 (2010)
17. Garcia, V., Debreuve, E., Nielsen, F., Barlaud, M. K-nearest neighbor search: Fast GPU-based implementations and application to high-dimensional feature matching. In: Proceedings of the 17th IEEE International Conference on Image Processing, pp. 3757–3760 (2010)
18. Barrientos, R.J., Gómez, J.I., Tenllado, C., Matias, M.P., Marin, M.: kNN query processing in metric spaces using GPUs. In: Euro-Par 2011 Parallel Processing, pp. 380–392 (2011)
19. Arefin, A.S., Riveros, C., Berretta, R., Moscato, P.: GPU-FS-kNN: A Software Tool for Fast and Scalable kNN Computation Using GPUs. PloS One **7**(8), e44000 (2012)
20. Sismanis, N., Pitsianis, N., Sun, X.: Parallel search of k-nearest neighbors with synchronous operations. In: Proceedings of the IEEE Conference on High Performance Extreme Computing, pp. 1–6 (2012)
21. Haghani, P., Michel, S., Cudré-Mauroux, P., Aberer, K.: LSH at large-distributed KNN search in high dimensions. In: Proceedings of the 11th International Workshop on Web and Database (WebDB'08) (2008)
22. Haghani, P., Michel, S., Aberer, K.: Distributed similarity search in high dimensions using locality sensitive hashing. In: Proceedings of the ACM 12th International Conference on Extending Database Technology: Advances in Database Technology, pp. 744–755 (2009)
23. Pan, J., Lauterbach, C., Manocha, D.: Efficient nearest-neighbor computation for GPU-based motion planning. In: Proceedings of the IEEE/RSJ International Conference on Intelligent Robots and Systems, pp. 2243–2248 (2010)
24. Pan, J., Manocha, D.: Fast GPU-based locality sensitive hashing for k-nearest neighbor computation. In: Proceedings of the 19th ACM SIGSPATIAL International Conference on Advances in Geographic Information Systems, pp. 211–220 (2011)
25. Datar, M., Immorlica, N., Indyk, P., Mirrokni, V.S.: Locality-sensitive hashing scheme based on p-stable distributions. In: Proceedings of the ACM twentieth annual symposium on Computational geometry, pp. 253–262 (2004)
26. Slaney, M., Casey, M.: Locality-sensitive hashing for finding nearest neighbors [lecture notes]. EEE Signal Process. Mag. **25**(2), 128–131 (2008)
27. Andoni, A., Indyk, P.: E2LSH 0.1 user manual (2005)
28. Lv, Q., Josephson, W., Wang, Z., Charikar, M., Li, K.: Multi-probe LSH: efficient indexing for high-dimensional similarity search. In: Proceedings of the 33rd international conference on Very large data bases, pp. 950–961 (2007)
29. Merrill, D., Grimshaw, A.: High Performance and Scalable Radix Sorting: A case study of implementing dynamic parallelism for GPU computing. Parallel Processing Letters **21** (2):245–272 (2011)
30. Shams, R., Kennedy, R.A.: Efficient histogram algorithms for NVIDIA CUDA compatible devices. Proceedings of the Int. Conf. on Signal Processing and Communications Systems (ICSPCS) **418–422** (2007)
31. Munro, J.I., Papadakis, T., Sedgewick, R.: Deterministic skip lists. In: Proceedings of the 3rd ACM-SIAM Symposium Discrete Algorithms, pp. 367–375 (1992)
32. Hervé, J., Matthijs, D., Cordelia, S.: Product quantization for nearest neighbor search. IEEE Trans. Pattern Anal. Mach. Intell. **33**(1), 117–128 (2011)
33. NVIDIA Corporation.: NVIDIA Visual Profiler, Version 5.0 (2013).
34. Dong, W., Wang, Z., Josephson, W., Charikar, M., Li, K.: Modeling LSH for performance tuning. In the Proceedings of the 17th Conference on Information and Knowledge Management, pp. 669–678 (2008)
35. Qiu, D., May, S., Nüchter, A.: GPU-accelerated nearest neighbor search for 3D registration. Lecture Notes in Computer Science **5815**:194–203 (2009)

Chapter 15
Soft Computing Methods for Big Data Problems

Shafaatunnur Hasan, Siti Mariyam Shamsuddin, and Noel Lopes

Abstract Generally, big data computing deals with massive and high-dimensional data such as DNA microarray data, financial data, medical imagery, satellite imagery, and hyperspectral imagery. Therefore, big data computing needs advanced technologies or methods to solve the issues of computational time to extract valuable information without information loss. In this context, generally, machine learning (ML) algorithms have been considered to learn and find useful and valuable information from large value of data. However, ML algorithms such as neural networks are computationally expensive, and typically, the central processing unit (CPU) is unable to cope with these requirements. Thus, we need a high-performance computer to execute faster solutions such graphics processing unit (GPU). GPUs provide remarkable performance gains compared to CPUs. The GPU is relatively inexpensive with affordable price, availability, and scalability. Since 2006, NVIDIA provides simplification of the GPU programming model with the Compute Unified Device Architecture (CUDA), which supports for accessible programming interfaces and industry-standard languages, such as C and C++. Since then, general-purpose graphics processing unit (GPGPU) using ML algorithms are applied on various applications, including signal and image pattern classification in biomedical area. The importance of fast analysis of detecting cancer or non-cancer becomes the motivation of this study. Accordingly, we proposed soft computing methods, self-organizing map (SOM) and multiple back-propagation (MBP) for big data, particularly on biomedical classification problems. Big data such as gene expression datasets are executed on high-performance computer and Fermi architecture graphics hardware. Based on the experiment, MBP and SOM with

S. Hasan (✉) • S.M. Shamsuddin
UTM Big Data Centre, Universiti Teknologi Malaysia, Skudai, Johor, Malaysia
e-mail: shafaatunnur@utm.my; mariyam@utm.my

N. Lopes
UTM Big Data Centre, Universiti Teknologi Malaysia, Skudai, Johor, Malaysia

CISUC, University of Coimbra, Coimbra, Portugal
e-mail: noel@ipg.pt

Y. Cai, S. See (eds.), *GPU Computing and Applications*,
DOI 10.1007/978-981-287-134-3_15

GPU-Tesla generate faster computing times than high-performance computer with feasible results in terms of speed and classification performance.

Keywords GPGPU • Big data • Soft computing • SOM • MBP • Biomedical classification problems

15.1 Introduction

The volume of data being produced is increasing at an exponential rate due to our unprecedented capacity to generate, capture, and share vast amounts of data. In this context, machine learning (ML) algorithms can be used to extract information from these large volumes of data. However, these algorithms are computationally expensive. Their computational requirements are usually proportional to the amount of data being processed. Hence, ML algorithms often demand prohibitive computational resources when facing large volumes of data. As problems become increasingly challenging and demanding (in some cases intractable by traditional CPU architectures), often tool kits supporting ML software development fail to meet the expectations in terms of computational performance. Therefore, the scientific breakthroughs of the future will undoubtedly be powered by advanced computing capabilities that will allow researchers to manipulate and explore massive datasets [1]. Somehow, the pressure is to shift development toward high-throughput parallel architectures, crucial for real-world applications. In this context, highly parallel and programmable devices such as GPU can be used for general-purpose computing applications [2]. GPUs can provide remarkable performance gains compared to CPUs. Moreover, they are relatively inexpensive with affordable prices, availability, and scalability. Over the last few years, the number of GPU implementations of ML algorithms has increased substantially [3]. However, most of the implementations are not openly shared. The lack of openly available implementations is a serious obstacle to algorithm replication and application to new tasks and therefore poses a barrier to the progress of the ML field [4]. By using CUDA architecture, an open-source GPU Machine Learning Library (GPUMLib) was developed by Lopes and Ribeiro [3]. The aim is to provide the building blocks for the development of efficient GPU ML software. GPUMLib offers several advantages such as being useful in adoption of soft computing methods particularly on the neural network algorithms and fast detection of errors. Moreover, most of the previous studies are focused on using artificial neural networks (ANNs) for pattern recognition [5–7]. Hence, we proposed soft computing algorithms for big data problems, particularly in biomedical area. The aim is to provide fast analysis in detecting the cancer from non-cancer based on the extraction of useful information in gene expression, protein profiling, and genomic sequence data. This study is also significant to women who have a high risk of ovarian cancer due to family or personal history of cancer [8]. The remainder of this paper is organized as follows: Sect. 15.2 discusses the previous studies on the development of soft computing methods such as ANN algorithms on graphics hardware. Section 15.3 provides explanation on GPUMLib implementation, particularly on SOM and MBP. Section 15.4 presents the experimental setup, followed by experimental results and discussion in Sect. 15.5. Finally, a conclusion of the study will be discussed in final section.

15.2 Related Work

Early studies of soft computing algorithm with graphics hardware implementation have been proposed in game console application and supervised and unsupervised artificial neural network (ANN) algorithms [6]. In 1998, Bohn started to implement SOM on computer graphics interface (CGI) workstation for computer graphic applications [9]. Later, Zhongwen et al. started to apply SOM algorithm with multipass method on commodity GPUs (ATI 9550 and NVIDIA 5700) and INTEL P4 2.4G for CPU computing [10]. Campbell et al. proposed a parameter-less SOM which eliminates the parameter of learning rate and neighborhood size [11]. Furthermore, SOM is also evaluated on GPU cluster to compute the scalability [12]. On the other hand, parallel implementation of SOM to observe the suitability for high-dimensional problem has been implemented by [13, 14]. In pattern classification, Kyoung-Su Oh and Keechul Jung applied multilayer perceptron (MLP) for text detection [5]. Prabhu proposed unsupervised SOM for pattern classifier [15]. Meanwhile, Gadjos et al. applied unsupervised SOM for outage database [16]. Subsequently, combination of supervised and unsupervised SOM for image segmentation was introduced by Faro et al. [7]. Moreover, Takatsuka et al. applied the Geodesic SOM on standard machine learning dataset [17]. Their experimental results suggested that the GPU speed performance is not significant for small datasets such as iris, but is considerable on larger datasets (ionosphere and torus). In medical area, preliminary studies focused mainly on detecting the cancer nodule and non-nodule based on medical imagery [18]. In addition, Lopes and Ribeiro proposed parallel BP and MBP for ventricular arrhythmias (VAs) in biomedical applications [19]. The aim of parallel MBP and BP is to equip fast detection of diseases which highly potential to sudden death.

Based on the previous study, there is still a lack of SOM-GPU implementation for high-dimensional pattern analysis particularly on biomedical area. This is due to most of the studies that proposed feature selection process to cater the nature of dataset problems. Furthermore, high-dimensional features and imbalance dataset have a great influence to the classification accuracy [20]. SOM is an algorithm for exploratory data analysis which provides mapping from high-dimensional features to low-dimensional features [21]. However, the distance calculation and searching for the best matching unit (BMU) generally increase greatly the computational cost. Hence, we proposed parallel implementation of SOM and MBP to speed up the computation time. Moreover, the SOM and MBP with GPUMLib implementation will be discussed in the next section.

15.3 GPU Machine Learning Library Implementation

In this study, we implement parallelism on soft computing approaches based on neural network (NN) algorithms, multiple back-propagation (MBP), and self-organizing map (SOM) using GPUMLib. MBP is an open-source algorithm built-in in GPUMLib [22]. Meanwhile, SOM algorithm is proposed in this study for the parallel implementation on the distance computation and BMU searching process. In the meantime, the parallelism on SOM algorithm uses the GPUMLib memory access and reduction frameworks. The GPUMLib memory access framework contains *HostArray*, *HostMatrix*, *DeviceArray*, *DeviceMatrix*, and *CudaArray* classes. The framework manages to allocate the memory on the host and device, transfer data between host to device and vice versa. In the reduction framework, the *MinIndex* kernel is designed to compute the minimum of an array and its corresponding index within the array. Both algorithms use batch training for parallel implementation and will be explained in Sects. 15.3.1 and 15.3.2 respectively.

15.3.1 Parallel Multiple Back-Propagation

MBP networks are designed based on multiple feed-forward architecture. They differ from standard BP networks as they integrate two networks designated by main and space networks. The main network contains selective activation neurons which determine their importance for the actual *stimuli* from the space network. Therefore, the selective activation neurons choose and respond to specific group of patterns based on the input presented to the main network. Consequently, the network response is fine-tuned according to the actual space localization features. The main network only calculates its outputs after space network outputs are evaluated. The implementation relies in five kernels: *FireLayer*, *FireOutputLayer*, *CalculateLocalGradient*, *CorrectWeights*, and *CorrectOutputWeights* which execute in each epoch [22]. Initially, *FireLayer* and *FireOutputLayer* kernels are launched by the host in order to determine the space and main network output. Consequently, the main network weights are adjusted using the parallelism of *CalculateLocalGradient*, *CorrectWeights*, and *CorrectOutputWeights* kernels. Finally, the space network weights are adjusted with *CorrectOutputWeights* kernel. In addition, an autonomous training system (ATS) is implemented to improve MBP result. The ATS train several MBPs to select an appropriate MBP network topology. As new MBP networks are trained, its performance is compared with the best MBP found so far. These results are then used to determine the number of hidden neurons of a new MBP and adjusted accordingly until the termination criterion is satisfied [23].

15.3.2 Parallel Self-Organizing Map

The SOM implementation, developed in this study, using the GPUMLib is executed on GPU (host and device) and CPU (host only). For better representation, the implementation is depicted in Fig. 15.1. Basically, the input data and the weights are initialized randomly on the host side. Meanwhile, the Best Matching Unit (BMU) searching is implemented on the device side. In this process, the memory is allocated for both sides (host and device) and also transfers from the host to the device (vice versa). For instance, the weights and input data function variables are defined in a *HostMatrix* (host side) and in a *DeviceMatrix* (device side). Next, the ComputeDistanceskernel $< < < \cdots > > >$, depicted in Fig. 15.2 is launched. This function is designed purposely to calculate the sum squared distance between the input data and weights, i.e., the *Euclidean* distance. Subsequently, the reduction framework, *MinIndex* Kernel is launched (See Fig 15.3). The reduction process synchronizes the threads, in order to find the minimum value of BMU (x, y). Consequently, the result of each block is written to global memory. The minimum values are copied back to the host for updating the weights. Hence, the looping process continues until the termination criterion is satisfied and finally displays the result. On the other hand, all the processes from read the input data to display output are fully executed on the host (CPU) implementation. The distance and BMU are computed on BestMatchingUnit()function, without transfer to the device (see Fig 15.1).

15.4 Experimental Setup

The dataset preparation and performance measurement for biomedical area are presented in Sects. 15.4.1 and 15.4.2, respectively.

15.4.1 Dataset Preparation

In this study, high-dimensional biomedical dataset including gene expression data, protein profiling data, and genomic sequence data that are related to classification is shown in Table 15.1. The leukemia training dataset consists of 38 bone marrow samples which categorize as 27 acute myeloid leukemia (ALL) and 11 acute lymphoblastic leukemia (AML), over 7,129 probes from 6,817 human genes. Also, 34 sample-testing data are provided, with 20 ALL and 14 AML [24]. The prostate cancer training set contains 52 prostate tumor samples and 50 non-tumors which are labeled as normal with 12,600 genes. While testing set that consist of 25 tumor and 9 normal samples [25], the proteomic patterns for ovarian cancer were generated by mass spectroscopy, which consists of 91 normal and 162 ovarian

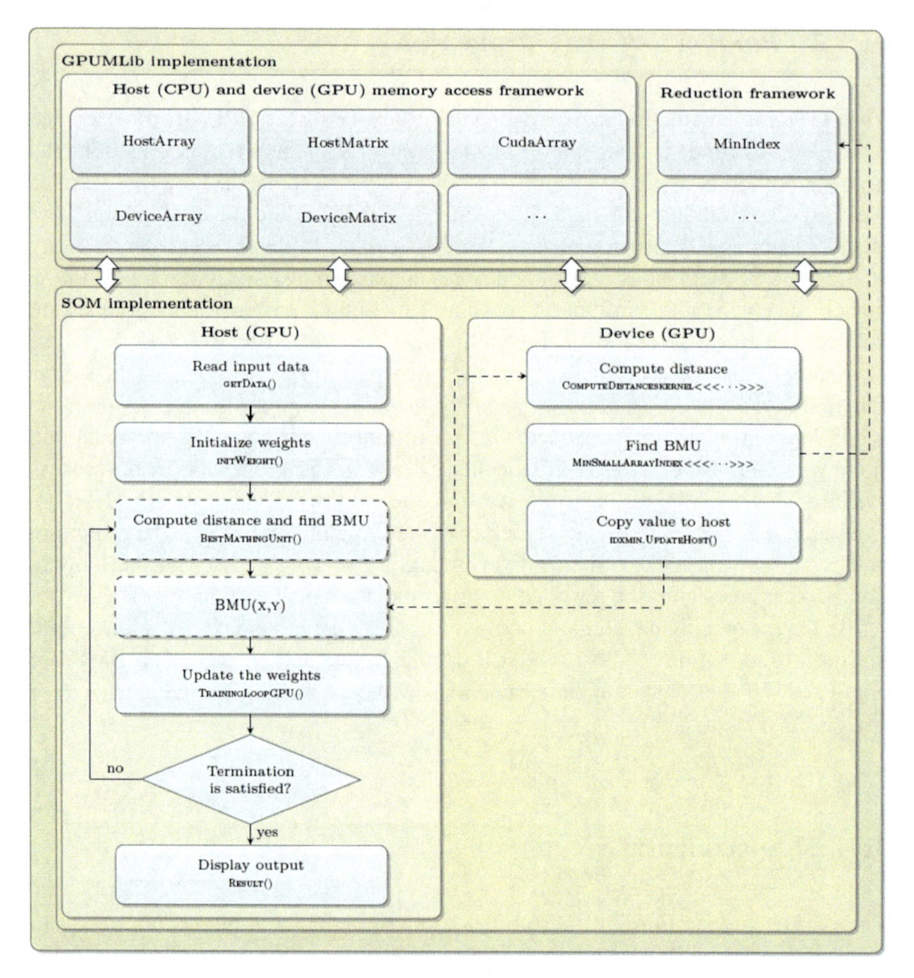

Fig 15.1 SOM with GPUMLib implementation on training the **host (CPU)** and **device (GPU)**

cancers. The raw spectral data of each sample contains 15,154 identities and 253 samples [8]. All datasets are normalized within the range of 0 to 1.

15.4.2 Performance Measurement

The performance measurement for classification task is shown in Table 15.3. The performance index is based on confusion matrix in Table 15.2, which TP, TN, FP, and FN correspond to true positive, true negative, false positive, and false negative, respectively. Generally, sensitivity is used to identify the portion of negative cases that are categorized as positive. Meanwhile, specificity determines portion of the

```
__global__ void ComputeDistancesKernel(float * inputData, float * weights, int
vector, int numberFeatures, float * distances) {
 extern __shared__ float sdist [];

 int i = blockIdx.x;
 int j = blockIdx.y;

 int w = i * gridDim.y + j; // weights have two dimensions

 float distance = 0.0;

 for (int feature = threadIdx.x; feature < numberFeatures; feature += blockDim.x) {
    float fdist = inputData[vector * numberFeatures + feature] -
                  weights[w * numberFeatures + feature];
    distance += fdist * fdist;
 }
 sdist[threadIdx.x] = distance;

 // reduction
 __syncthreads();

 for (int dist = blockDim.x; dist >= 2;) {
    dist /= 2;
    if (threadIdx.x < dist) {
      sdist[threadIdx.x] += sdist[threadIdx.x + dist];
    }
    __syncthreads ();
 }

 if (threadIdx.x == 0) {
    distances[w] = sqrt(sdist[0]);
 }
}
```

Fig 15.2 Launching a kernel to compute distances

```
  void KernelMinIndexes(cudaStream_t stream, int blocks, int blockSize, cudafloat *
  inputs, cudafloat * output, int * minIndexes, int numInputs, int * indexes) {
  MinSmallArrayIndex< blockSize ><<< blocks, blockSize, blockSize *
  (sizeof(cudafloat) + sizeof(int)), stream>>>(inputs, output, minIndexes, numInputs,
  indexes);
}
```

Fig 15.3 Launching a kernel to search the minimum value (http://gpumlib.sourceforge.net)

Table 15.1 Biomedical dataset (http://datam.i2r.a-star.edu.sg/datasets/krbd/)

No	Dataset	No. of samples	No. of features	Class name
1	Leukemia	72	7,129	ALL AML
2	Prostate cancer	136	12,600	Tumor Normal
3	Ovarian cancer	253	15,154	Tumor Normal

positive data that belong to negative cases. PPV and NPV are employed to calculate the average of positive (TP) and negative (TN) cases, respectively. High sensitivity, specificity, PPV, and NPV means the result is accurate with perfect scores 1, else the lowest score is 0. Furthermore, an average site performance (ASP) and

Table 15.2 Confusion matrix

Actual class	Predicted class		
		+(ve)	−(ve)
	+(ve)	TP	FN
	−(ve)	FP	TN

Table 15.3 Classification performance measurement

Performance measurement index	Abbreviation	Formula
Sensitivity	Sn	$\frac{TP}{(TP+FN)}$
Positive predictive value	PPV	$\frac{TP}{(TP+FP)}$
Specificity	Sp	$\frac{TN}{(TN+FP)}$
Negative predictive value	NPV	$\frac{TN}{(TN+FN)}$
Accuracy	ACC	$\frac{(TP+TN)}{(TP+TN+FP+FN)}$
Average site performance	ASP	$\frac{Sn+PPV}{2}$
Performance coefficient	PC	$\frac{TP}{(TP+FN+FP)}$

performance coefficient (PC) are being used to analyze the average of precision and recall performance in classification problems.

15.5 Experimental Result and Analysis

In this study, the SOM and MBP algorithms are executed on NVIDIA Tesla C2075 graphic hardware and Intel Xeon high-performance computer. Both algorithms are tested on high-dimensional biomedical datasets (leukemia, prostate cancer, and ovarian cancer). The SOM algorithm is set up for 1,000 iterations in three different sizes of mapping. While the MBP algorithm executes for 10,000 iterations using the autonomous training system (ATS), initially, the MBP generates 100 networks with one and two hidden layers. The biomedical datasets such as prostate cancer, ovarian cancer, and leukemia dataset are indicated as large, medium, and small feature dimensions. Meanwhile, the SOM mapping sizes (5×10, 10×10, and 10×15) are labeled as small, medium, and large, respectively. There are two sections of analyses, which are speed and classification analysis. In Sect. 15.5.1, the speed performance will be analyzed in terms of CPU and GPU elapsed time. For classification analysis in Sect. 15.5.2, the result will be based on the performance measurement index which was previously described in Table 15.3.

15.5.1 Speed Analysis

In this study, the aim of the analysis is to observe the capability of MBP and SOM algorithm using graphics hardware (GPU) on high-performance computer (CPU). In this experiment, the size of SOM mapping dimension is categorized as mapping $1 = 5 \times 10$, mapping $2 = 10 \times 10$, and mapping $3 = 10 \times 15$. While the number of hidden nodes is set to 100 for the first hidden layer, a total of 15 nodes are set for the second hidden layer (see Tables 15.4 and 15.5). Hence, large proportion size of mapping dimension, number of hidden nodes, iterations, and feature dimensions of the dataset generate slow computation times for both algorithms. Since the computational time depends on certain parameters, we evaluate both algorithms with similar datasets, number of nodes, and number of iterations. The SOM speed on GPU generates approximately three times faster than CPU for all datasets as depicted in Fig 15.4. Subsequently, MBP leukemia dataset produces significant performance with 27 times speed for 10,000 iterations on GPU (see Table 15.5 and Table 15.6). Meanwhile, MBP (100 nodes) generates 12 times more than SOM (size of mapping $= 10 \times 10$) for 1,000 iterations on CPU (see Table 15.4 and Table 15.6).

Table 15.4 SOM speed performance

Dataset	Performance evaluation	SOM result		
Leukemia	Max epoch	1,000		
	Size of mapping	**Mapping 1**	**Mapping 2**	**Mapping 3**
		5 × 10	**10 × 10**	**10 × 15**
	CPU time	356.436 s	533.726 s	974.418 s
	GPU time	115.441 s	207.574 s	301.79 s
	Speed	3.087603 x	2.57126 x	**3.228795 x**
Prostate cancer	Max epoch	1,000		
	Size of mapping	**Mapping 1**	**Mapping 2**	**Mapping 3**
		5 × 10	**10 × 10**	**10 × 15**
	CPU time	1,621.65 s	2,618.41 s	4,081.941 s
	GPU time	660.474 s	1,118.06 s	1,565.038 s
	Speed	2.455275 x	2.34192 x	**2.608206 x**
Ovarian cancer	Max epoch	1,000		
	Size of mapping	**Mapping 1**	**Mapping 2**	**Mapping 3**
		5 × 10	**10 × 10**	**10 × 15**
	CPU time	3,455.925 s	6,354.214 s	9,116.06 s
	GPU time	1,086.895 s	2,061.42 s	3,166.077 s
	Speed	**3.179631 x**	3.082445 x	2.879292 x

The bold values represent the best performance in terms of speed and classification analysis

Table 15.5 MBP-GPU (device) speed performance

Dataset	Performance evaluation	MBP result			
Leukemia	Max iteration	10,000			
	MBP network	**7129-100-1**		**7129-5-10-1**	
		Min	**Max**	**Min**	**Max**
	Iteration	146	163	230	10,000
	GPU time	8.892 s	9.928 s	**0.702 s**	30.654 s
Prostate cancer	Max iteration	10,000			
	MBP network	**12600-100-1**		**12600-5-10-1**	
		Min	**Max**	**Min**	**Max**
	Iteration	181	225	2,996	10,000
	GPU time	36.042 s	44.805 s	**23.678 s**	79.119 s
Ovarian cancer	Max iteration	10,000			
	MBP network	**15154-100-1**		**15154-4-10-1**	
		Min	Max	Min	Max
	Iteration	152	191	155	10,000
	GPU time	77.173 s	96.955 s	**2.075 s**	134.134 s

The bold values represent the best performance in terms of speed and classification analysis

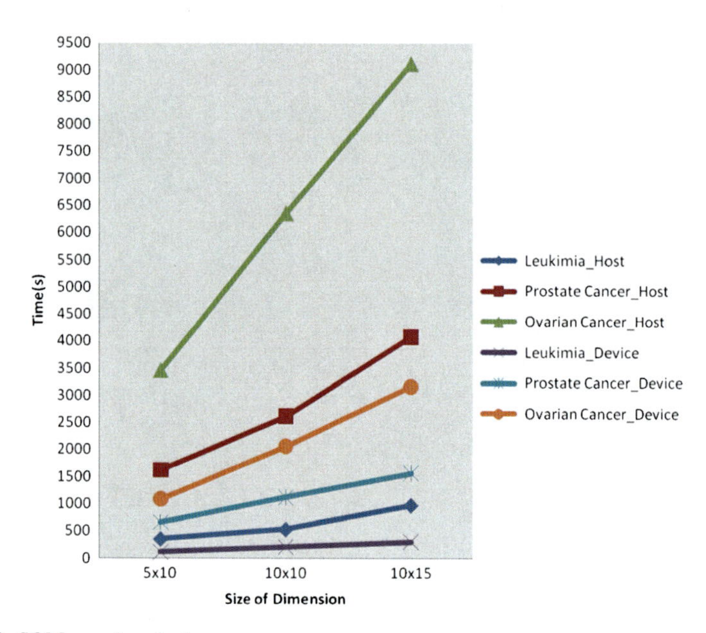

Fig 15.4 SOM speed analysis

Table 15.6 MBP-CPU (Host) speed performance in leukemia dataset

Performance evaluation	MBP result			
Iteration	1,000	10,000	1,000	10,000
MBP network	7129-100-1		7129-5-10-1	
CPU time	6,612 s	48,668 s	109 s	838 s

15.5.2 Classification Analysis

The classification analysis for SOM and MBP is shown in Table 15.7. The results are evaluated based on the percentage of *best*, *average*, and *worst* performance. MBP outperforms SOM in *best* result of all datasets. Otherwise, SOM generates significant result in *average* performance of prostate cancer dataset.

Table 15.7 Classification analysis

			SOM			MBP	
Dataset	Abbreviations	Average (%)	Worst (%)	Best (%)	Average (%)	Worst (%)	Best (%)
Leukemia	*Sn*	74.43	65.83	84.29	**78.95**	47.76	**91.18**
	PPV	72.73	66.67	78.79	**78.79**	48.49	**90.91**
	Sp	76.09	64.67	89.79	**79.32**	42.72	**89.27**
	NPV	62.53	56.80	67.37	**72.75**	42.28	**91.40**
	ACC	72.73	66.67	78.79	**78.79**	48.49	**90.91**
	ASP	73.58	66.25	81.54	**78.87**	48.12	**91.05**
	PC	56.12	49.10	63.08	**64.65**	32.58	**83.46**
Prostate cancer	*Sn*	**85.61**	75.78	91.64	74.20	17.83	**97.35**
	PPV	**79.41**	64.71	91.18	76.47	65.36	**97.06**
	Sp	**69.47**	53.63	84.19	65.11	6.42	**92.65**
	NPV	**85.48**	73.07	93.65	48.86	23.53	**98.94**
	ACC	**79.41**	64.71	91.18	76.47	23.53	**97.06**
	ASP	**82.51**	70.24	91.41	75.34	41.59	**97.21**
	PC	**68.89**	50.60	84.30	61.77	17.30	**94.41**
Ovarian cancer	*Sn*	83.23	81.26	86.42	**94.96**	90.90	**100**
	PPV	82.02	81.58	82.90	**94.74**	90.79	**100**
	Sp	83.07	79.93	92.94	**92.24**	88.16	**100**
	NPV	73.42	67.10	79.34	**95.42**	89.67	**100**
	ACC	82.02	81.58	82.90	**94.74**	90.79	**100**
	ASP	82.62	81.42	84.66	**94.85**	90.85	**100**
	PC	69.24	68.94	69.47	**90.12**	83.33	**100**

The bold values represent the best performance in terms of speed and classification analysis

Conclusion

In this study, we found that the results are proportionate to the mapping size of the SOM architecture and feature dimensions of the datasets. In other words, the larger the mapping size and feature dimensions, the slower the computation time for both CPU and GPU. This is due to ANNs' (SOM and MBP) parameters that depend on size of mapping (number of nodes), dataset feature dimensions, number of input samples, and termination criterion (number of iterations or convergence rate). Our findings are conformed to the findings conducted by [12, 14], i.e., larger mapping size will increase the memory transfer, thus, lower the computational time [14]. The current GPU parallel implement of the SOM algorithm performs three times ($3\times$) faster than the CPU, while the MBP is 27 times faster than the CPU. However, the SOM's speed could be improved with the parallelism on updating the weights. It is important for larger (big data) datasets that do not fit on the GPU memory, consists of devising methods, to choose a representative subset of the data. Alternatively, we can also create several maps for different data that could afterwards be merged together latter in a bigger map. Furthermore, the aim of SOM-GPUMLib implementation will be openly shared in the future.

Acknowledgments This work is supported by The Ministry of Higher Education (MOHE) under Long Term Research Grant Scheme (LRGS/TD/2011/UTM/ICT/03—4L805). The authors would like to thank Research Management Centre (RMC), Universiti Teknologi Malaysia (UTM) for the support in R & D, Soft Computing Research Group (SCRG) for the inspiration in making this study a success. The authors would also like to thank the anonymous reviewers who have contributed enormously to this work.

References

1. Hey, T., Tansley, S., Tolle, K. (eds.): The Fourth Paradigm: Data-Intensive Scientific Discovery. Microsoft Research, Redmond, WA (2009)
2. Owens, J.D., Houston, M., Luebke, D., Green, S., Stone, J.E., Phillips, J.C.: GPU computing. Proceedings of the IEEE **96**(5), 879–899 (2008)
3. Lopes, N., Ribeiro, B., Quintas, R. (2010): GPUMLib: A New Library to combine Machine Learning algorithms with Graphics Processing Units. In: HIS 2010, pp. 229–232
4. Sonnenburg, S., Braun, M.L., Ong, C.S., Bengio, S., Bottou, L., Holmes, G., LeCun, Y., Muller, K.-R., Pereira, F., Rasmussen, C.E., Ratsch, G., Scholkopf, B., Smola, A., Vincent, P., Weston, J., Williamson, R.C.: The need for open source software in machine learning. J. Mach. Learn. Res. **8**, 2443–2466 (2007)
5. Kyoung, S.H., Keechul, J.: GPU implementation of neural networks. Pattern Recognit. **37**(6), 1311–1314 (2004)
6. Meuth, R., Wunsch, D.C.: A survey of neural computation on graphics processing hardware. In: IEEE 22nd International Symposium on Intelligent Control, ISIC 2007. (2007)

7. Faro, A., Giordano, D., Palazzo, S.: Integrating unsupervised and supervised clustering methods on a GPU platform for fast image segmentation. In: 3rd International Conference on Image Processing Theory, Tools and Applications (IPTA) (2012)
8. Petricoin III, E.F., Ali, M.A., Ben, A.H., Peter, J.L., Vincent, A.F., Seth, M.S., Gordon, B.M., et al.: Use of proteomic patterns in serum to identify ovarian cancer. Lancet **359**(9306), 572–577 (2002)
9. Bohn, C.A.: Kohonen feature mapping through graphics hardware. In: Proceedings of 3rd International Conference on Computational Intelligence and Neurosciences (1998)
10. Zhongwen, L., Hongzhi, L., Zhengping, Y., Xincai, W.: Self-organizing maps computing on graphic process unit. In: 13th European Symposium on Artificial Neural Networks, Belgium, pp. 557–562 (2005)
11. Campbell, A., Berglund, E., Streit, A.: Graphics Hardware Implementation of the Parameter-Less Self-organising Map. In: Gallagher, M., Hogan, J., Maire, F. (eds.) Intelligent Data Engineering and Automated Learning - IDEAL, pp. 343–350. Springer, Heidelberg (2005)
12. McConnell, S., Sturgeon, R., Henry, G., Mayne, A., Hurley, R.: Scalability of Self-organizing Maps on a GPU cluster using OpenCL and CUDA. In: Paper presented at the Journal of Physics: Conference Series. (2012)
13. Platos, J., Gajdos, P.: Large data real-time classification with non-negative matrix factorization and self-organizing maps on GPU. In: International Conference on Computer Information Systems and Industrial Management Applications (CISIM) (2010)
14. Gajdoš, P., Platoš, J. (2013) GPU based parallelism for self-organizing map. In: Kudělka, M. et al. (eds.) Proceedings of the Third International Conference on Intelligent Human Computer Interaction (IHCI 2011), vol. 179, pp. 231–242, Prague, Czech Republic, August 2011. Springer, Heidelberg
15. Prabhu, R.D.: SOMGPU: an unsupervised pattern classifier on graphical processing unit. In: Evolutionary Computation, CEC 2008. IEEE World Congress on Computational Intelligence, pp. 1011–1018. (2008)
16. Gajdoš, P., Krátký, M., Bednár, D., Baca, R., Gono, R., Walder, J.: Efficient computation of SOM for outage database. In: ELNET 2011, 51 (2011)
17. Takatsuka, M., Bui, M.: Parallel Batch Training of the Self-Organizing Map Using OpenCL. In: Wong, K., Mendis, B.S., Bouzerdoum, A. (eds.) Neural Information Processing. Models and Applications, pp. 470–476. Springer, Heidelberg (2010)
18. Eklund, A., Dufort, P., Forsberg, D., LaConte, S. M.: Medical Image Processing on the GPU - Past, Present and Future. Medical Image Analysis. Elsevier (2013)
19. Lopes, N., Ribeiro, B.: Fast pattern classification of ventricular arrhythmias using graphics processing units. In: Proceedings of the 14th Iberoamerican Conference on Pattern Recognition (CIARP 2009), LNCS 5856, pp. 603–610. Springer, Heidelberg (2009)
20. Tanwani, A., Farooq, M. The role of biomedical dataset in classification. In: Combi, C., Shahar Y., Abu-Hanna, A. (eds.) Artificial Intelligence in Medicine, vol. 5651, pp. 370–374. Springer, Heidelberg (2009)
21. Kohonen, T.: Self-Organizing Maps. Springer Series in Information Sciences, vol. 30, 3rd edn (Extended Edition). Springer, Berlin (2001)
22. Lopes, N., Ribeiro, B.: GPU implementation of the multiple back-propagation algorithm. In: Corchado, E., Yin, H. (eds.) Intelligent Data Engineering and Automated Learning-IDEAL, pp. 449–456. Springer, Heidelberg (2009)
23. Lopes, N., Ribeiro, B.: A strategy for dealing with missing values by using selective activation neurons in a multi-topology framework. In: The 2010 International Joint Conference on Neural Networks (IJCNN), pp. 1–5. IEEE (2010)
24. Golub, T.R., Donna, K.S., Pablo, T., Christine, H., Michelle, G., Jill, P.M., Hilary, C., et al.: Molecular classification of cancer: class discovery and class prediction by gene expression monitoring. Science **286**(5439), 531–537 (1999)
25. Singh, D., Phillip, G.F., Kenneth, R., Donald, G.J., Judith, M., Christine, L., Pablo, T., et al.: Gene expression correlates of clinical prostate cancer behavior. Cancer Cell **1**(2), 203–209 (2002)

Chapter 16
Numerical Solution of BVP on GPU with Application to Path Planning

Lumír Janošek, Martin Němec, and Radoslav Fasuga

Abstract The problem of path planning in a virtual environment is a widely researched area, which finds application in fields such as robotics, simulations, and computer games. This article focuses on a comparison of numerical methods for solving partial differential equations with BVP on the GPU with NVIDIA CUDA, used in the path planning of virtual characters using the potential fields. The most commonly used methods for computing the potential fields on the GPU are compared in this article in terms of time consumption.

Keywords Path-planning • Agent • Iteration methods • Potential fields

16.1 Introduction

The original purpose of a graphic processing unit (GPU) was primarily for image data processing. Programming of graphical chips was not a simple matter. It was necessary to use an application programming interface (API) to access the graphic processor such as Direct3D® or OpenGL®. The release of NVIDIA CUDA in 2007 changed the approach to the programming of graphic processors [1].

This article focuses on a comparison of the implementation of iterative methods for solving partial differential equations on a GPU in the agent path-planning domain. This article is not intended to present new approaches, but only to show the differences in iterative methods implemented on the GPU, which are used in potential field-based path planning. In this article the most widely used methods for the generation of potential fields used for agent navigation are compared in terms of time consumption.

The problem of path planning is widely applied in areas such as robotics and computer games. Path finding can generally be understood as finding the optimal path from an arbitrary position in a virtual world to a goal. In practical applications, there is often the requirement that the methods must be able to find paths in real time. Currently, the A* algorithm is still widely used for path planning [2], falling among graph-oriented algorithms. An alternative to graph-oriented algorithms are

L. Janošek (✉) • M. Němec • R. Fasuga
Department of Computer Science, VŠB-Technical University, Ostrava, Czech Republic
e-mail: lumir.janosek.st@vsb.cz; martin.nemec@vsb.cz; radoslav.fasuga@vsb.cz

Y. Cai, S. See (eds.), *GPU Computing and Applications*,
DOI 10.1007/978-981-287-134-3_16

methods for path planning using the potential fields. These methods are traditionally used in robotics. The application of potential field-based path planning can also be found in computer games [3]. Application of the BVP path planning may not be limited just to 2D. In [4] a method for the new application of BVP path planning on the surface of a 3D object is presented.

The idea of BVP path planning is using the interplay between repulsion from obstacles and attraction to a target position to create the expected behavior. Potential fields are obtained from the class of partial differential equations (PDE) called the boundary value problem (BVP) [5]. BVP-based path planning can create realistic-looking complex humanlike behavior similar during the agent's movement toward to the goal. Implementation of the numerical solution of the BVP on the GPU then enables the application of these methods in multi-agent real-time applications [6].

This chapter is structured as follows: Sect. 16.2 summarizes the problems of BVP in the path-planning domain, Sect. 16.3 describes the iterative methods used for solving the partial differential equations, Sect. 16.4 presents the implementation of the listed methods on the GPU, Sect. 16.5 summarizes the achieved results during the implementation, and the final section presents our conclusion and future work.

16.2 Harmonic Potential Field

One of the most widely used methods for generating a potential field for agent navigation in a virtual environment is the numerical solution of a partial differential equation based on the boundary value problem (BVP). One of the first steps in this area was undertaken by Connolly and Grupen [7]. In their work they presented a method for the generation of potential fields, which do not have local minima. Such a local minimum may be the reason why the agent can end up trapped in local minima. In their work, Connolly and Grupen proposed a method for generating a potential field through a solution to the Laplace equation:

$$\nabla^2 u = 0, g(x, y) = \begin{cases} 1, \text{obstacle} \\ 0, \text{goal} \end{cases}, (x, y) \in \partial\Omega \tag{16.1}$$

called harmonic function. The property of the Laplace equation is that it does not present local minima. This property is based on the so-called maximum principle, which the Laplace equation satisfies [8].

Equation (16.1) is solved with preset values on the boundaries. This type of boundary condition is called the Dirichlet boundary condition in the terminology of the BVP given by $g(x,y)$. In the case of obstacle space, the potential values at the obstacles are preset to a higher value, while in the goal area the values are preset to zero. The resulting potential field is used to find the agent's path to the goal by gradient descent. Higher values of the obstacles repel the agent to prevent collision.

On the other hand, zero values of the goal create an attraction force. Because there is only one minimum defined in the goal area, there exists exactly one path from any point on the map to the goal [9].

16.3 Iterative Methods

In general, there exist two methods for solving the boundary value problem, classified as direct methods and iterative methods. Direct methods lead to an exact solution to the problem with the use of a finite sequence of operations. In contrast to direct methods are iterative methods, in which the solution is obtained by a number of iterations [10]. A typical procedure is to determine the initial solution, on the basis of which the new values are calculated. This procedure is repeated until the convergence reaches the desired solution. This is usually determined by some criterion of convergence.

The iterative solution of elliptic equations most commonly uses the following methods: Jacobi, Gauss-Seidel, or Successive Overrelaxation (SOR).

In the Jacobi method, the dependent variable at each grid point is solved using the initial values of the neighboring points or previously computed values [10]:

$$u_{i,j}^{(k+1)} = \frac{1}{4}\left[u_{i-1,j}^{(k)} + u_{i+1,j}^{(k)} + u_{i,j-1}^{(k)} + u_{i,j+1}^{(k)}\right]$$

where k denotes the values computed in the previous iteration and i, j denotes the grid point.

The Gauss-Seidel method is a modification of the Jacobi method. To compute the value of a dependent variable in the current iteration, the values from the previous and current iteration are used. This will certainly increase the convergence rate dramatically over the Jacobi method [10]. The iteration formula for the Gauss-Seidel method has the following form:

$$u_{i,j}^{(k+1)} = \frac{1}{4}\left[u_{i-1,j}^{(k+1)} + u_{i+1,j}^{(k)} + u_{i,j-1}^{(k+1)} + u_{i,j+1}^{(k)}\right]$$

where k denotes the values computed in the previous iteration, $k+1$ denotes the values computed in the current iteration, and i, j denotes the grid point.

Better convergence can be achieved with the Successive Overrelaxation (SOR) method. The main idea behind the SOR algorithm is to compute a better approximation to the true solution by forming a linear combination of the current updated solution $k+1$ and solution k from the previous iteration [11]. The iteration formula for SOR method is defined as:

$$u_{i,j}^{(k+1)} = (1-\omega)u_{i,j}^{(k)} + \frac{\omega}{4}\left[u_{i-1,j}^{(k+1)} + u_{i+1,j}^{(k)} + u_{i,j-1}^{(k+1)} + u_{i,j+1}^{(k)}\right] \quad (16.2)$$

where ω denotes the relaxation parameter and i, j denotes the grid point. The optimal value of ω should be in the range $1 < \omega < 2$. If $0 < \omega < 1$, this is so-called under-relaxation [12]. In the case of $\omega = 1$, the SOR algorithm is reduced to Gauss-Seidel.

16.4 Implementation

With access to today's NVIDIA CUDA-enabled GPU, it is possible to significantly accelerate the methods of numerical solution of elliptic equations using parallel implementation. With the parallel performance of the GPU, which is provided by the CUDA interface, it is possible to solve many complex computational problems with more efficiency than on the CPU. GPU is suitable for solving problems which require the parallel processing of large amounts of data.

Not all iterative methods for solving elliptic equations are suitable for implementation on the GPU. For parallel implementation and performance comparison of the numerical solution of elliptic equations on the GPU, the Jacobi, Jacobi Red-Black, and SOR Red-Black methods were chosen. The sequential implementation of the Gauss-Seidel uses two values from the current iteration and two values from the previous iteration to calculate the current cell. In the implementation of this method on the GPU, it is necessary to have some synchronization, which can lead to performance degradation [13]. Gauss-Seidel is an effective method for implementation on the CPU. Due to the need for synchronization, the Gauss-Seidel method is not best suited for parallel implementation on the GPU, and therefore was not taken into account for the implementation of iterative methods on the GPU.

As mentioned in the introduction, the methods presented in this article are focused on agent navigation in a virtual world. A virtual environment contains a number of obstacles, which the agent tries to avoid on the way to the goal. Before the start of the potential field calculation, it is necessary to discretize the virtual environment into a fixed homogeneous grid representation. Each grid cell (i,j) is associated with a small region of the real environment and maintains the potential value $u_{i,j}$, which holds information about whether the given cell is an obstacle or free space. Cells defined in place of the obstacles have the initial potential set to 1, while cells containing a goal have the potential value set to 0. Such a manner of setting the initial values corresponds to the Dirichlet boundary conditions [14].

With such a defined initial boundary condition, the values of all other cells are computed using a certain number of iterations. In order for the method to converge to the correct solution, a sufficient number of iterations must be specified. The number of iterations varies depending on the used method. One option of how to control the number of iterations is assessment of some convergence criteria based on which the calculation is terminated. Such criteria could be check of the error that occurred during the iterations, for instance. The iteration is terminated once the error is less than the given tolerance [11]. An alternative way is to specify a fixed number of iterations at the beginning of the algorithm. [12] shows that the required number of iterations can be determined by an analytical formula. The number of iterations r required to reduce the error by a factor 10^{-p}, for the Jacobi method, is defined as:

$$r \approx \frac{1}{2} p J^2 \tag{16.3}$$

J^2 denotes the number of grid points.

Using the Red-Black method in conjunction with the Jacobi method, it is possible to achieve certain optimization [11]. The Red-Black method divides grid points into odd and even, symbolically expressed by red-black coloring. The coloring of the grid points is done so that no point is directly adjacent to a point of the same color. The red point values from the previous iteration are utilized during the calculation of the values of the black points. This step is identical to the Jacobi iteration, applied to all black points. Updated black point values are used in the next step in the computation of the red points, which is identical to the Gauss-Seidel iteration. The Red-Black method is thus composed of one Jacobi iteration and one Gauss-Seidel iteration. As mentioned in the previous Sect. 16.3, the Gauss-Seidel method uses values computed in the previous iteration to compute the current values, thus significantly contributing to speeding up the convergence rate. The number of iterations for the Jacobi Red-Black method can then be defined practically as well as for the Gauss-Seidel method, for which it is defined as [12]:

$$r \approx \frac{1}{4} p J^2 \tag{16.4}$$

J^2 denotes the number of discrete grid points. The GPU implementation of the Red-Black methods uses two kernels, one for computation of the red points and one for computation of the black points. The number of black or red points on the y-axis of the grid is half. This can reduce the number of threads in each kernel on the y-axis by half. Reducing the number of threads leads to a certain optimization of the iterative process.

Compared to the Jacobi or Jacobi Red-Black, the SOR method leads to much faster convergence. As already stated, the SOR method uses the values from the previous iteration and the values from the current iteration to compute the current point, similarly as the Gauss-Seidel method, see (16.2). The parallel GPU implementation of the SOR method is enabled using the Red-Black ordering [15]. Updated values of the black points, i.e., values of the current iteration, are used to compute the red points. Updated values of the red points, i.e., values of the previous iteration, are used to compute the black points. The number of required iterations needed in order to reduce error by factor 10^{-p} is given by [12]:

$$r \approx \frac{1}{3} pJ \tag{16.5}$$

Comparing the number of iterations of the SOR method with the number of iterations of the Jacobi method (16.3) and Jacobi Red-Black (16.4), it is obvious that the optimal number of iterations of the SOR method is in the order of J, compared with J^2 of the Jacobi and Jacobi Red-Black method. The weak point of the SOR may be the choice of overrelaxation parameter ω. In [12] the following equation is stated, which can be used to estimate the overrelaxation parameter:

$$\omega \approx \frac{2}{1 + \frac{\pi}{J}}$$

In general, finding the correct value of ω is not an easy task. In many cases experimentation is the only possible way to determine the correct value of parameter ω.

16.5 Results

Implementation of the Jacobi, Jacobi Red-Black, and SOR Red-Black methods was compared in terms of time performance. These methods were tested on GeForce GTX 560 and GeForce GTX 670 graphics cards.

A map of static obstacles is copied into the device memory before the start of the actual iterative procedure. Since the obstacle map is read only, it is copied into the texture memory of the GPU before the calculation. The texture memory is optimized for a 2D spatial locality, so threads of the same warp that read texture addresses that are close together will achieve the best performance [16]. The map of obstacles only holds information about the position of the obstacles and walkable spaces. For this reason, the 8-bit data format was chosen for maximum reduction of the memory requirements.

In practical applications of these numerical methods in the field of path finding and agent navigation in a virtual environment, such as in [17], it is necessary to change the global obstacle map only in case of adding new obstacles or removing existing ones. Due to the individual approach to the implementation of the global obstacle map, the data transfers from the host to the device were not taken into account during the speed comparison of the methods.

Maps of different sizes were used to compare the speed of these methods. The resulting time difference of the method is shown in Fig. 16.1. The most optimal performance was achieved with SOR Red-Black when compared with the Jacobi and Jacobi Red-Black. For each method the number of iterations was determined based on equations (16.3) for Jacobi, (16.4) for Jacobi Red-Black, and (16.5) for SOR Red-Black.

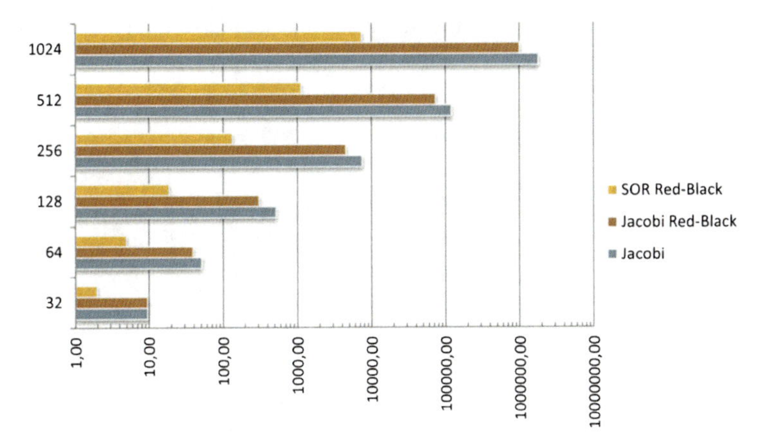

Fig. 16.1 Speed differences (in milliseconds) of the GPU computation of the Jacobi, Jacobi Red-Black, and SOR Red-Black methods. The comparison was made for input grid size $32^2 - 1024^2$

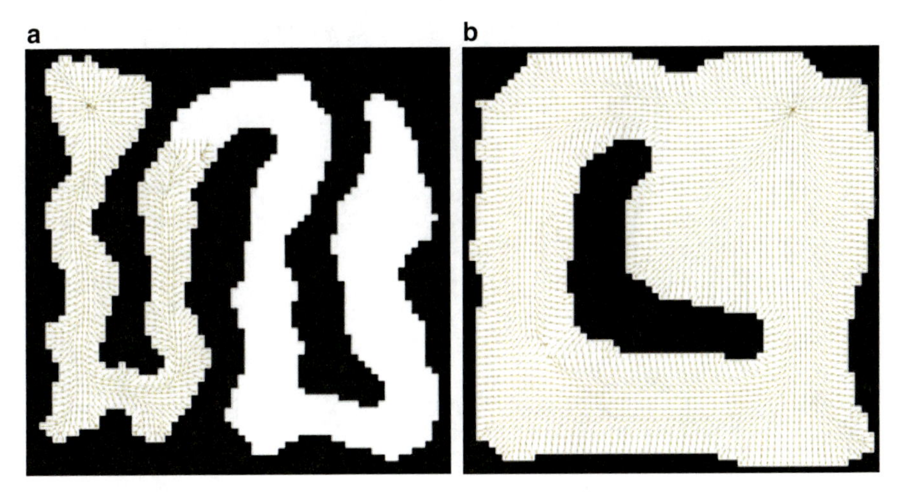

Fig. 16.2 Picture 1.2a shows the resulting gradient of the potential field. Picture 1.2b illustrates the failure of the calculation in confined space due to a lack of real number precision

Implementation of the tested methods was performed in the double-precision floating-point format. Potential field computation was tested in such obstacle configurations which simulated the cramped spaces. These configurations were often the cause of the loss of the potential value in locations too far from goal, because of insufficient accuracy of the real number. One such situation is illustrated in Fig. 16.2b. Values in this potential field were rounded to 1 due to insufficient accuracy of the real number. Final computation of the gradient cannot then be achieved in these cases. The potential field gradient illustrated on Fig. 16.2a and fig. 16.2b with size of 64^2 was computed using the Jacobi Red-Black method. The number of iterations required to obtain a valid solution was determined using equation (16.4).

Conclusion

In this chapter the implementations of the numerical methods for solving elliptic equations using CUDA with application on BVP path planning were compared. The Jacobi, Jacobi Red-Black, and SOR Red-Black methods were compared in terms of time complexity. Using the SOR Red-Black, we reached the fastest convergence, in comparison with Jacobi and Jacobi Red-Black. These methods were applied to the obstacle configuration simulating a real environment. It was shown that the configuration of obstacles simulating cramped spaces, such as underground caves, does not provide sufficient freedom for the convergence of methods. The information is lost due to insufficient accuracy of the real number during the convergence to the final potential field.

In [18] the methods of BVP path planning were combined with the Full Multigrid method, which solves elliptic equations using a hierarchical strategy. The hierarchical approach overwhelms the speed of convergence of the original SOR method.

In the previous section, an error caused by insufficient accuracy of the real number, leading to early rounding to 1, was described. One option of solving this problem is described in [19]. Future development of this work will focus on finding an alternative way to solving the problem with insufficient accuracy of the real number and to optimizing the convergence in cramped spaces. This would then allow the application of BVP path planning for space-limited interiors.

Acknowledgment This work was partially supported by the SGS in VSB Technical University of Ostrava, Czech Republic, under the grant No. SP2013/185.

References

1. Kirk, D.B., Hwu, W.-W.: Programming Massively Parallel Processors: A Hands-on Approach, 1st edn. Morgan Kaufmann Publishers Inc, San Francisco, CA (2010)
2. Cui, X, Shi, H.: A*-based pathfinding in modern computer games. IJCNIS **11**(1), 125–130 (2011)
3. Silveira, R., Fischer, L., Jos' e AntônioSalini F., Prestes, E., Nedel, L.: Path-planning for RTS games based on potential fields. In: Proceedings of the Third international conference on Motion in games, MIG'10, pp. 410–421. Springer, Heidelberg (2010)
4. Fischer, L., Fischer L.: Semi-automatic navigation on 3d triangle meshes using bvp based path-planning. In: 24th SIBGRAPI Conference on Graphics, Patterns and Images (Sibgrapi), pp. 33–40, (2011)

5. Marcelo, T., Idiart, M.A., Edson, P., Engel, P.M.: Exploratory navigation based on dynamical boundary value problems. J. Intell. Robotics Syst. **45**(2), 101–114 (2006)
6. Fischer, L.G., Silveira, R., Nedel, L.: Gpu accelerated path-planning for multi-agents in virtual environments. In: VIII Brazilian Symposium on Games and Digital Entertainment (SBGAMES), pp. 101–110, (2009)
7. Connolly, C.I., Grupen, R.A.: On the applications of harmonic functions to robotics. J. Robot. Syst. **10**, 931–946 (1993)
8. Strauss, W.A.: Partial Differential Equations: An Introduction. Wiley, New York, NY (1992)
9. Dapper, F., Prestes, E., Idiart, M.A.P., Nedel, L.P.: Simulating pedestrian behavior with potential fields. In: Proceedings of the 24th international conference on Advances in Computer Graphics, CGI'06, pp. 324–335. Springer, Heidelberg (2006)
10. Klaus, A.: Hoffmann and Steve T Chiang. Computational fluid dynamics vol.i - hoffmann.pdf. Int. J. Comut. Fluid. Dyn. **126**(2), 581–594 (2000)
11. Zhu, J.: Solving Partial Differential Equations on Parallel Computers. World Scientific Publishing Co. Inc., River Edge, NJ (1994)
12. Press, W.H., Teukolsky, S.A., Vetterling, W.T., Flannery, B.P.: Numerical Recipes 3rd Edition: The Art of Scientific Computing, 3rd edn. Cambridge University Press, New York, NY (2007)
13. Gomes, G.A.A.: Linear solvers for stable fluids: GPU vs CPU. In: 17th EncontroPortugues de ComputacaoGrafica (EPCG09), pp. 145–153 (2009)
14. Dapper, F., Prestes, E., Nedel, L.P.: Generating Steering Behaviors for Virtual Humanoids Using BVP Control. In: Proc. of CGI, pp. 105–114 (2007)
15. Konstantinidis, E., Cotronis, Y.: Graphics processing unit acceleration of the red/black SOR method. Concurr Comput. **25**(8), 1107–1120, (2012)
16. NVIDIA. CUDA C BEST Practices Guide (2012)
17. Fischer, L.G., Silveira, R., Nedel, L.: Gpu accelerated path-planning for multi-agents in virtual environments. In: VIII Brazilian Symposium on Games and Digital Entertainment (SBGAMES), pp. 101–110, (2009)
18. Silveira, R., e Silva, E.P., Jr., PorcherNedel, L.: Fast path planning using multi-resolution boundary value problems. In: IEEE/RSJ International Conference on Intelligent Robots and Systems, 18–22 October 2010, Taipei, Taiwan, pp. 4710–4715. IEEE (2010)
19. Renato, S., Fbio, D., Edson, P., Luciana, N.: Natural steering behaviors for virtual pedestrians. Vis. Comput. **26**(9), 1183–1199 (2010)

Chapter 17
Fast Multi-Keyword Range Search Using GPGPU

Amirul Abdullah, Amril Nazir, Mohanavelu Senapan, Soo Saw Meng, and Ettikan Karuppiah

Abstract Large organisations are constantly challenged by the need to handle big data. Big data sizes are a constantly moving target, as of 2013 ranging from a few dozen terabytes to many petabytes of data. The data is usually stored in very large databases that are often indexed off-line to enable the acceleration of on-line searches. More recently, the p-ary algorithm has been proposed to exploit the massively parallel architecture of graphics processors (GPUs) to substantially accelerate the search operations on such large databases. In this chapter we present a multi-keyword range search technique that efficiently exploits index data structures to search multiple text keywords in large databases. The multi-keyword range search is an extension of the p-ary algorithm which was originally developed by Kaldewey et al. We enhanced the p-ary algorithm to support multi-keyword search on GPGPU. We compare the performance in terms of response time, throughput and speed-ups between CPU and GPGPU implementations. The performance benchmarks demonstrated that our algorithm achieves up to 25× and 6× performance in terms of speed-up on Tesla K20c GPU card when compared to a single and multicore CPU implementations, respectively.

Keywords GPGPU • CUDA • GPU • P-ary • Multi-keyword search • Binary search

17.1 Introduction

Digital data explosion has exceeded petabytes and entered to zettabyte era. A large organisation has dire needs to analyse and interpret large data in meeting their business objectives. Such data are normally collected and stored in databases, and these may be either structured or unstructured. It is common that large volumes of data are indexed in order to facilitate searching and retrieval. For example, Google search engine constantly builds index of keywords to facilitate search of keywords on growing collection of compound and hyperlinked documents in the World Wide

A. Abdullah (✉) • A. Nazir • M. Senapan • S.S. Meng • E. Karuppiah
MIMOS Berhad, Kuala Lumpur, Malaysia
e-mail: amirul.abdullah@mimos.my; amrilnurman.nazir@mimos.my; mohanavelu.senapen@mimos.my; sm.soo@mimos.my; ettikan.karuppiah@mimos.my

Y. Cai, S. See (eds.), *GPU Computing and Applications*,
DOI 10.1007/978-981-287-134-3_17

Web. Such search engine typically contains an index, for instance, comprising text from a large number of uniform resource locators (URLs). However, database search typically involves long latency for the main memory access followed by small number of arithmetic operations, leading to ineffective utilisation of large number of cores and memory. This main memory access latency is difficult to be hidden due to irregular and unpredictable data accesses during information search and retrieval.

Recent approaches exploiting the massively parallel architecture of graphics processors (GPUs) to parallelize and accelerate search operations have achieved intriguing results. [1] has presented a novel parallel search (p-ary) algorithm for large-scale database index operations that scales with the number of processors and outperforms traditional thread-level parallel GPU and CPU implementations. The algorithm exploits the GPU by applying a "divide-and-conquer" strategy to speed up individual searches. The algorithm has shown to outperform conventional binary search on the GPU in terms of response time and throughput.

However, the original p-ary algorithm has several limitations. First, the resulting output of the matched value from the p-ary algorithm is returned in a non-predictable manner. The algorithm selects the matched value in a random fashion within a given range of characters from an index. As such, the returned value from a search operation is a non-deterministic. This poses a challenge when there is a need to find all possible matched values from a given range of matched values. For example, users may wish to determine all matched values resulting from a search operation to perform reduction operation, e.g. summation of a set of numbers. Hence, there is a need for multi-keyword range search to facilitate searching multi-keyword in index database.

Second, the p-ary algorithm is unable to identify the first occurrence and/or the last occurrence of matched values from the index database due to its non-deterministic characteristic. This can be a major limiting factor as it is often the case that the user needs to determine the index value of the first and the last occurrence from the resulting search operation to perform maximum and/or minimum arithmetic calculations or to count the total number of occurrences of the keywords from the database. Most real-life applications rely on this feature for operations like filtering documents by tags, counting words in documents and extracting links to related data.

In this chapter, we present multi-keyword range search algorithm that extends the original p-ary algorithm to address the limitations of the previously mentioned issues. The contributions of this chapter are as follows:

- We enhanced the original p-ary algorithm to capture the offset of the first and the last occurrence of the repetitive keyword matches.
- Our enhanced algorithm is able to search for multiple different keywords and return multiple ranges of search results.
- We present various optimisation techniques, which include data packing, memory coalescing and shared-memory optimisation techniques. Our experimental results show that enhanced algorithm achieves a notable 25× speed-up over a

single-core CPU implementation (1-thread CPU implementation). Similarly, our enhanced algorithm achieves 6× speed-up compared to 8-CPU-thread implementation.

The remaining of this chapter is organised as follows: Sect. 17.2 gives some background on keyword, binary and multi-keyword search and summarises some previous related work. Section 17.3 describes the implementations of multi-keyword range search methodologies on GPU architecture. Subsequently, Sect. 17.4 presents the results analysis on single-core and multicore systems of our GPU implementation. Final section summarises the future work and main conclusions of this work.

17.2 Background

Before we discuss the proposed algorithm in detail, we provide a review of selected search algorithms, namely exact match, binary search and P-ary, which play crucial role in understanding our proposed algorithm.

Searching for information is an indispensable component of our lives. Rapid growth of available text in unstructured data (e.g. docx, txt files, etc.) and structured data (i.e. relational databases) increases the need for ordinary users to search such information. The size of these data can be very large at the scale or terabytes and petabytes. For example, it was reported that Facebook performs text searches on 250 petabytes of data on regular basis [2]. Hence, having the ability to quickly analyse petabytes of data at an affordable cost is indispensable.

Traditionally, major RDBMS (e.g. MySQL, Oracle, etc.) provided full-text search capabilities that enable string searches on structured data on a database. However, as the number of user increases and the size of database increases, the speed of the searching is important where it may cost time or money to organisations and end users. Furthermore, large database solutions (e.g. K data, Vertica, Netezza) are very expensive. On the other hand, GPU-based solution provides a scalable and affordable solution.

17.2.1 Keyword Search

We define keyword search as the string searching problem which looks for all occurrences of a string *str* of length *strlen* in another text of length *textlen*. The goal is to search for one or more occurrences of a string or pattern in large text databases. The earliest fast exact string search algorithms include [[3], [4], [5]], etc. Traditionally, these algorithms were implemented on CPU architecture. Recently, there has been great attention paid to GPU acceleration of string matching using GPU CUDA. [6] present an overview of CUDA implementation of the Boyer-Moore,

Knuth-Morris-Pratt, Horspool and Quick-Search string search implementations on the GPU architecture. They have shown that these algorithms in average achieved up to 18× speed-up on the GPU architectures. [7] also reported some other search algorithms such as FASTA and BLAST that were implemented on the GPU architecture achieving similar speed-ups.

17.2.2 Binary Search

Binary search is one of the fast search algorithms which perform searching in a sorted list of data and based on divide-and-conquer strategy [8, 9]. However, the main disadvantage of the binary search is that it imposes the requirement that the list be sorted. Binary search works by storing the starting value (index 0) and the end value (index N-1). Next, it compares values of the keyword with the intended data at index N/2. If the value of the keyword is same as the value of the data, then it will return the value. Otherwise, if the value of the keyword is less than the value of the data, a new ending value is provided. On the other hand, if the value of the keyword is larger than the value of the data, a new starting value will be provided. A comparison will be made between keyword and the middle value of these two new points. This process will be repeated until a match is found. Figure 17.1 illustrates an example of binary search. However, the time taken to perform the binary search depends on the way the data is sorted. For example, if the keyword to be searched is of the largest value, the search operation is more efficient if it traverses out in descending order (best-case performance). However, in this particular scenario, if the searching is traversed in an ascending order list, the algorithm gives a worst-case performance [10].

Kaldewey et al. [1] presented a binary search implementation in GPU where multiple thread blocks are created and spawned to perform search on multiple keywords simultaneously. However, the algorithm employs a blocking approach where a thread needs to wait for the other running thread within same block to finish its search before a new search can be spawned. This is somewhat inefficient resulting in low resource utilisation especially when the number of keywords is less than the number of used threads. Moreover, the algorithm has a limiting factor whereby when the same pivots are being used for all running threads, these threads are quickly diverging as each thread is assigned a different key search [11]. As a result, it is not amenable to caching or coalescing. Moreover, contention may likely occur due to large amount of small memory access.

17.2.3 Multi-keyword Search (P-ary Search)

Multi-keyword or p-ary searching algorithm uses divide-and-conquer strategy with a complexity of $\log_p(n)$, where p is the number of parallel threads. Similar to binary

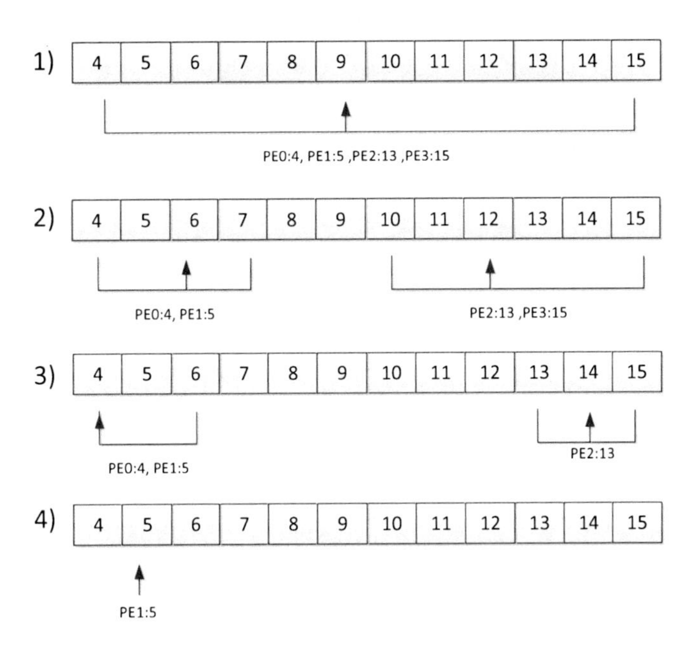

Fig. 17.1 Binary search example

search, it requires input data in sorted format [8]. The algorithm takes advantage of SIMD architecture in GPU by dividing the keywords amongst many blocks and performs searching in parallel manner using threads and/or processing elements (PEs) in shared memory. The p-ary is different from binary search where each thread block is used to search for the same keyword [1]. Each thread will have certain range to search within the data, with its own starting and ending points. The threads will compare both points to determine either the search keyword is within that range or not. A new range will be assigned to the thread if there is a possibility that a keyword match can be found in any one of the given ranges. These processes will continue until the keyword is found. The advantages of this algorithm are due to its ability to spawn multiple threads to leverage the GPU gather operations while having data in the memory to be coalesced [11]. The method also scales with increasing number of threads.

Figure 17.2 shows the example of p-ary implementation using 4 threads that are spawned to search same keywords. Each thread (PE) is spawned to search for the same keyword "11". As it can be observed, the algorithm takes fewer steps to complete the search in comparison to binary search, which would require additional steps. In the next section, we will describe the implementation of our p-ary enhancement and techniques that we employ to support fast multi-keyword range search.

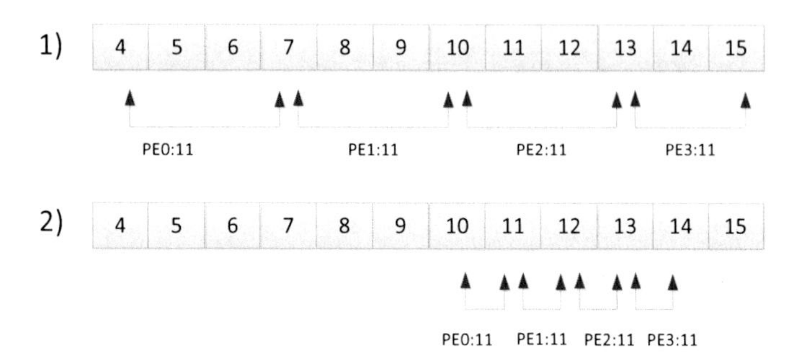

Fig. 17.2 P-ary search algorithm

17.3 Implementation

In this section, we describe our implementations of multi-keyword range search methodologies on GPU architecture. Our aim is to provide the capacity for searching a keyword in an array of strings where it could have repeated data in the array. The output should return the start and the end of the keyword in the input array when a match is found. To achieve this, we leverage both the brute force search and binary search to obtain the index of repeated data element that is found. First, we must ensure that the datasets are sorted before the p-ary search can be applied. Example of the scenario is illustrated in Fig. 17.3 where the keyword "eeee" is searched in an array of sorted input data where the keyword occurs at index 5 to 7.

Contrast with the implementation of [1], we serialise the p-ary algorithm to ensure that the output from p-ary algorithm always returns the index of the first and the last occurrence of the keyword in the dataset. In this way, we eliminate the race condition problem which could occur in the original p-ary implementation. In our implementation we assign each thread to handle one keyword. In the original p-ary implementation, four or more threads are assigned for each keyword. Instead, we only employ one thread per keyword in order to return the first and last occurrence of keywords. Issuing more threads to search the same keyword has the effect of increasing memory synchronisation, which we aim to avoid. Furthermore, by assigning only one thread per keyword, we can achieve higher load utilisation since more tasks can be assigned to one thread while reducing the communication between many different threads. Moreover, employing one thread to handle one keyword consumes less shared memory in the GPU. This gives more room for the GPU shared memory to store large number of keywords.

Figure 17.4 shows the kernel pseudocode of our implementation in GPU. Lines 1 to 12 search within the input array for specific keyword. We make use of two offset variables, namely, *offset* and *offset_rev* in which these two variables are used to store the offset of the first occurrence and last occurrence separately. Next, lines 13 to 14 perform exact match of the keywords against the datasets and store matched data index as result.

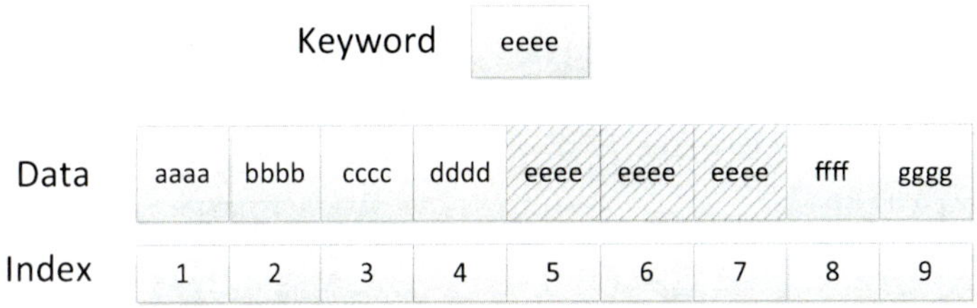

Keyword eeee

Data	aaaa	bbbb	cccc	dddd	eeee	eeee	eeee	ffff	gggg
Index	1	2	3	4	5	6	7	8	9

Fig. 17.3 Searching for the range of repeated data

```
while (range > granuality){
  old_offset = offset
  old_range = range
  range = range/granuality

  for(i=0; i<granuality)
   temp = temp + ((cache_key-data) >0)
   temp_rev = temp_rev + ((cache_key-data) <0)

  offset = old_offset + temp*range
  offset_rev = old_offset_rev + temp*range
 }
for(i=0;i<Gran){
 if(!(cache_key-(data+offset+i))
   result = offset +i
 if(!(cache_key-(data+offset+i))
   result = offset +i
}
```

Fig. 17.4 P-ary search pseudocode

We made few additional optimisations to the original p-ary algorithm to increase speed-up and throughput significantly. These optimisation techniques include data packing, memory coalescing, shared-memory optimisation and CUDA occupancy optimisation. In the next subsection, we will discuss each of these techniques in detail.

17.3.1 Data Packing

Reading data from device memory is costly and incurs large overhead due to large number of GPU cycles before data being received by GPU threads. Even if GPU provides wide memory bus, loading small data such as single *char* (1 byte) each time will underutilise the capability of the GPU memory bus and may lead to overhead to the application. CUDA provides availability to load different sizes of data from global memory such as 1 byte, 4 bytes or 8 bytes. Since our implementation involves text string which has long array of *char*, it is much efficient to read

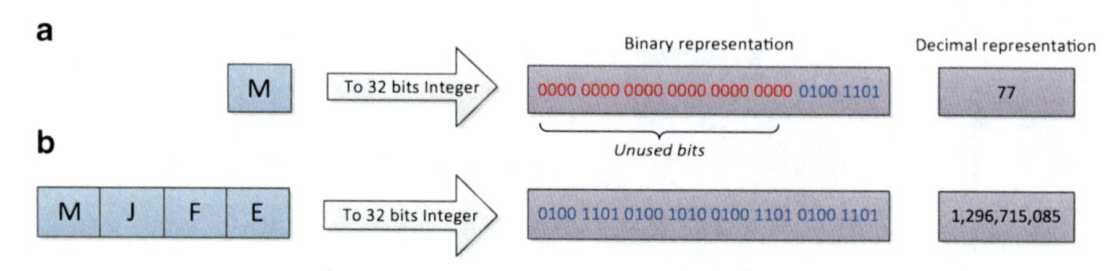

Fig. 17.5 Data packing technique

multiple *char* at once to increase the memory bandwidth efficiency. In achieving this, we present some optimisation techniques.

First, we combine multiple small data structures to form a larger data structure, in which it can be fitted into one instruction call. The idea is to fill in as much data in one instruction call so that more data can be fetched with less overhead. In achieving this, we pack text data into common size of bits such as 4 bytes or 8 bytes. This method can significantly reduce the waiting time for the GPU to fetch instruction when compared to the original method of accessing 1 byte at one time.

Second, the text data is compressed in the form of integers rather than characters and/or strings. This is because transferring data in the form of integers is more efficient compared to transferring strings of characters. Moreover transferring raw text data to the GPU will result in low memory throughput within a given cycle. To avoid the low throughputs, we therefore convert text data into a series of integers. For example, converting four characters into one integer will significantly increase the amount of data read within given cycle. In this case, the GPU can fetch 32 bits at one time from the global memory while text data will occupy 8 bits of memory at a time.

Figure 17.5a illustrates how packing one character into an integer incurs less space and will not fully occupy the extra bits available in one integer. Figure 17.5b further shows an example of packing four characters into one integer. Packing four characters into one integer can significantly reduce the memory calling time by 25 % compared to calling one character at a time. One point to take into account is that converting a character into an integer incurs the big-endian little-endian effects. Hence, the integer needs to imitate as a text string.

17.3.2 Memory Coalescing

Uncoalesced memory accessing can become a bottleneck for GPU implementation due to overhead of uneven pattern of reading data from the global memory, which takes considerable amount of cycles. To maximise the usage of global memory bandwidth, we aim to minimise the number of bus transactions in the GPU so that GPU will be busy spending more time for computations rather than reading data. In achieving this, we have to coalesce the memory accesses so that each thread can

Before :	A_1	A_2	A_3	B_1	B_2	B_3	C_1	C_2	C_3
After :	A_1	B_1	C_1	A_2	B_2	C_2	A_3	B_3	C_3

Fig. 17.6 Example of the text data rearrangement to increase coalescing

fetch the data within the same block of memory that is being read. Bingsheng He et al. [12] have implemented similar technique to gain speed-up by aligning the data to be coalesced in order to increase memory locality for database operations.

Since p-ary algorithm is built upon on B-tree data structure search, it is very likely that it will access data in non-coalesce pattern. To make the data more coalesced, we advocate that the datasets are to be allocated in an array of structure manner. This will increase chances the likelihood that any thread that reads the data from the global memory being stored in the L2 cache of the GPU. This approach will significantly reduce the overhead of miss caching.

Figure 17.6 shows an example of how an array of text is rearranged in the GPU to reduce cache miss. As it can be observed, the datasets are firstly arranged where each character in a single element of the datasets is placed next to each other. After the rearrangement, each of the first character from each element of the datasets is positioned continuously one by one. In this way, there is more likeliness for memory that needs to be fetched by threads is within the cache.

17.3.3 Shared Memory

Accessing shared memory in the GPU uses less cycle when compared to accessing global memory. To achieve maximum performance, it is important to minimise redundant accesses to global memory whenever possible. Whenever possible, one should store data in the shared memory, especially when we need frequent access to the data for specific operations. Hence, our approach for optimisation is to store the keyword into shared memory since the algorithm requires frequent access to the keyword when doing comparison against the datasets. This method will substantially reduce cycles needed to read the keyword. However, using too much shared memory will also decrease the total occupancy level. Hence, determining the right amount of shared memory to be allocated is equally important. From our observations, in achieving at least 50 % occupancy, best performance can be obtained by using 128 bytes of shared memory with each block comprising of 128 threads. Furthermore, each thread is allocated with 4 bytes of shared memory.

17.4 Experimental Evaluation

In our research, we have implemented the algorithm in GPU and CPU. Experimental evaluation was conducted to evaluate performance between CPU and GPU implementations in terms of response time, speed-up and throughput. We will compare the performance of the algorithms between GPU and CPU by using two Quad-Core Intel® Xeon E5506 (2.13 GHz) with a total of 8 cores and 12 GB of RAM for CPU and Tesla K20c for GPU. The API we use to parallelise CPU implementation is OpenMP, and for GPU we use CUDA 5.0. All the experiments is carried out in Windows environment and using Visual Studio 2010 Premium as our IDE. The data generated for all the experiments are randomly distributed but sorted in alphabetical order. We measure the execution time as the total time for task to run on the GPU which includes the data transfer and communication time between CPU and GPU. Each experiment is repeated three times and an average is obtained. We define "datasets" as the list of input text data where it is structured in a way that each row has the same length and in sorted order. The word "query" is defined as list of text keywords to be searched in given datasets. The "text length" is defined as length of each row on both the query and datasets. A sample of datasets or query is shown in Fig. 17.7. In this figure, each row has 32 characters and the data is completely random. Both query and datasets used in this experiment will look similar as these.

17.4.1 *Response Time*

We have identified three different experiment parameters to demonstrate the influence of these parameters on performance. For each experiment, we vary the variable of the text length, the number of queries and the number on datasets. It is to be expected that these three parameters will have significant effect on the performance. This experiment also will show the performance scalability for GPU compared to CPU when we increase the value for some of these parameters. First, we vary the number of queries with 5 million, 10 million and subsequently

index	Data
1	AACBRMWOLDKMSMOTFKDOSLAMSIWEKDIF
2	ABHSDKFMTKGLCVRIQPDNFNVOZPXMDKSA
3	ACDKMEMFEPQPZVEYRTPWMBHCKDLSPWMK
4	BAITPWYPERWAKZORTKMDMSSLFORPWLLV
.	.
.	.
.	.
n-1	ZAROPHPREKWLSWOEIRMAXMZPDMRMFITK
n	ZZAPDKLFDKGOTIRTKTMHJKSQIUEYTRPZ

Fig. 17.7 Sample of input data or query

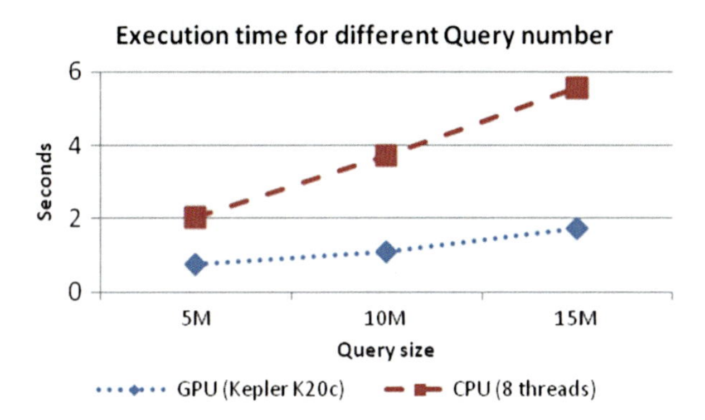

Fig. 17.8 Execution of GPU and CPU for 5 million, 10 million and 15 million queries

15 million queries whereby each element comprises 32 characters and a query has datasets totalling 15 million rows. The result is shown in Fig. 17.8.

As it can be observed, as the number of queries increases, the execution time also increases for both GPU and CPU. However, we can observe that the execution time for multicore CPU implementation with 8 threads incurs higher execution time when compared to GPU implementation. This indicates that GPU implementation is suitable for large queries as more threads are spawned to do multiple search operations at the same time. As the number of query size increases, we can observe a steady increase in execution time due to the overhead of traversing large data for both CPU and GPU implementations. As expected, GPU implementation incurs less execution time when compared to the CPU implementation.

Interestingly, we can see that performance gap between CPU and GPU widens as we increase query size. This shows the GPU implementation scales well when compared to the CPU implementation, which has a problem of scaling as the query size increases. The CPU implementation cannot scale simply because only a maximum of eight threads can be deployed at any time, and since each query is handled by one thread, we can see this as a limiting factor for the CPU to scale. Nevertheless, we conjecture that accessibility to more CPU cores will most likely result in better performance since more threads will be spawned to handle additional queries.

Next, we examine the impact of performance on the dataset size. We fix the query size to be 15 million where each element comprises 32 characters. By fixing the query size and the number of characters, we can observe the effects of increasing datasets. Fig. 17.9 presents our result. Interestingly, we can observe that execution time for CPU increases almost exponentially. The CPU implementation clearly cannot scale when dataset size is increased. The best plausible explanation for the poor performance is most likely due to low memory throughput in CPU while the GPU has high memory throughput. Hence, more threads in warps can fetch from L2 cache at a given time which allows higher throughput and efficiency.

Finally, we investigated the performance between the CPU and GPU for different text lengths of 8, 16 and 32 characters long. The result is shown in Fig. 17.10. From the graph, we can observe that increasing text length will also increase the

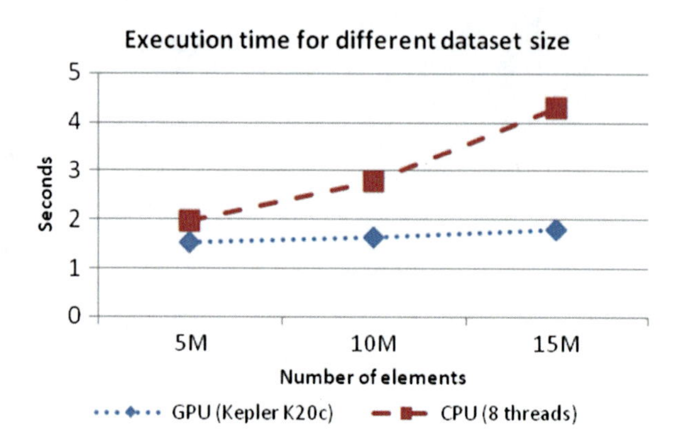

Fig. 17.9 Execution of GPU and CPU for 5 million, 10 million and 15 million queries

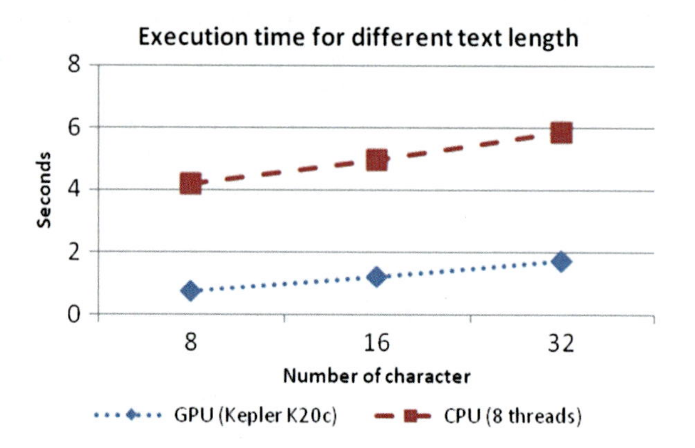

Fig. 17.10 Execution of GPU and CPU for 8, 16 and 32 characters of text length

computation time for both CPU and GPU. It can be seen that the execution time of the CPU and GPU increases almost at the same rate as the number of characters increases. In fact, we observe a linear increase for both CPU and GPU as the number of characters increases. Nonetheless, GPU implementation is still showing faster performance when compared to the CPU implementations for different text length.

Based on our previous experiment results, we have examined the performance difference between CPU and GPU implementations when increasing dataset size, query size and text length. For all cases, we have observed that the GPU implementation provides tremendous improvement of execution under all parameters and workloads. This is due to the parallelism nature of search algorithm that effectively utilises all GPU cores for accelerating search operations. Each query is executed independently from one another, and this enables the GPU to spawn more threads to handle many queries simultaneously without compromising performance.

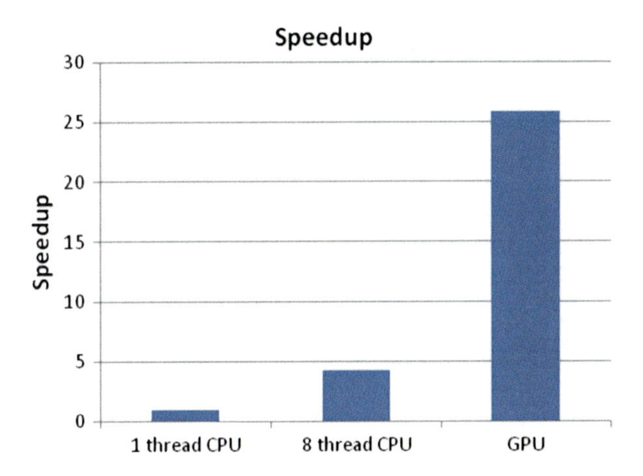

Fig. 17.11 Speed-up average for GPU and CPU implementation

17.4.2 Speed-Ups

Next, we will look at the speed-up gain between GPU and CPU. Figure 17.11 shows the speed-ups achieved by different configurations of CPU and GPU implementations. We first compare the differences of speed-up between 1-thread CPU, 8-thread CPU and GPU Nvidia K20c implementations. To examine the speed-up, we tested the GPU and CPU with 15 million queries against 15 million rows of datasets where each dataset consists of 32 characters.

Since we are running the experiments on 2-quad-core machine, we spawned 8 threads to fully utilise the 8-core machine (spawn 1 thread per CPU core). As can be observed, the 8-thread CPU implementation achieves approximately 4× speed-up in comparison to a single-core implementation (1-thread CPU). We can observe reasonable gains from exploiting the maximum number of physical CPU cores.

However, the GPU implementation achieves far greater speed-up to 25× than a single-core implementation (1-thread CPU). When comparing to 8-core CPU implementation, the GPU implementation achieves a notable 6× speed-up. This is due to the highly parallel structure of GPU that effectively enables up to 2,048 maximum threads per SM to be deployed on our Kepler GK110 card. One interesting observation is that accessing data is deemed to be a major bottleneck due to small local memory and caches in the GPU. However, the ability for the GPU to execute 32 threads in a single warp at any one time makes it possible for the GPU to effectively hide memory latency by executing other warps while current warps are in idle.

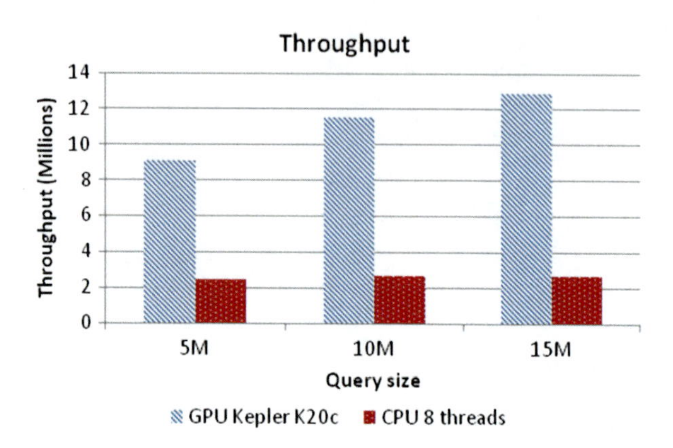

Fig. 17.12 Throughput vs. query size for GPU and CPU

17.4.3 Throughput

Throughput is defined as the amount of keywords searched at a given response time of the kernel execution. Figure 17.12 shows the throughput we calculated for GPU Kepler K20c and 8-thread CPU for 5, 10 and 15 million queries with each query comprising 32 characters. We can observe that the GPU throughput is relatively very high when compared to CPU implementation, which is almost up to 4 times higher. Besides that, we can see that the throughput of GPU increases as query size increases. This shows that the throughput is not yet saturated for this amount of query size. This is due to the fact that our latest Kepler K20c GPU card is able to process huge amount of threads; hence it is possible to achieve even higher throughput. For the CPU implementation, we can see that the throughput is relatively low in comparison to GPU, and as we increase the query size, the throughput is maintained. This is because the CPU has limited number of processors to process the query; thus throughput is saturated when it reaches to a certain amount of query size.

Conclusion and Future Work
In this chapter, we present multi-keyword range search algorithm that extends the p-ary algorithm to address some of the limitations of the original p-ary algorithm. The aim is to further optimise the performance for large-scale index search operations. Our multi-keyword range search algorithm is able to capture the offset of the first and the last occurrence of the repetitive keyword matches, while maintaining its fast performance by employing some of the optimisation techniques which include data packing, memory coalescing and shared-memory optimisation techniques. As a result, our enhanced algorithm is able to search multiple keywords and return multiple range search results. From our experimental results, our enhanced algorithm achieves a notable

(continued)

25 times speed-ups over a single-core CPU implementation, whereas it achieves six times speed-up over 8-thread multicore CPU implementation.

In future work, we aim to further optimise our implementation by minimising the non-coalesced memory accesses on the GPU. In achieving this, we seek to rearrange the datasets based on the B-tree data structure. The B-tree data structure will help enforcing next data sequence to reside closer within the GPU memory, which in turn provides efficient reading and writing large blocks of data. This allows more datasets to be fetched in the cache at any one time, which can substantially increase the performance. Besides that, we plan to implement our algorithms on Many Integrated Core (MIC)/Intel® Xeon Phi™ Coprocessor that has 60 CPU cores and compare these against GPGPU performance.

Acknowledgement The research was done under Joint Lab, NVIDIA-HP-MIMOS GPU R&D and Solution Center. This is the first GPU solution centre in South East Asia. Funding for the work came from MOSTI, Malaysia. The authors would like to thank Prof. Simon See and Pradeep Gupta from NVIDIA for the supports. We also would like to acknowledge assistance and help provided by Zakiah Zulkefli from Universiti Sains Malaysia during her internship at MIMOS Berhad.

References

1. Kaldewey, T., Hagen, J., Blas, A.D., Sedlar, E.: Parallel search on video cards. In: HotPar'09: Proceedings of the First USENIX conference on Hot topics in parallelism, p. 9. (2009). http://portal.acm.org/citation.cfm?id=1855591.1855600
2. Souders, S.: High-performance web sites. CACM, pp. 36–41 (2008)
3. Boyer, R.S., Strother Moore, J.: A fast string searching algorithm. Commun. ACM **20**(10), 762–772, (1977). 10.1145/359842.359859
4. Knuth, D.E., Morris, J.H., Pratt, V.R.: Fast pattern matching in strings. SIAM J. Sci. Comput. **6** (2), 323–350 (1977)
5. Karp, R.M., Rabin, M.O.: Efficient randomized pattern-matching algorithms. IBM J. Res. Dev. 249–260 (1987). doi:10.1147/rd.312.0249
6. Kouzinopoulos, C.S., Margaritis, K.G.: String matching on a multicore GPU using CUDA. In: Proceedings of the 2009 13th Panhellenic Conference on Informatics, pp. 14–18. (2009). doi:10.1109/pci.2009.47
7. Vouzis, P.D., Sahinidis, N.V.: GPU-BLAST: using graphics processors to accelerate protein sequence alignment. Bioinformatics 182–188 (2011). http://bioinformatics.oxfordjournals.org/content/27/2/182.abstract
8. Bentley, J.L.: Multidimensional binary search trees used for associative searching. Commun. ACM. **18**(9), 509–517, (1975). doi:10.1145/361002.361007
9. Nievergelt, J.: Binary search trees and file organization. In: ACM Computing Surveys (CSUR), pp. 195–207 (1974)
10. Yao, A.C-C.: Should tables be sorted. J. ACM **28**(3), 615–628 (1981). doi: 10.1145/322261.322274
11. Hwu, W.W: GPU Computing Gems Jade Edition. Elsevier Science. (2011). http://books.google.com.my/books?id=LsNVFUnzcVMC

12. He, B., Yang, K., Fang, R., Lu, M., Govindaraju, N., Luo, Q., Sander, P.: Relational joins on graphics processors. In: Proceedings of the 2008 ACM SIGMOD international conference on Management of data (SIGMOD '08), pp. 511–524. ACM, New York, NY. (2008). doi:10.1145/1376616.1376670
13. Gregg, C., Hazelwood, K.: Where is the data? Why you cannot debate CPU vs. GPU performance without the answer. In: 2013 I.E. International Symposium on Performance Analysis of Systems and Software (ISPASS), pp. 134–144 (2011)
14. Fang, W., Lau, K.K., Lu, M., Xiao, X., Lam, C.K., Yang, P.Y., He, B., Luo, Q., Sander, P.V., Yang, K.: Parallel Data Mining on Graphics Processors. Technical Report HKUST-CS08-07, Hong Kong University of Science and Technology (HKUST) (2008)
15. Lustig, D., Martonosi, M.: Reducing GPU Offload Latency via Fine-Grained CPU-GPU Synchronization. In: HPCA, pp. 354–365 (2013)
16. Che, S., Boyer, M., Meng, J., Tarjan, D., Sheaffer, J.W., Skadron, K.: A performance study of general-purpose applications on graphics processors using {CUDA}. J. Parallel Distrib. Comput. 1370–1380 (2008). http://www.sciencedirect.com/science/article/pii/S0743731508000932. General-Purpose Processing using Graphics Processing Units

Index

Y. Cai, S. See (eds.), *GPU Computing and Applications*,
DOI 10.1007/978-981-287-134-3

D

E

R

S

T

U

V

W

Z